Sleep 1980

Fifth European Congress on Sleep Research, Amsterdam, September 2–5, 1980

Sleep 1980
Circadian Rhythms, Dreams, Noise and Sleep, Neurophysiology, Therapy

Editor: *W. P. Koella*, Basel

94 figures and 69 tables, 1981

S. Karger · Basel · München · Paris · London · New York · Sydney

European Congresses on Sleep Research

Sleep. First European Congress on Sleep Research, Basel 1972
    Editors: W.P. Koella, P. Levin, Basel
    XVI + 612 p., 148 fig., 87 tab., 1973. ISBN 3–8055–1604–5
Sleep 1974. Second European Congress on Sleep Research, Rome 1974
    Editors: P. Levin, W.P. Koella, Basel
    XIV + 526 p., 100 fig., 44 tab., 1975. ISBN 3–8055–2069–7
Sleep 1976. Third European Congress on Sleep Research, Montpellier 1976
    Editors: W.P. Koella, P. Levin, Basel
    XIV + 490 p., 143 fig., 53 tab., 1977. ISBN 3–8055–2663–6
Sleep 1978. Fourth European Congress on Sleep Research, Tîrgu-Mureş 1978
    Editors: L. Popoviciu, B. Aşgian, G. Badiu, Tîrgu-Mureş
    XVI + 794 p., 331 fig., 94 tab., 1980. ISBN 3–8055–0778–5

National Library of Medicine, Cataloging in Publication
    Sleep 1980. 5th European Congress on Sleep Research, Amsterdam 1980.
    Sponsored by the European Sleep Research Society.
    Editor: W.P. Koella. – Basel; New York: Karger, 1981.
    Consists of the proceedings of the 5th European Congress on Sleep Research.
    1. Sleep Disorders – congresses 2. Sleep – congresses I. Koella, Werner Paul, 1917–, ed.
    III. European Congress on Sleep Research IV. European Sleep Research Society
    W3 SL632 DNLM
    ISBN 3–8055–2045–X

Drug Dosage
    The authors and publisher have exerted every effort to ensure that drug selection and dosage set forth in this text are in accord with current recommendations and practice at the time of publication. However, in view of ongoing research, changes in government regulations, and the constant flow of information relating to drug therapy and drug reactions, the reader is urged to check the package insert for each drug for any change in indications and dosage and for added warnings and precautions. This is particularly important when the recommended agent is a new and/or infrequently employed drug.

All rights reserved.
    No part of this publication may be translated into other languages, reproduced or utilized in any form or by any means, electronic or mechanical, including photocopying, recording, microcopying, or by any information storage and retrieval system, without permission in writing from the publisher.

© Copyright 1981 by S. Karger AG, P.O. Box, CH-4009 Basel (Switzerland)
    Printed in Switzerland by Effingerhof AG, Brugg
    ISBN 3–8055–2045–X

Contents

Fifth European Congress on Sleep Research. Organizing Committee and Acknowledgements XII
Preface XIII

Part I. Symposia

A. Sleep Deprivation and Depression
Chairman: *R.H. van den Hoofdakker*, Groningen

van den Hoofdakker, R.H.; Elsenga, S. (Groningen): Clinical Effects of Sleep Deprivation in Endogenous Depression 2
Zander, K.J.; Lorenz, A.; Wahlländer, B.; Ackenheil, M.; Rüther, E. (München): Biogenesis of the Antidepressive Effect of Sleep Deprivation 9
Gillberg, M.; Åkerstedt, T. (Stockholm): Sleep Deprivation in Normals – Some Psychological and Biochemical Data from Three Studies 16
Uhde, T.W.; Post, R.M.; Ballenger, J.C.; Cutler, N.R.; Jimerson, D.C.; Weitzman, E.D.; Bunney, W.E., Jr. (Bethesda, Md./Charlottesville, Va./Bronx, N.Y.): Circadian Rhythm and Sleep Deprivation in Depression 23
Wehr, T.A.; Wirz-Justice, A. (Bethesda, Md./Basel): Internal Coincidence Model for Sleep Deprivation and Depression 26
List of References of Symposium A 34
Addresses of Authors 39

B. Circadian Rest-Activity and Sleep-Wake Rhythms
Chairman: *A.A. Borbély*, Zürich

Borbély, A.A. (Zürich): Introduction 40
Groos, G.A. (Leiden): The Suprachiasmatic Nuclei as a Central Pacemaker Timing the Rest-Activity Cycle 42

Chouvet, G.; Emptoz, H.; Burtschy, B.; Routhier, H.L.; Valatx, J.L. (Lyon/Villeurbane/Paris): Circadian Rhythms of Sleep-Waking Cycle: A Genetic Approach 51
Eastman, C. (Chicago, Ill.): Circadian Rhythms of Rats at the Limits of Entrainment and a Phase-Shift Model of 'Spontaneous Internal Desynchronization' in Humans. 59
Wirz-Justice, A.; Wehr, T.A. (Bethesda, Md./Basel): Uncoupling of Circadian Rhythms in Hamsters and Man 64
Schulz, H. (München): Sleep Onset REM Episodes in Depression 72
List of References of Symposium B 79
Addresses of Authors 84

C. Sleep and Emotional Stress

Chairmen: *D. Schneider-Helmert*, Königsfelden; *P. Visser*, Amsterdam

Schneider-Helmert, D.; Visser, P. (Königsfelden/Amsterdam): Introduction 85
Bastiaans, J. (Leiden): The Psychosomatic Approach of Sleep Disturbances 86
De Koninck, J. (Ottawa): Dream and Mood Regulation in Stressful Situations 91
Horne, J.A. (Loughborough): Sleep Deprivation, Stress and Sleep Function 95
Åkerstedt, T.; Gillberg, M. (Stockholm): Sleep, Stress and Recuperation 98
Cluydts, R.J.G.; Visser, P. (Brussels/Amsterdam): Emotional Stress, Mood State and Sleep 102
Schneider-Helmert, D. (Königsfelden): Clinical and Conceptual Aspects of Sleep and Emotional Stress 107
List of References of Symposium C 114
Addresses of Authors 117

Part II. Workshops

A. Periodic Hypersomnia

Chairmen: *M. Billiard*, Montpellier; *B. Roth*, Prague

Billiard, M. (Montpellier): Introduction 120
Roth, B.; Nevšímalová, S. (Prague): The Clinical Picture of Periodic Hypersomnia. A Study of 38 Personally Observed Cases 120
Billiard, M. (Montpellier): The Kleine-Levin Syndrome 124
Hishikawa, Y.; Iijima, S.; Tashiro, T.; Sugita, Y.; Teshima, Y.; Matsuo, R.; Kaneda, H. (Osaka): Polysomnographic Findings and Growth Hormone Secretion in Patients with Periodic Hypersomnia 128
Montplaisir, J.; de Liry, J.L.; Dardenne, A. (Montreal): Hypersomnia and Manic Depressive Illness 133
List of References of Workshop A 138
Addresses of Authors 140

Contents

B. Hypnotics and Insomnia Models in Animal and Man
Chairmen: *R. Scherschlicht, W.P. Koella*, Basel

Koella, W.P. (Basel): Introduction	141
Oswald, I. (Edinburgh): Testing an Hypnotic in Man	142
Scherschlicht, R.; Marias, J.; Schneeberger, J.; Steiner, M. (Basel): Model Insomnia in Animals	147
Haefely, W. (Basel): Mechanism of Action of Benzodiazepines	155
Koella, W.P. (Basel): Side Effects of Today's Hypnotics and the Hypnotic of the Future	158
List of References of Workshop B	164
Addresses of Authors and Discussants	168

C. Special Symposium on Dreams
Chairmen: *D. Lehmann, I. Strauch*, Zürich

Strauch, I. (Zürich): Introduction	169
Lehmann, D.; Koukkou, M. (Zürich): Dream Formation in a Psychophysiological Model: the State-Change Theory	170
Foulkes, D. (Atlanta, Ga.): Dreams and Cognitive Development	174
Salzarulo, P.; Cipolli C. (Paris): Memory Processing of the Mental Sleep Experience	178
Kramer, M. (Cincinnati, Ohio): The Function of Psychological Dreaming: a Preliminary Analysis	182
List of References of Workshop C	186
Addresses of Authors	189

D. Chronobiology, Shiftwork and Sleep
Chairman: *O. Benoit*, Paris

Åkerstedt, T.; Torsvall, L. (Stockholm): Age, Sleep and Adjustment to Shiftwork	190
Foret, J.; Benoit, O. (Paris): Individual Factors, Sleep Characteristics and Circadian Evolution of Body Temperature	195
Knauth, P.; Kiesswetter, E.; Bruder, S.; Romberg, H.P.; Rutenfranz, J. (Dortmund): Day Sleep during the Morning and during the Afternoon between Experimental Night Shifts	198
Zulley, J. (Andechs): Comparison of Long and Short Sleep Durations in Free-Running Sleep-Wake Cycles	202
Benoit, O. (Paris): Concluding Remarks	206
List of References of Workshop D	207
Addresses of Authors	210

Contents   VIII

E. Traffic Noise, Sleep and Performance
Chairmen: *A. Kumar*, Amsterdam; *A. Muzet*, Strasbourg

Kumar, A. (Amsterdam): Introduction 211
Muzet, A.; Ehrhart, J.; Eschenlauer, R.; Lienhard, J.P. (Strasbourg): Habituation and Age Differences of Cardiovascular Responses to Noise during Sleep............. 212
Griefahn, B.; Gros, E.; Kauth, H. (Düsseldorf): Noise and Sleep in the Home: General Methodology.. 215
Jurriëns, A.A. (Delft): Noise and Sleep in the Home: Effects on Sleep Stages 217
Vallet, M.; Gagneux, J.M.; Blanchet, V. (Bron): Noise and Sleep at Home: Stage Changes and Arousals.. 220
Wilkinson, R.T. (Cambridge): Effects of Traffic Noise upon Sleep in the Home. Subjective Report, EEG, and Performance the Next Day........................... 225
Hofman, W.F.; Kumar, A.; van Diest, R. (Amsterdam): Noise and Sleep in the Home: Effect on Heart Rate 228
Williams, H.L. (Oklahoma City, Okla.): Comments on Traffic Noise and Sleep 231
List of References of Workshop E 232
Addresses of Authors 234

Part III. Special Lectures

Corner, M.A.; Mirmiran, M.; Bour, H. (Amsterdam): On the Role of Active (REM) Sleep in Ontogenesis of the Central Nervous System 236
Foulkes, D. (Atlanta, Ga.): Dreams and Dream Research 246
Passouant, P. (Montpellier): La narcolepsie 258

Part IV. Free Communications and Posters

A. Pharmacology and Endocrinology of Sleep

Benedek, G.; Obál, F., Jr.; Bari, F.; Obál, F. (Szeged): The Effect of Atropine on the Hypnogenic Action of Basal Forebrain Stimulation.......................... 272
Spiegel, R. (Basel): Increased Slow-Wave Sleep in Man after Several Serotonin Antagonists.. 275
Wauquier, A.; Van Den Broeck, W.A.E.; Niemegeers, C.J.E. (Beerse): On the Antagonistic Effects of Pimozide and Domperidone on Apomorphine-Disturbed Sleep-Wakefulness in Dogs 279
Depoortere, H. (Paris): Sleep Induction and Central Effects of Some α-Adrenoceptor Agonists 283
Miettinen, M.V.J. (Helsinki): α-Adrenergic Function and Sleep in Kittens 287
Lanfumey, L.; Adrien, J. (Paris): Effects of a Noradrenergic Agonist on Sleep in the Rat . 290
Kafi, S.; Gaillard, J.-M. (Chêne-Bourg): Pre- and Postsynaptic Effect of Yohimbine on Rat Paradoxical Sleep 292

Contents

Helojoki, M.; Putkonen, P.T.S. (Helsinki): α_2-Adrenoceptive Inhibition of Avian Paradoxical Sleep 294
Depoortere, H.; Lefèvre-Borg, F.; Cavero, I. (Paris): Electroencephalographic Studies on Clonidine and Para-Aminoclonidine, a Potent Peripherally Acting Imidazoline ... 297
Blois, R.; Monnier, M.; Tissot, R.; Gaillard, J.-M. (Chêne-Bourg): Effect of DSIP on Diurnal and Nocturnal Sleep in Man 301

B. Neurophysiology and Biochemistry of Sleep

Bowker, R.M. (Galveston, Tex.): The Awakening of the Sleeping Ponto-Geniculo-Occipital Wave 304
Gnirss, F.; Schneider-Helmert, D.; Schenker, J. (Königsfelden): The Role of Biological Rhythms versus Sleep-Wake Regulation in the Determination of Nocturnal Arterial Hypotension 307
Kobayashi, T.; Saito, Y.; Endo, S.; Tsuji, Y.; Okuno, H. (Tokyo) Fluctuation of REM Activity as Analyzed by REM Momentum 311
Obál, F., Jr.; Benedek, G.; Józsa, C.; Obál, F. (Szeged): Insomnia after Olfactory Tubercle Lesion in Cats 314
Mirmiran, M.; Corner, M.A. (Amsterdam): Polyneuronal Discharges in the Cerebral Cortex of Developing Rats during Sleep 318
Bour, H.L.; Corner, M.A. (Amsterdam): Bioelectric Activity of Brainstem Reticular Neurons during Sleep and Waking in Free Moving Developing Rats 321
Declerck, A.C.; Wauquier, A.; Sijben-Kiggen, R.; Pans-Raedts, M. (Heeze): A Central and Central-Temporal 3 c/s Rhythm Occuring during REM and Stage 2 of the Non-REM Sleep in Humans 324
Kahn, J.P.; Adrien, J. (Paris): Selective Paradoxical Sleep Deprivation in 6-OHDA-Lesioned Kittens 328
Zamboni, G.; Perez, E.; Parmeggiani, P.L. (Bologna): cAMP Concentration in the Hypothalamus and Cerebral Cortex of the Rat in Wakefulness and Sleep 331

C. Physiology and Phenomenology of Sleep

Roy, J.C.; Freixa i Baqué, E.; Delerm, B. (Lille): Electrodermal Activity as a Function of Sleep Stages in the Cat 334
Weber, U.; Haller, H.; Boenke, R.; Kendel, K. (Lahr): Nocturnal Sweat Secretion during the Dissociated Sleep Stages 337
Shapiro, C.M.; Moore, A.T. (Edinburgh): Circadian Heat Transfer 340
Hasan, J.; Alihanka, J. (Turku): Construction of REM-NREM Sleep Hypnograms from Body Movement Recordings 344
Horne, J.A. (Loughborough): Is Human Sleep for Body Restitution? 348

D. Sleep, Perception and Memory

Mirmiran, M.; Van den Dungen, H. (Amsterdam): Influence of a Complex ('Enriched') Environment on the Sleep-Waking Patterns of Developing Rats 351
Baylor, G. W. (Montreal): Dreams as Problem Solving. A Computer-Aided Dream Diary 354
Kramer, M.; McQuarrie, E.; Bonnet, M. (Cincinnati, Ohio): Problem-Solving in Dreaming: an Empirical Test 357
Cipolli, C.; Salzarulo, P.; Calasso, E.; Maccolini, S.; Pani, R.; Tuozzi, G. (Bologna/Paris): Recall and Retrieval Processes Related to Stage II Sleep 361
Empson, J.A.C.; Hearne, K.M.T.; Tilley, A.J. (Hull): REM Sleep and Reminiscence 364
Tilley, A.J.; Empson, J.A.C. (Cambridge/Hull): Picture Recall and Recognition following Total and Selective Sleep Deprivation 367
Seitz, R.; Bernhard, J.; Burkhardt, R.; Faeh, M.; Moser, U.; Strauch, I. (Zürich): Reliability of a Structural Dream Analysis Method 370
Heynick, F. (Eindhoven): Freud and Chomsky: Our Linguistic Capacities during Dreaming 373
Strauch, I.; Soukos-Valavani, I. (Zürich): Differential Aspects of Self-Rated Sleep Latencies .. 377

E. Physiopathology of Sleep and Sleep Disorders

De Graaf, W.; Poelstra, P.A.M.; Visser, P. (Amsterdam): An Epidemiological Sleep Survey in a General Practice of a Middle-Sized Town 380
Partinen, M.; Putkonen, P.T.S. (Helsinki): Sleep Habits and Sleep Disorders in 2,537 Young Finnish Males 383
Hayashi, Y.; Yoshida, R.; Watanabe, H.; Endo, S. (Tokyo): All-Night Polygraphies for Healthy Aged People (2nd Report) 386
Billiard, M.; Michel, F.B.; Bertrand, A.; Milane, J.; Passouant, P. (Montpellier): Excessive Daytime Somnolence in Patients with Chronic Respiratory Failure Showing Prolonged Episodes of Severe Oxygen Desaturation during REM Sleep 389
Garma, L.; Bouard, G.; Benoit, O. (Paris): Age and Intervening Wakefulness in Chronic Primary Insomnia 391
Nevšímalová, S.; Roth, B.; Ságová, V.; Paroubková, D.; Horáková, A. (Prague): Clinical and Polygraphic Studies of Sleep Drunkenness 394
Montplaisir, J.; Malo, J.L.; Walsh, J.; Monday, J. (Montreal): Nocturnal Asthma: Sleep and Dream Analysis 397
Billiard, M.; Touchon, J.; Passouant, P. (Montpellier): Sleep Apneas and Mental Deterioration in Elderly Subjects 400
Nogues, B.; Monod, N.; Samson-Dollfus, D. (Rouen): A Comparative Study of the Frequency of Sleep Apnea during the First Part of the Night and the Whole Night in Infants 2–12 Months of Age 403
Porter, J.M.; Horne, J.A. (Loughborough): Exercise and Sleep Behaviour: a Questionnaire Approach 406
Porter, J.M.; Horne, J.A. (Loughborough): Bed-Time Carbohydrate Ingestion and Sleep 408

Contents XI

Schneider-Helmert, D. (Königsfelden): Interpretation of Night Terrors in Adults Based upon Observations in the Dark 411
Putkonen, P.T.S.; Bergström, L. (Helsinki): Clonidine Alleviates Cataplectic Symptoms in Narcolepsy 414
Schneider-Helmert, D.; Gnirss, F.; Schoenenberger, G.A. (Königsfelden): Effects of DSIP Applications in Healthy and Insomniac Adults 417
Schneider-Helmert, D.; Gnirss, F.; Schenker, J. (Königsfelden): Successful Treatment of Insomnia by Interval Therapy with L-Tryptophan 421
Velasco, M.; Velasco, F.; Romo, R.; Perez, M.A. (Mexico City): Effect of We-941 (a New Thienodiazepine) on Sleep Patterns 424
Steinberg, R.; Nedopil, N.; Rüther, E. (München): Long-Term Effects of L-Tryptophan and Oxprenolol in Hyposomnia 426

F. Psychiatry, Neurology, Sleep Deprivation

Declerck, A.C.; Lustenhouwer, A.W. (Heeze): Evaluation of Automatic Sleep Spindle Detection in Epileptic Patients 429
Landau-Ferey, J.; Benoit, O.; George, B. (Paris): Night Sleep after Recovery from Brainstem Traumatic Injury 431
Shimizu, T.; Sugita, Y.; Iijima, S.; Teshima, Y.; Hishikawa, Y. (Osaka): Sleep Study in Patients with Spinocerebellar Degeneration and Related Diseases 435
Coenen, A.M.L.; Van Hulzen, Z.J.M.; Van Luijtelaar, E.L.J.M. (Nijmegen): Effects of Different Paradoxical Sleep Deprivation Treatments on Locomotor Activity in Rats 438
Smirne, S.; Franceschi, M.; Bareggi, S.R.; Comi, G.; Mariani, E.; Mastrangelo, M. (Milan): Sleep Apneas in Alzheimer's Disease 442
Touchon, J.; Besset, A.; Billiard, M.; Passouant, P. (Montpellier): Data Compiled through the Study of Sleep Concerning the Prognosis of Petit Mal 445

G. Methodology of Sleep Studies. Automatic Analysis

Campbell, K.B. (Ottawa): Principles of the Automatic Pattern Recognition of Human Sleep Waveforms 448
Chouvet, G.; Odet, P.; Etienne, J.P.; Chemarin, P.; Valatx, J.L.; Pujol, J.F. (Lyon): Estimation of the Circadian Sleep-Wake Cycle with a Doppler Radar 452
Glatt, A.F. (Basel): Computer Analysis of the EEG in Infrahuman Species 456
Chouvet, G.; Buda, C.; Janin, M.; Odet, P.; Pujol, J.F. (Lyon): An Automatic Sleep Classifier for Cat 459

Author Index 463

Fifth European Congress on Sleep Research

Amsterdam, September 2–5, 1980

Organizing Committee

P. Visser, Chairman, Amsterdam; *H.A.C. Kamphuisen*, Vice-Chairman;
J. Snel, Treasurer; *M. Veenman-van der Linden*, Secretary;
J. Bastiaans, Leiden; *D.D. Breimer*, Leiden; *A. Bruzova*, Amsterdam;
M.A. Corner, Amsterdam; *R.H. van den Hoofdakker*, Groningen;
W.P. Koella, Basel; *A. Kumar*, Amsterdam; *D. Schneider-Helmert* Königsfelden;
A.H.M. Schoenmakers, Tilburg

Acknowledgements

The organizing committee of the Fifth European Congress on Sleep Research is deeply grateful to both the University of Amsterdam and the Hoffmann-La Roche company, whose advise and support greatly facilitated the successful organization of the sleep congress in Amsterdam.

Through the initiative of *A. Kumar*, the E.S.R.S. was able to organize financial support covering the travel expenses of 24 young scientists who presented communications at the congress. The generous gifts of the Environment Research Programme of the Commission of the European Communities and Beckman Instruments Nederland B.V. are acknowledged with appreciation. The committe hopes that support for young scientists, granted for the first time during this congress, will become a policy for all future meetings of the E.S.R.S.

Preface

This volume – Sleep 1980 – contains the material presented during the Fifth European Congress on Sleep Research, held at the Ian Swammerdam Institute, Amsterdam, from September 2 to 5, 1980. The congress, organized by Prof. *Visser* with the fine support of his Organizing Committee, as on earlier occasions offered a number of symposia and workshops and in addition left ample space and time for the presentation of Free Communications and Posters.

The compilations of the symposia and workshops were accepted to be included in this volume without the advice and critique of the Scientific and Publications Committee; so were the manuscripts of a number of 'Special Lectures'. However, for the first time, the 'mini-papers' of the Free Communications and Posters were scrutinized and screened by this Committee, resulting in the elimination of contributions judged to be either too long and/or too deficient in content or style (or both). As happened before, a number of authors presenting papers or posters evidently decided of their own not to submit a manuscript for publication.

It remains to thank all the contributors to this volume and to the organizers of this Congress for a fine piece of work without which the publication of this volume would not have been possible. We have to thank too Miss *Jacqueline Baud* for her never tiring help with the editorial work. And, we thank, as on earlier occasions, Mr. *T. Karger* and Miss *Anke Rogal* of S. Karger AG for their excellent support and cooperation.

Basel, May 1981 *W.P. Koella*

Part I. Symposia

A. Sleep Deprivation and Depression

Chairman: *R.H. van den Hoofdakker,* Groningen
Participants: *M. Gillberg,* Stockholm; *K.J. Lander,* Munich; *T.W. Uhde,* Bethesda, Md.; *A. Wirz-Justice,* Basel

Clinical Effects of Sleep Deprivation in Endogenous Depression[1]

R.H. van den Hoofdakker, S. Elsenga, Groningen, The Netherlands

Introduction

The main goal of this symposium is to attempt to explain the therapeutic effects of sleep deprivation (SD). Our role will be the presentation of empirical data for which any explanation must account.

Schulte [71, 72] was the first to mention that SD might be therapeutically effective in endogenous depression. This inspired *Pflug and Tölle* [52, 53] to perform their pioneering study in which they attempted to provide these fragmentary observations with a more systematic empirical basis. These authors measured the difference in depressive symptomatology before and after SD in endogenously and neurotically depressed patients. Their conclusions were that SD caused at least some improvement in every patient in the endogenous group and a large improvement in this group taken as a whole, whereas the neurotic group reacted less, and more variably. For obvious reasons, these conclusions gave rise to a considerable amount of research.

This review will focus on the role of SD as a treatment variable. From an experimental point of view, two general problems have to be dealt with in reviewing the literature. First, there is the problem of the reliability and the validity of the measures used in the experiments. Do reports provide the necessary psychometric information, or must acceptable standards be taken for granted? By and large, the latter is, unfortunately, the case. Second,

[1] For references see compound list on page 34.

there is the problem of the validity of designs and procedures; in other words, do the designs and procedures rule out alternative explanations? SD is a drastic intervention. Effects attributed to SD may, in fact, be due to other factors inherent in the experimental set-up. Expectations as to the outcome may be elicited either in the subjects themselves, or in the people around them, i.e. nurses, observers, etc. The SD procedure is, in general, accompanied by a drastic change in the attention paid to the subjects. The sleepless night involves more than just deprivation of sleep; group sessions, walks, dinners, etc. during the night may also have therapeutic effects. Last but not least, psychopharmacological treatment before, during, or after SD may play a major role in the outcome.

In view of these latter considerations, the question of the role of SD as a treatment variable, will be subdivided into two more specific questions: (1) What is the effect of SD per se on the symptoms of endogenous depression? (2) What is the relationship between SD and other potentially important treatment variables?

The Effects of SD per se

Only a small number of studies have dealt with the role of SD per se. *van den Burg and van den Hoofdakker* [11] subjected 10 hospitalized endogenously depressed patients to 2 SDs separated by 1 night of sleep. No drugs were given for at least 3 days before the experiment. In this study subjects were not highly motivated, and nurses were sceptical. Ratings of the effects were made by blind psychiatrists and non-blind nurses. A slight, though statistically significant improvement was found after SD. As a rule, relapse followed subsequent sleep (fig. 1). Comparable results were obtained by *Post et al.* [60], who applied 1 SD to 19 hospitalized subjects with primary affective disorders, as defined by *Spitzer et al.* [78]. The patients were drug-free for at least 10 days prior to treatment. Ratings were done by 14 patients themselves and by non-blind nurses. However, both were told that the procedure was related to the study of urine and CSF. 10 patients showed some improvement, usually lasting 1 day, whereas 9 showed slight changes or felt worse. *Gerner et al.* [24] reported on 25 hospitalized patients with primary affective disorder who were subjected to one SD. 11 of these were included in the study of *Post et al.* [60]. These patients were free of medication for at least 3 weeks. Ratings were made by patients (as far as possible) and by nurses who were informed that the purpose of the study was to assess the

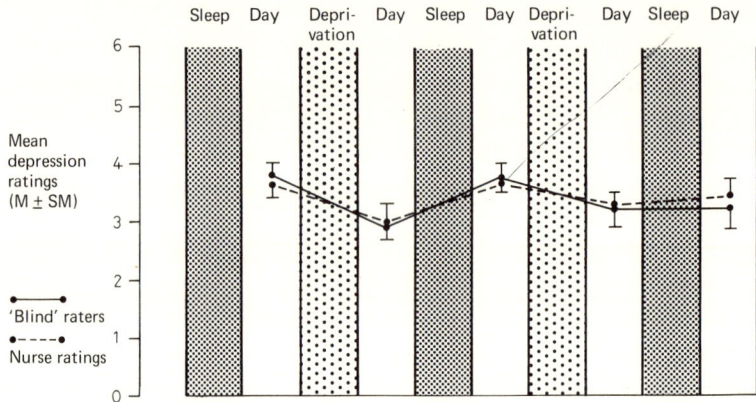

Fig. 1. Means of mean daily depression ratings on days after SD and after nights of sleep. The more depressed a patient, the higher the score.

effects of SD on chronobiological variables. Some decrease in depression was found. Improvement was followed by relapse, as indicated by the fact that ratings on the day after subsequent sleep were not significantly different from those on the pre-treatment day. The data here are somewhat difficult to evaluate. A positive response was observed in 15 of the 25 patients studied. The definition of response, however, hampers a precise estimation of the relevance of SD, as it may include minor and possibly irrelevant changes. The last study to be mentioned is that of *Larsen et al.* [40]. 26 patients were deprived, 19 of whom were endogenously depressed, and 7 of whom showed non-endogenous depression. 4 subjects received diazepam, thioridazine, chlorprothixene and lithium; the others were drug-free for at least 36 h. Only the data on the endogenous group are presented; it is not clear whether the above-mentioned 4 drug-treated patients belonged to this group. Information regarding instructions, expectations and bias of patients, nurses and raters is lacking. 6 subjects reacted slightly positively. The duration of the effects is not mentioned.

What conclusions can be drawn from these data? A serious handicap in the assessment of the effects is the impossibility of fully controlling subject bias because double-blind placebo experiments cannot be carried out. Although the patients may have been blind to the purpose of the intervention, they certainly were not blind to the nature of it. Many of them perceive lack of sleep as detrimental to their mental health. Another problem is that the effects of instructions—such procedures as deceiving or encouraging–induce positive, negative or neutral pre-experimental sets. The extent to which these instructions may influence the results is hard to determine as long as this is not investigated. Still, another not easily controllable factor is the substitution of a troublesome lonely period of wakefulness in bed for the non-specific therapeutic factor of the company and the attention of nurses. Apart from these partially inevitable weaknesses, the studies show some strength regarding the control of confounding factors. The great majority of the results have been established under drug-free conditions, while no specific psychotherapeutic measures were taken. In at least three of these investigations an attempt was made to control the expectations and attitudes of subjects and raters. All things considered, the conclusion must be drawn that SD per se probably induces some transient antidepressant effect in endogenously depressed patients.

Relationship between SD and Other Treatment Variables

Very little can be definitively said about this subject because of the small amount of systematic research which has been carried out. The conclusion which may be derived from the large number of studies in which SD was combined with other measures, including antidepressant medication, are, at best, speculative. At this point we shall consider in detail those studies involving medication.

Most studies make use of a one group pre-post test design. The treatment factors, however, are frequently not kept equal for all subjects. Either the number of SDs is not the same, or the subjects are treated with different antidepressant drugs, or both the number of SDs and the drug treatment vary. Attribution of a specific antidepressive effect to any one factor or to a specific combination of factors is therefore impossible [6, 14, 52, 53, 65, 68, 79, 82].

Nevertheless, certain inferences regarding the effectiveness of the combination of SD and antidepressant drugs emerge from certain studies. *Buddeberg and Dittrich* [9] examined a number of depressives, 12 of whom were endogenously depressed. These last ones received drugs, supposedly antidepressants. Interview ratings and global self-ratings showed significant improvement after 1 SD, whereas self-rated 'vitality' scores did not. *Rudolf and Tölle* [65] also observed improvement as indicated by interview ratings after 1 SD in endogenous depressives (n = 29), the majority of whom received tricyclics. Self-rating scores, however, did not show significant amelioration of mood. A third study notable in this context is that of *Schmocker et al.* [70]. Interview ratings as well as self-ratings showed significant improvement in a group of 30 depressed patients with unspecified diagnoses. All patients were treated with tricyclics; many were also treated with neuroleptics or sedatives.

A cautious conclusion of these studies may be that the administration of antidepressant drugs has no detrimental influence on the immediate effects of one SD. As to the long-term relationship between tricyclics and SD, some indications are provided by the aforementioned study of *Buddeberg and Dittrich* [9]. These authors report a significant increase of the self-rated 'vitality' on the day after recovery sleep relative to the baseline day. The data presented by *Philipp* [56] also indicate that antidepressants might prolong the effects of total SD by preventing relapse after recovery sleep. More elucidating on this point are the findings of *Loosen et al.* [41]. Two groups of 8 primary depressives were treated with either clomipramine alone or with one SD, followed by clomipramine treatment starting on the day after SD. The first group showed a slow remission in contrast to the second, in which a prompt, long-lasting remission was obtained. It is notable that in this study no relapse after the recovery night was observed. In other words, no relapse after the subsequent sleep-night was observed. Thus, antidepressants are likely to prevent the relapse found when SD is used without these drugs. This is not to say that this influence of antidepressant drugs has been proven. Conclusive results can only be obtained in further controlled studies. Secondly, the possibility of placebo effects needs to be taken into account.

Elsenga and van den Hoofdakker [19, 20] compared the effects of the combination of SD and clomipramine with those of either the combination of SD and placebo or clomipramine without SD. Three groups of 10 endogenous depressives were examined over a period of 2 weeks, with the SD groups receiving 4 SDs. All patients were drug-free for at least 3 days. The administration of drugs was double-blind; the effects were measured by means of self-ratings and ratings by blind psychiatrists and non-blind nurses. SD was introduced as a part of the treatment plan. No specific other therapeutic actions were undertaken. Only the results for the depression variable (Von Zerssen) are mentioned here. In figure 2 the average of morning and afternoon self-ratings are shown; figure 3 presents the analogous observer rating data obtained at the same time. The major finding is that the combination of SD and clomipramine induces a development of mood over time which is significantly different from that produced by clomipramine alone. The latter does not differ significantly from SD plus placebo, indicat-

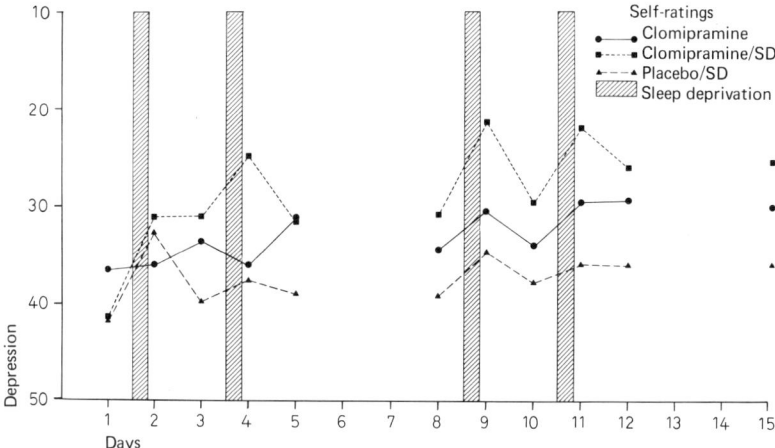

Fig. 2. Mean depression self-ratings for the clomipramine, clomipramine/SD and placebo/SD-treated groups on 11 experimental days.

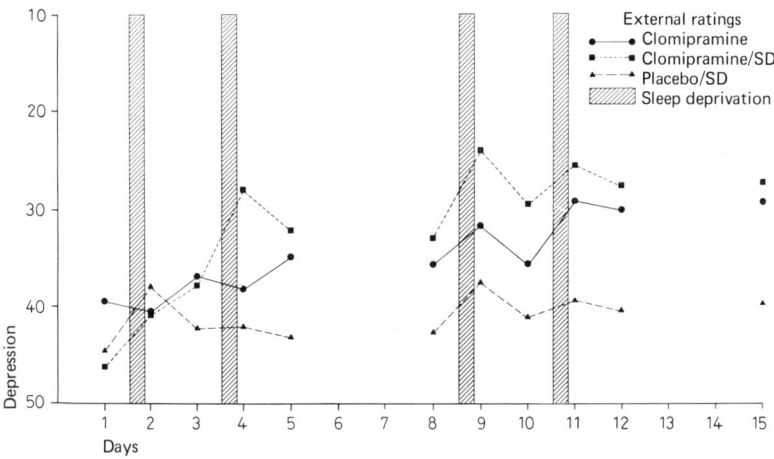

Fig. 3. Mean depression ratings as rated by external raters for the clomipramine, clomipramine/SD and placebo/SD groups on 11 experimental days.

ing that this last combination is not an alternative to a classical treatment with clomipramine alone. Further, it may be concluded that it is the specific interaction of SD, the antidepressant drug and time, rather than any of these factors alone, which is responsible for the beneficial results.

As shown earlier, the effect of SD per se is usually short-lived. The same is found for the combination of SD and placebo. A close resemblance can be seen between the effects of both combinations – SD plus placebo and SD plus clomipramine – after the first SD. After the first recovery night, however, there is no relapse in the clomipramine/SD group, in contrast to the placebo/SD group. Thus, clomipramine acts in such a way as to prevent the relapse found after the first recovery night. This property of prevention seems to be present irrespective of whether clomipramine is introduced before or after SD [41].

One further observation should be noted, namely that a large variability in depression between days appeared to exist in the group treated with clomipramine only, especially in the afternoon. Whether this is due to the natural course of endogenous depression or to the drug is unknown. In any case, not immediately occurring effects which are sometimes attributed to SD may, in fact, represent the therapeutic effects of antidepressant medication and/or the natural course of the disorder.

Final Remarks

Little attention has been paid to the generalizability of the effects of total SD, both in this review and in the literature. Numerous questions remain to be answered. What, for example, are the best psychopathological or biological predictors of the effect? Are the effects specific for total SD, or can they be produced by various manipulations of the sleep-wake cycle? Is clomipramine the only drug which may effectively be combined with SD, or are other antidepressants or other classes of drugs comparable in effect? Few of these questions has yet been answered on a valid, empirical basis. The greatest progress appears to have been made with regard to the generalizability of the effects to manipulations of the sleep-wake cycle other than total SD. There is evidence that partial SD [56, 69], selective REM SD [81], partial recovery sleep [5], and phase shifting of the sleep-wake cycle [85b] may produce beneficial effects. However promising these and other lines of research may be, the hard core of the empirical knowledge consists of the fact that total SD very probably has therapeutic effects in endogenously depressed patients and that clomipramine may prevent the relapse after the first recovery night. These are data for which any explanation must account.

Sleep 1980. 5th Eur. Congr. Sleep Res., Amsterdam 1980, pp. 9–15
(Karger, Basel 1981)

Biogenesis of the Antidepressive Effect of Sleep Deprivation [2]

K.J. Zander, A. Lorenz, B. Wahlländer, M. Ackenheil, E. Rüther,
München, FRG

Introduction

Sleep deprivation (SD) [11, 51], partial sleep deprivation during the second half of the night [62] and REM deprivation [80] have been demonstrated to be effective as therapy for major depression in 30–70% of the patients. The effect of a single SD appears to be short but can be considerably prolonged by the repetition of the procedure without [40, 96] and with [94] accompanying antidepressive pharmacotherapy.

The primary meaning of SD was thought to lie in the relationship between the symptomatology of depression, the disturbance of endogenous rhythms – and the special features of the SD effects [55]. Thus, the pathological circadian temperature profile in depression may [54] or may not [24] normalize after SD.

However, the most important findings regarding the effects of SD have been established in biochemical research. Before SD, MHPG excretion in urine [43] and in liquor [24] is higher in responders (R) than in nonresponders (NR).

Only few studies deal with the polygraphic recording of the effects of SD [50]. During depression, more patients complain of sleep disturbances

[2] For references see compound list on page 34.

than can be proven by sleep polygraphy [59]. The patterns of objective sleep disturbance in major depression have been characterized by a shortened REM latency [39] and by an 'insomnia profile': reduced total sleep time (TST), slow-wave sleep (SWS), fragmentation of sleep continuity by waking stages, and early morning awakening [31]. In this investigation to be checked at the same time were: the symptomatology of depression, EEG sleep polygraphy and chemical data before and after SD in order to test for mechanisms which are concerned with the therapeutic effects of a 'one-night' SD and repeated SD.

The possible prognostic value of any analysed parameter was of special interest. In this report the main interest is focused on the physiological data.

Methods

11 patients aged from 35 to 64, with a unipolar depression according to *Spitzer's* [77] criteria, have been subjected to repetitive SD after having been drug-free for at least 4 weeks. The pretreatment period to be analysed lasted for 6 days. Each SD lasted 36 h and was repeated 6 times with intermittent breaks of 1–4 days. In 2 patients there were no breaks between each 2 SDs. In responders a follow-up SD was performed 10–15 days after the therapeutic period. Psychometric analysis was performed by means of the Hamilton rating scale of depression, the AMP–3 rating scale and self-rating scales: self-adjustment scale of depression by v. Zerssen [13], 100-mm rating scale and MMPI. The cut-off score for a response to a 1st SD was: (1) a reduction of 10 points in the Hamilton score (HAM) and (2) a decrease of 25% in the v. Zerssen self-rating score. Final response (R) was defined by complete remission, i.e. ready for release from hospital.

Blood pressure, pulse frequency and sublingual temperature were measured every 3 h during the day and during SD. Plasma cortisol measured by standard RIA, noradrenaline and adrenaline monitored by the COMT method [46] were investigated during an experimental, standard stress procedure [1] before and after SD.

The sleep-waking polygraphy was recorded continuously by telemetry (Intrinsic 8F8) for 12–85 hour periods (night- and day-record of adaptation, baseline, SD) during the first and the 6th and also during the follow-up SD-period. Scoring of sleep stages was performed according to Rechtschaffen and Kales [17], and of waking stages, according to Rüther et al. [66].

Results

The mean HAM of depression before the SD therapy period was 34 ± 8; the mean self-rating score (Bf-S) was 32 ± 11 in the morning. 1 patient dropped out early. 9 of 10 patients improved at any given time during the SD therapy. Suicidal behavior and subjective sleep disturbances were

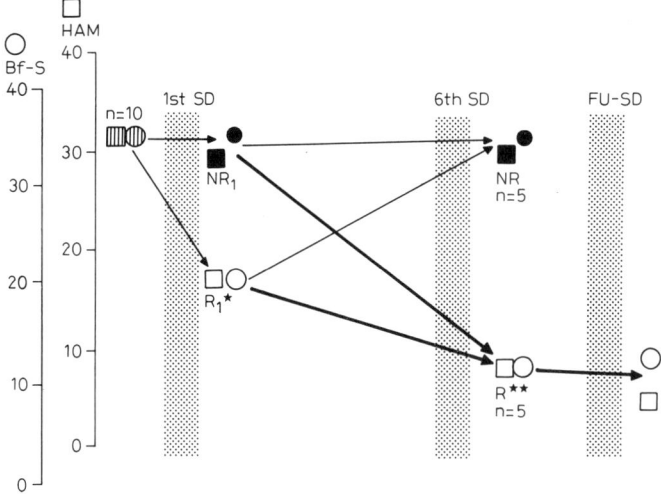

Fig. 4. One-night sleep deprivation (1st SD) and 6 times repeated SD. HAM = Hamilton score of depression; Bf-S = v. Zerssen self-rating score of depression; NR_1 = nonresponder to 1st SD, NR = nonresponder to repeated SD (black); R_1 = responder to 1st SD, R = responder to repeated SD (white). *,**: $p < 0.05$ (Man-Whitney U test). FU-SD = follow-up SD of R after 2 weeks.

significantly reduced in all patients. The mean HAM decreased from 35 to 22 after the 1st SD and to 15 after the 6th SD. The self-rating score (32 at 8 h and 36 at 17 h before test) fell significantly to 30 (8 h) and 25 (17 h) after the 6th SD. 7 of 10 patients decreased in the self-rating score by at least 25% after the 1st SD. Four different response patterns to repetitive SD were obtained (fig. 4): (1) Patients who improved already after the 1st SD (fig. 5); (2) no initial improvement but continuous amelioration during therapy; (3) improvement after the 1st SD, but subsequent relapse into depression, and (4) no essential improvement during repetitive SD.

The analysis of the data refers primarily to the subgroups, 5 R and 5 NR.

REM Latency. The mean REM latency of all patients before SD was 77 min, decreasing in both (R/NR) subgroups during the therapeutic period. The analysis of single data revealed that only 3 patients accidentally had a normal REM latency. It was shortened in 4 patients and enlarged also

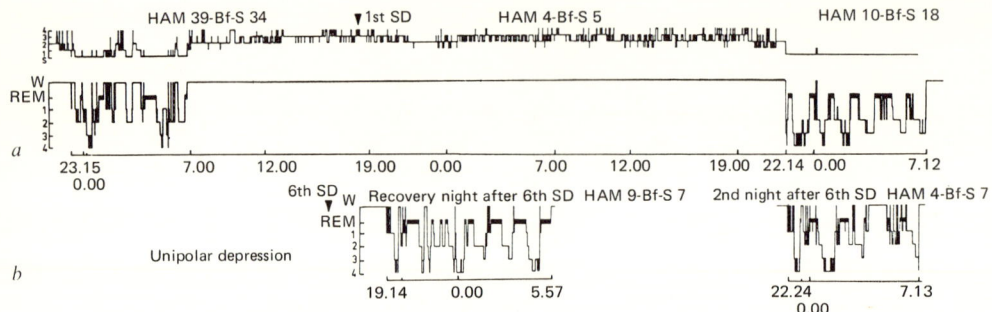

Fig. 5. a Scoring of stage awake 1–4 (upper graph): Motor activity increased after good response to SD with maximum in the evening [21]. Scoring of sleep stages (lower graph): Insomnia sleep profile while depressed. Recovery night after good response to SD with normalized sleep profile; REM latency remains shortened. *b* Sleep profile of 2 recovery nights after 6 times repeated SD. The sleep profile is again disturbed in remission.

in 4 patients. The largest intraindividual variability occurred in 1 patient with 9 min and 189 min.

Sleep Stages. The stage awake decreases after the 1st and the 6th SD and complementary TST and SWS increases after each SD. The increase of both after the last SD is less pronounced than after the 1st SD. The conditions tend to relapse to baseline conditions within few days after the 6th SD and after the follow-up SD in the R group. The REM rebound after the 1st and 6th SD occurs as expected significantly in all patients. The REM% of sleep period time (SPT) is 18% and thus slightly below the percentage expected for normals in comparable age [89 c].

Analysis of Responder-Nonresponder Differences. Only the R group demonstrates severe sleep disturbance. The fragmentation of the sleep by stage awake throughout the night is most apparent in the early improved and stabilized patients (fig. 6). Thus, R have significantly more stage awake than NR during the baseline conditions. During the therapeutic period the R reduce the waking time from 93 min (baseline) to 49 min, whereas NR are practically unchanged (30–27 min). TST of R is by 120 min (mean) shorter than TST of NR (484 min during baseline sleep). R and NR are significantly different in TST during the whole therapeutic period and TST increases significantly after each SD. SWS is predominant in NR before and

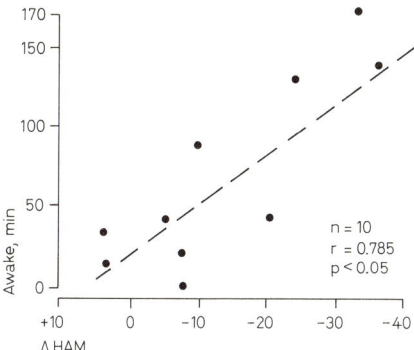

Fig. 6. Correlation between time of stage 'awake' (Rechtschaffen-Kales) of the total sleep period and the decrease in Hamilton score (Δ HAM) before and after repeated SD. Spearman rank correlation (r).

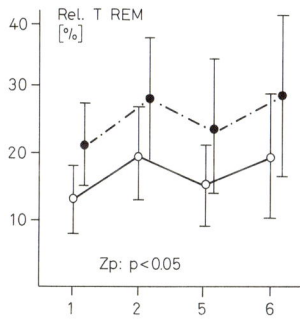

Fig. 7. Percent of REM sleep (Rel. T REM) before (1) and after (2) 1st sleep deprivation (SD) and before (5) and after (6) a 6th SD. ○ = responder; ● = nonresponder, n = 4. Group differences are not significant, changes in the time course (Zp) are significant (anova $p < 0.05$).

after the 1st SD; however, after the 6th SD the difference between R and NR is minimalized. Stage REM in R is 14% (of SPT) and in NR 19% during the 1st and 6th baseline night. REM rebound is significant in both groups after 1st and 6th SD (fig. 7). 2 days after the 6th SD REM% remains enhanced in both groups (R 21%, NR 25%). 3 of 4 patients being much improved had no REM rebound after the follow-up SD.

The circadian temperature profile was disturbed in both groups, the NR having a significantly higher mean body temperature than R. In both groups the temperature decreased in the mean during the therapy and tended to be reinstated to a physiological circadian profile. A similar but less pronounced effect occurred with plasma cortisol which tended to decrease during the therapeutic period at 8 and 17 h in both groups.

Preliminary analysis of the further biochemical and neurophysiological data revealed no further trends in group differences.

Analysis of the First-Night Response. The results of the polygraphic analysis refer mainly to the R/NR-dichotomization, whereas the EEG pattern of the first-night response is much less impressive concerning the differences between the subgroups R_1 and NR_1. However, after the 1st SD the NR_1 responders are shown to have significantly increased in 8-h cortisol plasma levels (NR_1 28.55 ± 2.5 μg/100 ml; R_1 18.4 ± 2.7 μg/100 ml). Also this NR_1 subgroup showed no noradrenergic or adrenergic reaction to experimental stress [1]. In contrast, R_1 seemed to have high baseline levels of catecholamines, and also these patients' responses were clearly adrenergic to experimental stress.

Discussion

The insomnia-like sleep pattern of a depressive patient may have a positive predictive value for response to SD. However, the number of patients in this investigation is yet too small to allow a certain hypothesis. The analysis trend of the data could be summed up as follows: (1) Patients who have severe sleep disturbances during depression (and less predominantly during good health) have a better chance to profit therapeutically by repeated sleep deprivation than those patients with a normal polygraphic sleep profile, even when complaining of reduced sleep quality. (2) During depression, SD procedure, but also during remission there is an apparent variability of REM latency, being frequently but not regularly shortened as also in other conditions [42]. Thus, the role of a shortened REM latency in depression remains to be elucidated. (3) There is no convincing relationship between the REM suppressive effect of antidepressive pharmacotherapy [26] and the REM-sleep i.e. REM rebound during SD therapy. Up until the end of a therapeutic period REM sleep increases identically after each SD. 3 of 4 good responders showed a 'negative' REM rebound when monitored in a good state during follow-up SD. In this condition, a mechanism also influenced by TCA may be activated by SD.

The biological factors influenced by SD are of different relevance as far as the therapeutical SD effect on depression is concerned. Thus, only when sleep is in fact disturbed by depression may SD considerably influence both depression and sleep. The repeated SD may possibly have a trigger function to start remission. There is no clear relationship between pathological endogenous rhythms in general and depressive disease.

The results of this study dealing with the mechanisms of a *one*-night SD are in accordance with earlier evidence that it's therapeutical effects may be related to sufficient resources of central catecholamine release, above all to norepinephrine (NE) [43]. However, only a limited number of severely depressed patients seem to be able to activate the NE system [24]. The early responding patients of this study provide high peripheral catecholamines which are mobilized by experimental stress. Although there is no certain relationship between peripheral and central catecholaminergic reactions [97], the high cortisol plasma values after the first SD may indicate that the blocking central NE system is less activated in early NR_1 than in one-night responders R_1 [35]. However, the relationship between response to SD and cortisol has been found to be opposite by other investigators [24]. Some of the early ameliorated patients relapsed during the further therapeutic periods and, in those patients who finally improved, peripheral catecholamines then tended to be reduced basically and after experimental stress, and to differ not much from the NR conditions. It may be concluded that the NE activation is only responsible for early and short-time effects of SD on depression.

Acknowledgement

We would like to thank Dipl.-math. *W. Hoebel* and Dr. *E. Pöppel* of MEDIS-Institute of GSF in Munich (FRG), and Dr. *R. Engel*, Psychology Department of the Psychiatric Clinic of the University, Munich, who provided statistical assistance. Dr. *Wolferstätter* and Dr. *Wagner* provided data for this investigation.

Sleep Deprivation in Normals – Some Psychological and Biochemical Data from Three Studies[3]

M. Gillberg, T. Åkerstedt, Stockholm, Sweden

In the literature on the anti-depressive effects of sleep deprivation, remarkably few references have been made to the relatively large amount of research on sleep deprivation and circadian rhythms in normals. Here data from three such experiments will be presented.

All three experiments were performed with variations of the constant conditions design described by *Conroy and Mills* [15]. The idea behind constant conditions is that environmental influences are controlled by being allowed to occur in equal 'portions' during identical time intervals. In the present three experiments time intervals, or 'modules', were 3 h in experiments 1 and 2, and 2 h in experiment 3. The experiments had in common that standardized food (sandwiches) and drink (mineral water) was served once every module. Measurements and other activities were also on the 2- or 3-h schedule. Clocks were not allowed, neither were radio, newspapers or telephone calls. The laboratories were also isolated from daylight and external noise. Care was also taken to change experimentors in a seemingly unsystematic fashion. And, perhaps most important, the subjects were deprived of sleep in all three experiments.

The Three Experiments

In *experiment 1*, 32 subjects, officers in the age range 40–60 years, were kept awake for 72 h. Results from this experiment have been published elsewhere [22, 23]. Here data on self-ratings, body temperature and urinary excretion of adrenaline and noradrenaline have been extracted. *Experiment 2* was performed on 12 subjects, 20–29 years old. They were kept awake for 64 h. Data have been published earlier [3, 4, 49]. Self-ratings, body temperature, urinary excretion of adrenaline, noradrenaline, cortisol and melatonin are presented here. Plasma

---

[3] For references see compound list on page 34.

levels of thyroid hormones (T3 and T4) were also evaluated, however, only on three occasions: before the vigil, after 48 h without sleep and 5 days after the experiment. *Experiment 3* demanded 36 h of continuous wakefulness. 6 subjects between 29 and 45 years of age participated. Except for the thyroid hormones the variables chosen were essentially the same as in experiment 2.

It should be mentioned that, in spite of the constant condition routines, there were important methodological differences between the three experiments. The group sizes differed somewhat between experiments and so did the types of task the subjects had to perform. These differences make direct comparisons between experiments sometimes hazardous. On the other hand similarities of the results in spite of these differences, perhaps point to reliable findings.

The rating scales used were of the visual analogue type with the extreme ends anchored with 'very much' and 'not at all', respectively. Altogether, 13 rating scales are presented here. All of them were not used in all three experiments. For this paper the scales have been a posteriori divided into four groups: 'activation', 'stress', 'negative mood' and 'positive mood'. This grouping was made without any statistical background.

Results

To give an overview, the results are presented in tables that simply state the following: if variables exhibit a circadian variation, when an eventual peak (or trough) occurs and whether there were significant changes due to 1 or 2 nights without sleep and the direction of the change.

Table I. Summarized results on self-ratings of *activation* variables (xxxx = not measured)

Rating scales	Experiment number	Circadian rhythm?	Trough (T) or peak (P) at	Significant effect of sleep deprivation (decrease –, increase +)	
				1 night	2 nights
'Performance capacity'	1	xxxx	xxxx	xxxx	xxxx
	2	yes	0800–1100 hours (T)	yes (–)	yes (–)
	3	yes	0500–0700 hours (T)	yes (–)	xxxx
'Tired'	1	no	–	yes (+)	yes (+)
	2	yes	0800–1100 hours (P)	yes (+)	yes (+)
	3	yes	0300–0500 hours (P)	no	xxxx
'Alert'	1	no	–	yes (–)	yes (–)
	2	yes	0500–0800 hours (T)	yes (–)	yes (–)
	3	yes	0300–0500 hours (T)	yes (–)	xxxx
'Energetic'	1	xxxx	xxxx	xxxx	xxxx
	2	yes	0800–1100 hours (T)	no	no
	3	yes	0300–0500 hours (T)	yes (–)	xxxx

Table II. Summarized results on self-ratings of *'stress'* variables (xxxx = not measured)

Rating scales	Experiment number	Circadian rhythm?	Trough (T) or peak (P) at	Significant effect of sleep deprivation (decrease –, increase +)	
				1 night	2 nights
'Stressed'	1	no	–	yes (–)	yes (–)
	2	no	–	no	no
	3	no	–	no	xxxx
'Tense'	1	xxxx	xxxx	xxxx	xxxx
	2	no	–	no	no
	3	no	–	no	xxxx
'Restless'	1	xxxx	xxxx	xxxx	xxxx
	2	no	–	no	no
	3	no	–	no	xxxx

Self-Ratings

'Activation' contains the scales 'performance capacity', 'tired', 'alert' and 'energetic' (table I). Circadian rhythms existed in all scales in experiments 2 and 3, but not in experiment 1. The peaks (or troughs) occurred in the early morning. In the majority of variables there was a marked effect of sleep deprivation, indicating lowered activation.

'Stress'. Scales were: 'stressed', 'tense' and 'restless'. As can be seen in table II, no circadian rhythms existed. Increased stress with increasing sleep deprivation was the case only in experiment 1, but this experiment was also characterized by a rather 'stressful' environment. But the overall impression is that sleep deprivation is not experienced as stress in the normal sense.

'Negative Mood'. Under this heading the following items are presented: 'depressed', 'uneasy' and 'anxious' (table III). No circadian rhythms appeared in the scales 'depressed' and 'anxious'. Only in experiment 1 there was an increase in depression due to sleep deprivation.

Positive Mood'. The scales were 'stimulated', 'high spirits', and 'elated' (table IV). Results differed somewhat between experiments, but the overall impression is a decrease in positive mood.

One observation that we have made during our sleep deprivation studies is that many subjects are talkative and almost euphoric in the morning after their recovery sleep. In exper-

Table III. Summarized results on self-ratings of *'negative mood'* variables (xxxx = not measured)

Rating scales	Experiment number	Circadian rhythm?	Trough (T) or peak (P) at	Significant effect of sleep deprivation (decrease –, increase +)	
				1 night	2 nights
'Depressed'	1	no	–	no	no
	2	no	–	yes (+)	no
	3	no	–	no	xxxx
'Uneasy'	1	xxxx	xxxx	xxxx	xxxx
	2	yes	0800–1100 hours (P)	no	no
	3	yes	0300–0500 hours (P)	no	xxxx
'Anxious'	1	xxxx	xxxx	xxxx	xxxx
	2	no	–	no	no
	3	xxxx	xxxx	xxxx	xxxx

Table IV. Summarized results on self-ratings of *'positive mood'* variables (xxxx = not measured)

Rating scales	Experiment number	Circadian rhythm?	Trough (T) or peak (P) at	Significant effect of sleep deprivation (decrease –, increase +)	
				1 night	2 nights
'Stimulated'	1	xxxx	xxxx	xxxx	xxxx
	2	no	–	yes (–)	yes (–)
	3	no	–	no	xxxx
'High spirits'	1	no	–	yes (–)	no
	2	no	–	yes (–)	yes (–)
	3	yes ?	0300–0500 hours (T)	no	xxxx
'Elated'	1	xxxx	xxxx	xxxx	xxxx
	2	yes ?	0800–1100 hours (T)	no	no
	3	no	–	no	xxxx

iment 2, where this effect could be tested, it was found that ratings of 'elated' and 'high spirits' were significantly higher after the recovery night than after the baseline night. (And also higher than during the last hours of the vigil.) This change in mood could very well reflect the feeling of satisfaction and liberation when an ordeal is over, but we have the feeling that it is of a 'different' quality. Furthermore, subjects have spontaneously reported that the positive mood persisted a day or so after the recovery day. *Gerner* et al. [24] also report similar findings. They found that 'non-responding' depressed patients and the normal controls who reacted 'worst' to sleep deprivation were positively affected by the recovery night. This is contrary to what has been reported on 'responding' depressed patients [cf. 11].

Our results regarding self-ratings are similar to those found by other experimentors, namely increased fatigue and sleepiness, while scales and items measuring stress, depression and the like show no changes or unclear patterns [cf. 7, 16, 45].

Hormones and Body Temperature (table V)

Urinary excretion of adrenaline showed circadian rhythms in all three experiments with the usual nighttime trough between 0200 and 0500 hours. In experiments 2 and 3 weak effects of sleep deprivation were seen. In noradrenaline no circadian rhythm or effect of sleep deprivation could be traced in any of the experiments. For cortisol a marked circadian rhythm with a peak in the morning and no effect of sleep deprivation was seen (except in experiment 1, where cortisol was not measured). The same very marked peak and lack of sleep deprivation effect was found in melatonin with the peak being somewhat earlier, i.e. 0300–0800 hours. Regarding the thyroid hormones, the vigil significantly increased both T3 and T4. 5 days

Table V. Summarized results on biochemical variables and body temperature (xxxx = not measured)

Variable	Experiment number	Circadian rhythm?	Trough (T) or peak (P) at	Significant effect of sleep deprivation (decrease –, increase +)	
				1 night	2 nights
Adrenaline	1	yes	0200 hours (T)	no	no
	2	yes	0200–0500 hours (T)	yes (+)?	no
	3	yes?	0300–0500 hours (T)	yes (+)?	xxxx
Noradrenaline	1	no	–	no	no
	2	no	–	no	no
	3	no	–	no	xxxx
Cortisol	2	yes	0800–1100 hours (P)	no	no
	3	yes	0600–0800 hours (P)	no	xxxx
Melatonin	2	yes	0500–0800 hours (P)	no	no
	3	yes	0300–0500 hours (P)	no	xxxx
Body temperature	1	yes	0630–0930 hours (T)	no	no
	2	yes	0200–0500 hours (T)	no	no
	3	yes	0300–0500 hours (T)	no	xxxx

after the vigil levels had returned to baseline. The increase in thyroid hormones was interpreted as response to an increased need for energy.

Finally, body temperature showed in all three experiments a regular, marked sinusoidal variation with early morning troughs and virtually no effect of sleep deprivation.

The hormones show, perhaps surprisingly, little effect of sleep deprivation. It seems that sleep deprivation cannot be regarded as a kind of stressor but the increase in thyroid hormones suggests that it at least is a taxing experience.

Comments

If one wants to give a simplified picture of how normal subjects are affected by sleep deprivation it must be that they get tired and want to go to sleep. Their possible mood changes are hidden by this sleepiness.

According to *Wever* [89 b] and *Aschoff* [2] the effects of sleep deprivation on circadian rhythms are that overt rhythms persist, but sometimes with diminished amplitudes. They also state that there is a slight phase delay of the rhythms, e.g. the minimum of body temperature occurs somewhat later in the early morning during sleep deprivation. These effects can to some extent be seen also in our data. It is however difficult to study the effect of sleep deprivation on circadian rhythms because keeping subjects awake during a sufficiently long time without extreme and disturbing measures is methodologically complicated.

Our investigations of how normals react to sleep deprivation give disappointingly few hints to the anti-depressive effect on depressed patients. Firstly because the patients seem to differ in some important aspect from normal subjects regarding their sensitivity to sleep deprivation. One theory is that depressed patients have disordered or desynchronized circadian rhythms and that sleep deprivation acts as some kind of synchronizer [52, 55]. *Wehr* et al. [85 b] demonstrate in an elegant way that an important factor behind depressive illness is that certain rhythms, e.g. the rhythm of REM sleep, might be phase-advanced. By advancing the sleep period 6 h a clear anti-depressive effect was found, in spite of no sleep deprivation occurring during the experimental period.

Secondly, the data on depressions have been collected under different and often not very controlled experimental conditions. Our experiments have had rather special, controlled designs that are perhaps difficult to achieve with depressed patients. We would nevertheless like to suggest that the constant condition type of experiments are tried on depressed patients. As circadian rhythms undoubted play an important role both in the pathology and therapy of depression, such detailed studies would perhaps provide valuable hints. Comparisons could be made with constant condition studies of normals in order to trace the specific differences between the rhythms of depressives and normals.

In the literature on sleep deprivation and depression it is often described that patients sometimes experience a 'critical period' during the night and early morning [65] and a rather sudden improvement in connection with that. It would therefore be interesting to follow continuously some variables, e.g. body temperature and hormones in plasma, during these critical hours in order to find some physiological correlate of this sudden change of mood.

Finally, we know of no experiment with depressives where the vigil has been extended over more than 1 night without sleep. 2, or perhaps 3, consecutive nights without sleep would give the experimentor possibility to follow more than one circadian cycle. This would also make it possible to follow an eventual 'circadian rhythm of the therapeutic effect': in cases where the maximal effect is reached during the early morning after a night without sleep, what would happen if the vigil is extended over the next night?

Conclusion

Circadian rhythms of normal subjects persist remarkably well during sleep deprivation. Slight phase delays and effects on amplitudes are sometimes noted. The main effects are however changes that can be attributed to a marked increase of sleepiness. Inferences of the effects of sleep deprivation on circadian rhythms demand very strict control of the environment and the experimental setting.

Circadian Rhythm and Sleep Deprivation in Depression

T.W. Uhde, R.M. Post, Bethesda, Md.; *J.C. Ballenger,* Charlottesville, Va.; *N.R. Cutler, D.C. Jimerson,* Bethesda, Md.; *E.D. Weitzman,* Bronx, N.Y.; *W.E. Bunney, Jr.,* Bethesda, Md., USA

Introduction

The occurrence of an acute antidepressant response to 1 night's SD in medication-free depressed patients and an equally robust relapse following 1 night's recovery sleep provide a unique paradigm for examining mechanisms underlying affective illness and its amelioration. We have found several clinical and biological markers which are associated with increased antidepressant response to SD. The SD response is more likely to occur in patients with: higher baseline ratings of depression [60]; classical endogenous illness; a sleep structure typical of endogenous depression [18]; lower atypicality ratings of depression [73]; a reducing profile of average evoked response to somatosensory threshold testing [8]; lower baseline CSF HVA and higher CSF MHPG [24]. Diurnal variations in temperature rhythm remain normal in depressed patients who respond to SD, but nonresponders show a blunting of the temperature rhythm.

We restrict this report to our preliminary findings of the study of two bipolar II patients previously known to be excellent sleep deprivation responders, and two normal volunteers. The purpose of this ongoing research is twofold: (1) to determine the clinical efficacy of repeated sleep deprivation in medication-free bipolar patients, and (2) to document the secretory pattern of cortisol, growth hormone, and 3-methoxy-4-hydroxy-phenylethyleneglycol (MHPG) during the sleep-wake cycle and during and following SD.

---

[4] For references see compound list on page 34.

Subjects and Method

2 patients had Research Diagnostic Criteria for Major Depressive Disorder, bipolar type II. *Patient 161* was a 45-year-old woman with a 15-year history of major mood swings rapidly cycling on a weekly to monthly basis. Depressive phases were characterized by either severe psychomotor retardation or agitation. She had profound hypophagia, weight loss, and constipation. During depressive periods she perceived herself as helpless, hopeless, and worthless. She had frequent suicidal thoughts. *Patient 164* was a 50-year-old woman with a 5-year history of major mood swings occurring every 1–2 months. Unlike patient 161, this woman lacked any significant neurovegetative signs. Although she had a disordered sleep structure, her appetite, weight control, and motor behavior were always normal. Her major symptoms were cognitive in nature. She complained of an inability to concentrate and 'make decisions', especially in the morning hours. She had intense feelings of self-reproach and reported decreased interest in social contacts. She did, however, continue to perform skills of daily living and maintain the quality of her interpersonal relationship. Both patients had hypomanic episodes characterized by flight of ideas, grandiose ideation, social intrusiveness and sexual provocativeness.

Normal volunteers were 2 male college students, 20 and 21 years of age. Both patients were sleep deprived for 1 night on at least three separate occasions, as previously described [24]. Patients 161 and 164 had SD on nights 1, 9, 12, and 1, 8, 10, 13, respectively.

Patients and volunteers had an indwelling venous catheter inserted and 20-min continuous blood sampling for cortisol, growth hormone (HGH), and MHPG was obtained before, during, and after the first SD period. Samples were obtained, therefore, throughout the sleep-wake cycle, during SD, and during the period of mood improvement following SD. In addition, we also obtained samples through a spontaneous switch into hypomania in patient 161.

Cortisol and HGH were measured by radioimmunoassay and MHPG was determined by a modified mass spectrometric technique [33].

Results

Repeated Sleep Deprivation. Both patients had dramatic mood improvements following the first night of SD. Nurse-rated scores of depression decreased from baseline means of $12.5 \pm$ SEM 0.3 and 8.8 ± 0.4 to 2.3 ± 0.3 and 3.2 ± 0.7 following SD in patients 161 and 164, respectively. Decreases in self-rated measures of depression were equally robust. In both patients repeated SD resulted in tolerance to and delay of the antidepressant effects following SD. In patient 164 later attempts to maintain or capture the SD response by combining SD with a 6-h phase advance or a 6-h phase delay were unsuccessful. In addition, repeated SD prior to the expected switch into the depressive phase was also unsuccessful in patient 161.

Endocrine and MHPG Rhythms. Both patients demonstrated baseline cortisol hypersecretion, complete loss or blunting of cortisol diurnal rhythm, alterations in sleep-related HGH secretion, and failure to restore normal en-

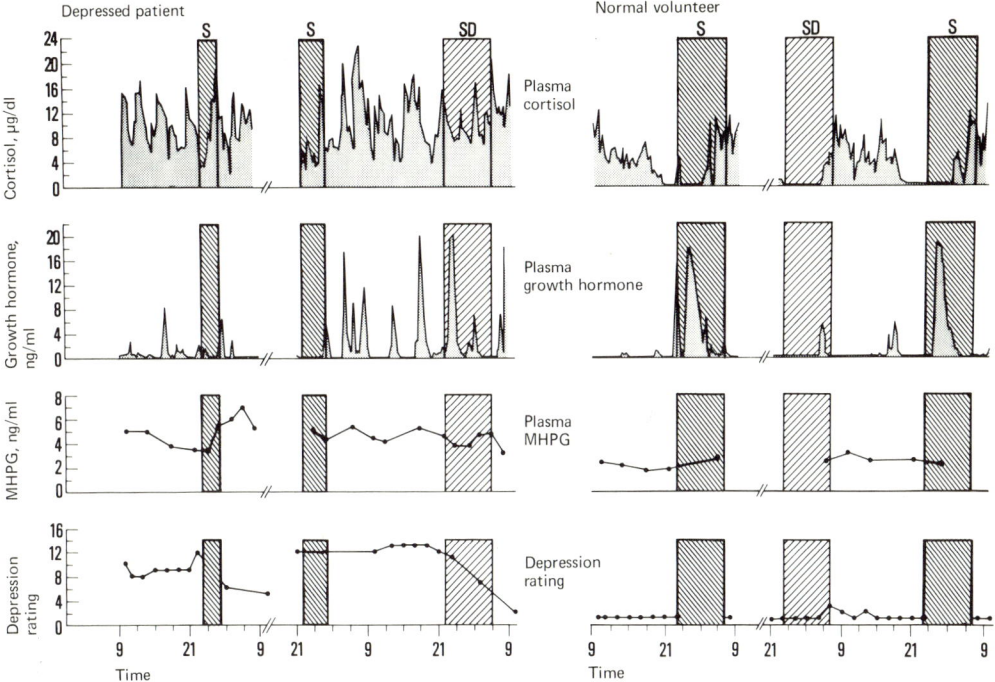

Fig. 8. Effects of sleep (S), spontaneous mood improvement (a.m. following first night sleep), and sleep deprivation (SD)-induced mood improvement on cortisol hypersecretion, growth hormone dysrhythmia, and plasma MHPG in a bipolar II depressed patient.

docrine patterns following SD. These findings were most prominent in patient 161, who had a more characteristic endogenomorphic depression and higher baseline depression ratings than patient 164, and are presented in figure 8. As illustrated, the marked clinical response to SD was not associated with a reduction in cortisol hypersecretion, and multiple peaks of HGH occurred throughout the day in contrast to normal sleep-related secretion. Also, plasma-free MHPG increased during a spontaneous mood switch at 0200 hours the first night, but not during SD-induced mood improvement.

Discussion

The antidepressant response of medication-free patients to a single SD is typically transient, usually lasting 1 day. Despite initial dramatic responses, repeated SD was an ineffective treatment in our 2 bipolar patients resulting in the delay of and tolerance to the mood improvement. More sustained antidepressant effects to SD have been reported in patients concurrently receiving antidepressant medication. Further studies are required to determine whether the combination of antidepressant pharmacotherapy and SD may provide special treatment advantages, i.e. more rapid drug response and/or capture of the SD response.

The dramatic 'off-sets' of depression with SD and 'onsets' following sleep suggest that mood alterations might be associated with sleep-related neurotransmitters or peptides which are inhibited by SD and stimulated by sleep. Our preliminary data suggest that disordered cortisol (hypersecretion and blunting of diurnal rhythm) and growth hormone patterns (multiple secretory episodes and uncoupling from sleep-related secretion) in depression are not immediately normalized by SD and that studies of other systems are indicated.

Sleep 1980. 5th Eur. Congr. Sleep Res., Amsterdam 1980, pp. 26–33
(Karger, Basel 1981)

Internal Coincidence Model for Sleep Deprivation and Depression [5]

Thomas A. Wehr, Bethesda, Md., USA;
Anna Wirz-Justice [6], Basel, Switzerland

Sleep deprivation can lead to a rapid, albeit short-term amelioration of depressive symptomatology [51, 63; for review, see *van den Hoofdakker,* this symposium]. Rebound sleep (even very short naps) after therapeutic sleep deprivation results in immediate return of depressive symptoms. Partial sleep deprivation in the second half of the night is as efficacious as total sleep deprivation [56, 69]. These observations, which have been replicated in

---

[5] For references see compound list on page 34.
[6] Visiting Fellow of the Swiss Foundation for Biomedical Research.

many independent studies, led us to carry out an experiment whereby sleep per se was not deprived, but sleep was shifted to an earlier period in the day: this phase advance of the sleep period induced clinical remission (fig. 11) [85b]. These various manipulations of the sleep-wake cycle in depressive patients suggest that sleep may interact with a sensitive phase of the circadian system to produce and sustain depression, and that being awake at this sensitive phase, in the second half of the night [63], induces remission [87, 88]. This paper presents an internal coincidence model for sleep and depression based on a two circadian oscillator model of affective illness, that may be of heuristic value in understanding therapeutic sleep deprivation.

A Circadian Rhythm Phase-Advance Hypothesis of Depression

The formal properties of the human circadian system have been described in terms of a strong oscillator driving the circadian rhythms in, for example, rapid-eye-movement (REM) sleep propensity, body temperature, and cortisol secretion, that is coupled to a weak oscillator driving the sleep-wake cycle [89b]. Under normal entrained conditions there is a stable phase relationship between the two oscillators. This is schematically illustrated in figure 9A. The normal phase relationships between the two circadian oscillators and their overt rhythms can be temporarily disturbed by phase-shift experiments and rapid transmeridian travel. A similar shift occurs when normal subjects are isolated from time cues [17, 89b, 98]. Both delay phase shifts or jet lag (from east to west) and free-running conditions result in a relative phase-advance of the strong oscillator that controls REM sleep, temperature and cortisol with respect to the sleep-wake cycle (fig. 9B). This means that sleep onset, which normally occurs some hours after the temperature maximum, is shifted later until it almost coincides with the temperature minimum. There is an associated change in sleep architecture, in particular the temporal distribution of REM sleep: REM latency is shorter, first REM episodes are longer, early REM% is increased [17, 98].

Depressive patients, in general, also have short REM latency, long first REM episodes, more REM sleep in the first third of the night [25, 27, 29, 31, 39, 76, 81]. These changes can be interpreted in terms of an abnormal temporal distribution of REM sleep analogous to that described occurring in normals sleeping late [81] or free-running (fig. 9B) [17, 98], and could result from a phase-advance of the circadian rhythm in REM sleep propensity,

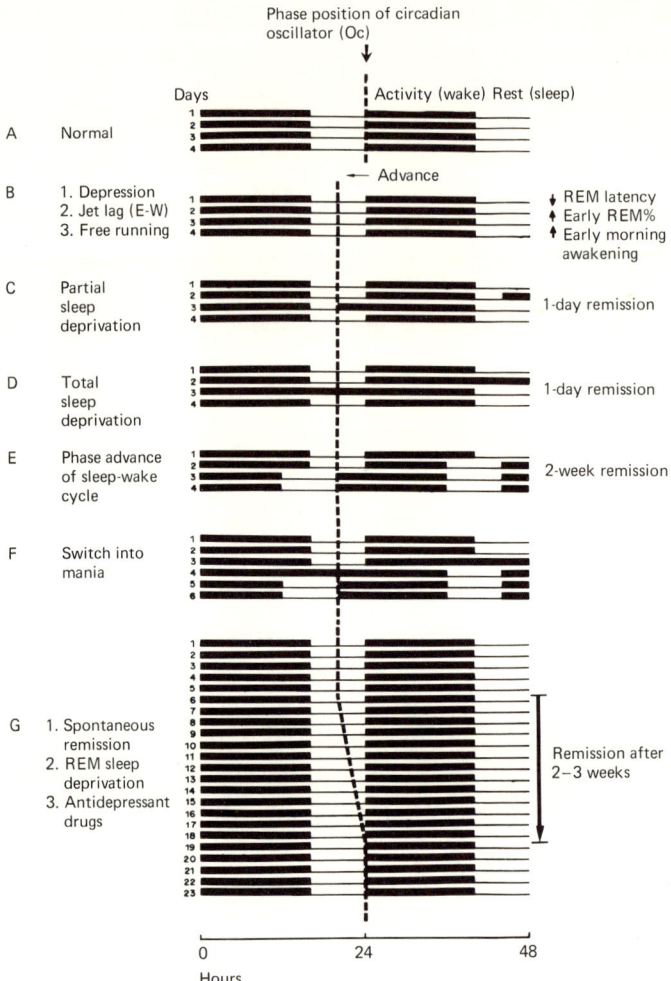

Fig. 9. Schematic representation of a two circadian oscillator model of depression. The activity (wake)-rest (sleep) cycle is delineated by a conventional double plot of black bars representing activity, and thin black lines representing rest for each 24-h day. The strong circadian oscillator (Oc) that drives the rhythms of REM sleep propensity, temperature, and cortisol is delineated by a vertical dotted line. This represents the phase position (e.g. the time of maximum propensity for REM sleep) of this oscillator from day to day with respect to the sleep-wake cycle. The examples A–G are hypothetical formulations of the relative phase position of these two circadian oscillators under different conditions, in particular manipulations of the sleep-wake cycle, and are explained in detail in the text.

such that its maximum occurs near the beginning instead of near the end of sleep [50, 76, 81, 85b]. This analogy is more explicitly detailed in figure 10.

Since the circadian rhythm in REM sleep propensity appears to be closely coupled to that of body temperature and cortisol secretion, evidence for a phase advance in the latter two rhythms would tend to support the circadian phase-advance interpretation of the REM sleep abnormalities. Some (but not all) studies of temperature and cortisol rhythms, as well as

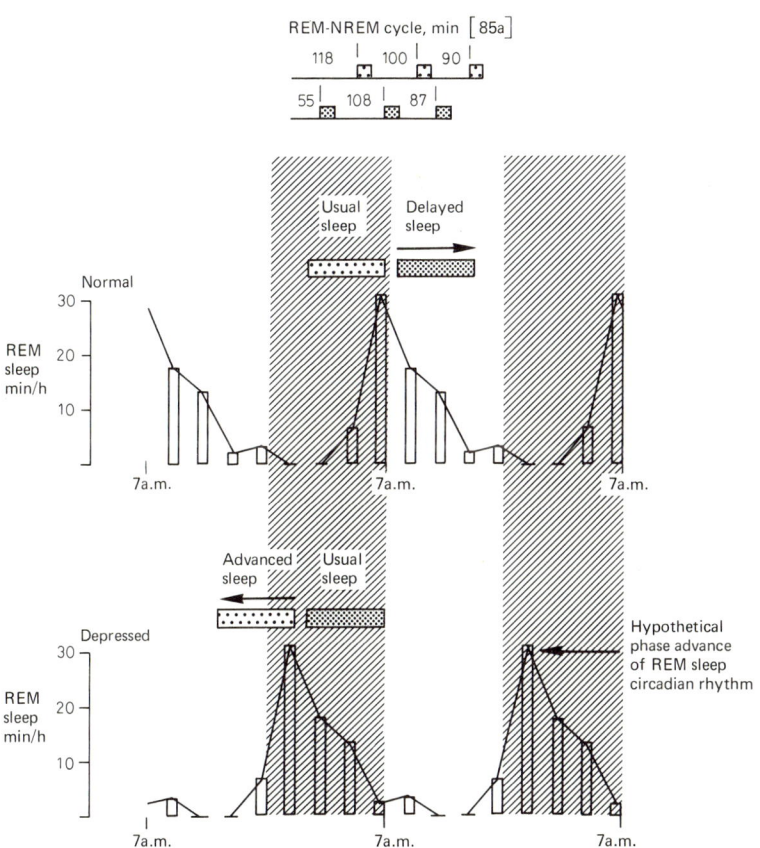

Fig. 10. Schematic representation of the hypothetical phase advance of REM sleep propensity with respect to the sleep-wake cycle in depression. The black sleep periods result in sleep EEG patterns as typified in the upper REM-NREM cycle: the sleep periods marked with a cross result in sleep EEG patterns as typified in the lower REM-NREM cycle. Data are taken from refs [85a, 89a].

many studies of neurotransmitter metabolites and other parameters, indicate that the phase-position of circadian rhythms is abnormally advanced in depressives [all studies reviewed in 86, 88].

Manipulation of the Sleep-Wake Cycle in Depression

A possible causal relationship between an abnormally early phase position of a number of circadian rhythms and psychopathology can be tested in a number of ways. In figure 9C–E the three manipulations of the sleep-wake cycle known to have antidepressant effects are expressed in terms of the phase relationships of the two circadian oscillators. Partial sleep deprivation in the second half of the night can be considered as 'advancing wake' to a 'normal' phase-position with respect to the strong oscillator. Total sleep deprivation can also be considered in this light (although here deprivation of sleep per se cannot be distinguished from phase shifting): both usually result in only a single day's remission. Phase advance of the sleep-wake cycle also advances 'wake' to a 'normal' phase position with respect to the strong oscillator, but here there is no deprivation of sleep. Thus, phase advance of the sleep period can be continued on subsequent nights without deleterious effects, and in this case remission lasts about 2 weeks. This is the time required after 'jet lag' for the strong oscillator to gradually adjust to the shifted sleep schedule (and presumably reestablish the preexisting depressive phase disturbance) (fig. 11).

A common feature of these three procedures that may be related to their clinical efficacy is that they shift 'wake' to an earlier time of day. If we consider that a critical phase of the REM sleep-temperature-cortisol circadian rhythm which is normally associated with the first hours of waking, is advanced into the last hours of sleep (fig. 12), then we can hypothesize that depression occurs in susceptible individuals when sleep interacts with this sensitive phase, and that 'wake' causes a temporary remission.

Besides the therapeutic efficacy of partial sleep deprivation in the second half of the night, there are other clinical observations that support the idea that this particular consequence of the depressive phase advance is pathogenic. First, switches into and out of depression tend to occur most often during this critical circadian phase [10, 12, 27, 31, 50, 64, 74, 75]. Second, depression is most severe just after and least severe just before this sensitive phase each day (the classical pattern of diurnal variation in depressive mood) [21, 28, 30, 44, 83, 84].

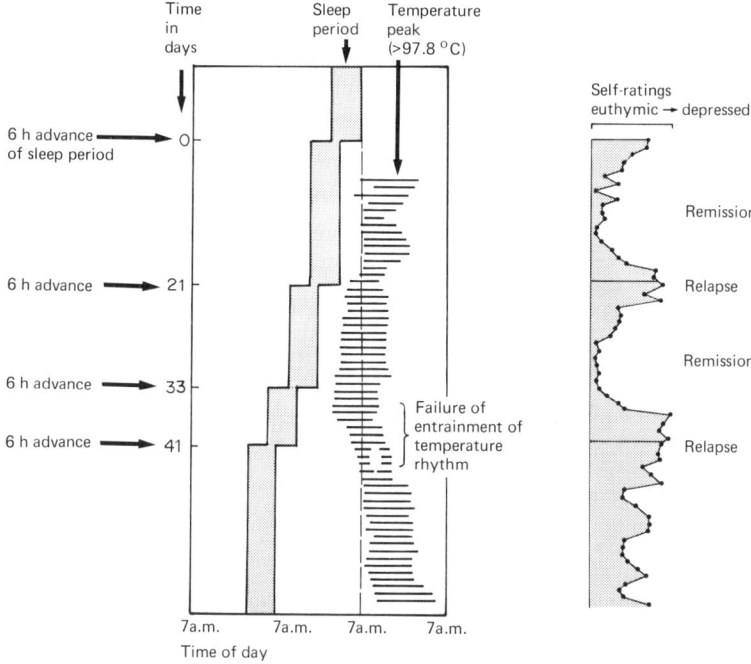

Fig. 11. Phase advance of the sleep-wake cycle in a manic-depressive woman. The timing of the sleep period and the four sequential phase advances are shown together in a double plot with the circadian temperature rhythm: the time of day when the temperature exceeded 97.8°F are indicated by one horizontal bar for each day. The right-hand diagram plots the daily self-rating scores (100 mm line) as the patient moves from depression to euthymic states with subsequent relapse and remission. It can be seen that the first two phase advances were successful in inducing a 2-week remission: the third phase advance was undertaken before relapse occurred in the hope of a prophylaxis, but relapse did take place with a concomitant failure of entrainment of the circadian temperature rhythm.

This concept of an internal coincidence of a critical circadian phase and sleep or wake is analogous to the internal coincidence model proposed to underlie seasonal switches in animal behaviour [57, 58], in which critical circadian phase-relationship is established between internal oscillators.

Figure 9F represents the sleep-wake cycle of depressive patients when they spontaneously switch out of depression into mania. At these times we found patients typically experience one or more 48-hour sleep-wake cycles and a subsequent advance of the time of waking. We have documented such patterns of total insomnia in a large number of patients [85b, 87], and it is no new observation that manic patients appear to require very little or even

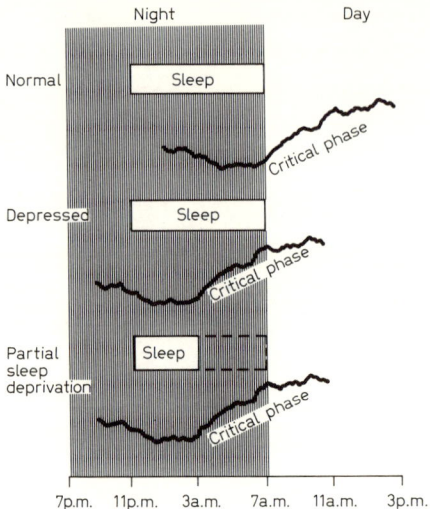

Fig. 12. Diagram of the internal coincidence model. In the normal case, sleep occurs towards the minimum of the circadian temperature rhythm, and the postulated 'critical phase' occurs during wake. In the phase-advance model of depression, this critical phase has moved into the last hours of sleep, i.e. depressive patients go to sleep on the ascending limb of the temperature rhythm. Partial sleep deprivation may act by imposing 'wake' on this critical phase.

no sleep. In some of these patients an imposed total sleep deprivation could also induce switches out of depression (usually transient). There is an obvious similarity between the 48-h sleep-wake cycles associated with spontaneous switches out of depression into mania (fig. 9F) and the total sleep deprivation treatment that can induce switches out of depression (fig. 9D). The clinical effect implies that the spontaneous total insomnia may be causally important in the process responsible for the switch, and may also depend on the interaction of a sleep-sensitive switch mechanism with a critical early-morning circadian phase.

Manipulation of Other Circadian Rhythms in Depression

Advancing the time of the sleep period relative to the REM sleep-temperature-cortisol circadian rhythm is one of two possible ways of correcting the depressive phase disorder. The other approach would be to delay the phase position of the REM sleep-temperature-cortisol circadian rhythm relative to the sleep period (fig. 9G). Since these rhythms are driven by the strong oscillator, it will therefore be less easily phase-shifted than the controlling oscillator of the sleep-wake cycle (this difference is the basis of jet lag). As with jet lag, correction of the phase-position of these rhythms might require 1–2 weeks. The process of spontaneous remission could be considered as a slow reestablishment (through whatever mechanism) of 'normal' phase relationships between the two oscillators. Selective REM sleep deprivation [81] has been reported to be an effective antidepressant but with a slow onset of therapeutic effect similar to that found with thymoleptic drugs. In responders, REM sleep deprivation tends to correct their abnormally early temporal distribution of REM sleep. Whether this effect occurs via a phase delay of the REM sleep oscillator, as schematically represented, is conjectural. The third long-term antidepressant modality is chronic treatment with drugs such as monoamine-oxidase inhibitors, tricyclics, and lithium. Preliminary animal experiments indicate that these drugs can delay the phase position of various circadian rhythms, in particular of those neurotransmitter receptors which have been the focus of investigations of depressive disorders and their treatment [36, 47, 90, 91]. Total sleep deprivation has no such effect on phase position [95]. Furthermore, these drugs may delay phase position by lengthening the intrinsic period of the driving oscillator [32, 34, 38, 91–93]. These results are in the direction required of drugs used to treat depression by a phase-advance hypothesis of the illness. The issue of specificity and selectivity of these effects on the circadian system remains to be addressed.

This circadian model of affective illness attempts to bring together clinical observations and the present knowledge of the normal physiology of the human circadian system [86, 88]. The phase-advance hypothesis of depression can integrate the classical symptoms of endogenous depression, such as early morning awakening and diurnal variation in mood, epidemiological phenomena such as seasonality and cyclicity of the illness, and disturbances in REM sleep and endocrine function. It also provides a model for understanding the effects of sleep deprivation and chronic antidepressant drugs [88]. It thus explicitly points to clinical experiments to test the hypothesis as well as new treatment modalities based on direct manipulation of the circadian system.

List of References of Symposium A

1 Ackenheil, M.; et al. Catecholamine response to short-term stress in schizophrenic and depressive patients, in Usdin, Kopin, Barchas, Catecholamines: basic and clinical frontiers. Proc. 4th Catecholamine Symp., pp. 1937–1940 (Pergamon Press, Oxford 1979).
2 Aschoff, J.: Features of circadian rhythms relevant for the design of shift schedules. Ergonomics 21: 739–754 (1978).
3 Åkerstedt, T.; Fröberg, J.E.: Sleep and stressor exposure in relation to circadian rhythms in catecholamine excretion. Biol. Psychol. 8: 69–80 (1979).
4 Åkerstedt, T.; Fröberg, J.E.; Friberg, Y.; Wetterberg, L.: Melatonin excretion, body temperature and subjective arousal during 64 hours of sleep deprivation. Psychoendocrinology 4: 219–225 (1979).
5 Bemmel, A.L. van; Hoofdakker, R.H. van den: Maintenance of therapeutic effects of total sleep deprivation by means of limitation of subsequent sleep. A pilot study. Acta psychiat. scand. (in press).
6 Bhanji, S.; Roy, G.A.: The treatment of psychotic depression by sleep deprivation. A replication study. Brit. J. Psychiat. 127: 222–226 (1975).
7 Bohlin, G.; Kjellberg, A.: Self reported arousal during sleep deprivation and its relation to performance and physiological variables. Scand. J. Psychol. 14: 78–86 (1973).
8 Buchsbaum, M.S.; Gerner, R.H.; Post, R.M.: The effects of sleep deprivation on averaged evoked response in depressed patients and normals. Biol. Psychiat. (in press).
9 Buddeberg, C.; Dittrich, A.: Psychologische Aspekte des Schlafentzugs. Arch. Psychiat. Nervenkrankh. 225: 249–261 (1978).
10 Bunney, W.E., Jr.; Murphy, D.L.; Goodwin, F.K.: The switch process in manic-depressive illness. Archs gen. Psychiat. 27: 295–302 (1972).
11 Burg, W. van den; Hoofdakker, R.H. van den: Total sleep deprivation on endogenous depression. Archs gen. Psychiat. 32: 1121–1125 (1975).
12 Burton, R.: The anatomy of melancholy, vol. 1, p. 171 (Dutton, New York 1961).
13 CIPS Internationale Skalen for Psychiatrie, Berlin (1977).
14 Cole, M.G.; Müller, H.F.: Sleep deprivation in the treatment of elderly depressed patients. J. Am. Geriat. Soc. 24: 308–313 (1976).
15 Conroy, R.T.W.L.; Mills, J.N.: Human circadian rhythms (Churchill, London 1970).
16 Cutler, N.R.; Cohen, H.B.: The effect of one night's sleep loss on mood and memory in normal subjects. Compreh. Psychiat. 20: 61–66 (1979).
17 Czeisler, C.A.; Zimmerman, J.C.; Ronda, J.M.; Moore-Ede, M.C.; Weitzman, E.D.: Timing of REM sleep is coupled to the circadian rhythm of body temperature in man. Sleep 2: 329–346 (1980).
18 Duncan, W.C.; Gillin, J.C.; Post, R.M.; Gerner, R.H.; Wehr, T.: Relationship between EEG patterns and clinical improvement in depressed patients treated with sleep deprivation. Biol. Psychiat. (in press).
19 Elsenga, S.; Hoofdakker, R.H. van den: Sleep deprivation and clomipramine in endogenous depression; in Popoviciv, Asgian, Badiv, Sleep 1978, pp. 625–629 (Karger, Basel 1980).
20 Elsenga, S.; Hoofdakker, R.H. van den: Clinical effects of sleep deprivation and clomipramine in endogenous depression (submitted for publication).

21 Fleck, U.; Kraepelin, E.: Über die Tagesschwankungen bei Manisch-Depressiven. Kraep. Psychol. Arb. 7: 213–253 (1922).

22 Fröberg, J.E.; Karlsson, C.-G.; Levi L.; Lidberg, L.: Psychobiological circadian rhythms during a 72-hour vigil. Försvarsmedicin 11: 192–201 (1975).

23 Fröberg, J.E.; Karlsson, C.-G.; Levi, L.; Lidberg, L.: Circadian rhythms of catecholamine excretion, shooting range performance and self-ratings of fatigue during sleep deprivation. Biol. Psychol. 2: 175–188 (1975).

24 Gerner, R.H.; Post, R.M.; Gillin, J.C.; Bunney, W.E. Jr.: Biological and behavioral effects of one night's sleep deprivation in depressed patients and normals. J. psychiat. Res. 15: 21–40 (1979).

25 Gillin, J.C.; Mazure, C.; Post, R.M.; Jimerson, D.; Bunney, W.E. Jr.: An EEG sleep study of a bipolar (manic-depressive) patient with a nocturnal switch process. Biol. Psychiat. 12: 711–718 (1977)

26 Gillin, J.C.: The relationship between changes in REM sleep and clinical improvement in depressed patients, treated with amitriptyline. Psychopharmacology, Berlin. 59: 267–272 (1978)

27 Gillin, J.C.; Duncan, W.; Pettigrew, K.D.; Frankel, B.L.; Snyder, F.: Successful separation of depressed, normal and insomniac subjects by sleep EEG data. Archs gen. Psychiat. 36: 85–90 (1979).

28 Graw, P.; Hole, G.; Gastpar, M.: Tagesschwankungen bei hospitalisierten depressiven Patienten während der Depression und in gesunden Zeiten. Arch. Psychiat. Nervenkrankh. 228: 329–339 (1980).

29 Gresham, S.C.; Agnew, W.F.; Williams, R.L.: The sleep of depressed patients. Archs gen. Psychiat. 13: 503–507 (1965)

30 Hall, P.; Spear, F.G.; Stirland, D.: Diurnal variation of subjective mood in depressive states. Q. Psychiat. 38: 529–536 (1964).

31 Hartmann, E.: Longitudinal studies of sleep and dream patterns in manic-depressive patients. Archs gen. Psychiat. 19: 312–329 (1968).

32 Hofmann, K.; Gunderoth-Palmowski, M.; Wiedenmann, G.; Engelmann, W.: Further evidence for period lengthening effect of Li$^+$ on circadian rhythms. Z. Naturf. 33c: 231–234 (1978).

33 Jimerson, D.C.; Markey, S.P.; Oliver, J.O.; Kopin, I.J.: Simultaneous measurement of plasma 3-methoxy-4-hydroxyphenylglycol (MHPG) and 3,4 dihydroxyphenylglycol (DHPG) by gas chromatography-mass spectrometry (GCMS) (submitted for publication).

34 Johnsson, A.; Engelmann, W.; Pflug, B.; Klemke, W.: Influence of lithium ions on human circadian rhythms. Z. Naturf. 35c: 503–507 (1980).

35 Jones, M.T.; et al.: Secretion of corticotropin releasing hormone in vitro. Frontiers in neuroendocrinology, vol. 4 (Raven Press, New York, 1976).

36 Kafka, M.S.; Wirz-Justice, A.; Naber, D.; Wehr, T.A.: Circadian acetylcholine receptor rhythm in rat brain and its modification by imipramine. Neuropharmacology (in press, 1981).

37 Kollar, E.J. et al.: Stress in subjects undergoing sleep deprivation. Psychosom. Med. 28: 101–113 (1966).

38 Kripke, D.F.; Whyborney, V.G.: Lithium slows rat circadian activity rhythms. Life Sci. 26: 1319–1321 (1980).

39   Kupfer, D.J.: REM latency: a psychobiologic marker for primary depressive disease. Biol. Psychiat. *11:* 159–174 (1977).
40   Larsen, J.K.; Lindberg, M.L.; Skovgaard, B.: Sleep deprivation as treatment for endogenous depression. Acta psychiat. scand. *54:* 167–173 (1976).
41   Loosen, P.T.; Merkel, V.; Amelung, V.: Combined sleep deprivation and clomipramine in primary depression. Lancet *ii:* 156–157 (1976).
42   Maier-Ewert, K.; et al.: Drei narkoleptische Syndrome. Nervenarzt *46:* 624 (1975).
43   Matussek, N.; et al.: MHPG excretion during sleep deprivation in endogenous depression. Neuropsychobiology *3:* 23–29 (1977).
44   Middlehoff, H.D.: Tagesrhythmische Schwankungen bei endogenen Depressionen im symptomfreien Intervall und während der Phase. Arch. Psychiat. Nervenkrankh. *209:* 315–339 (1967).
45   Murray, E.J.; Williams, H.L.; Lubin, A: Body temperature and psychological ratings during sleep deprivation. J. exp. Psychol. *56:* 271–273 (1958).
46   Müller, T.: Radioenzymatische Simultanbestimmung von Adrenalin und Noradrenalin im Plasma. Arzneimittel-Forsch. *28:* 1304 (1978).
47   Naber, D.; Wirz-Justice, A.; Kafka, M.S.; Wehr, T.A.: Dopamine receptor binding in rat striatum: ultradian rhythm and its modification by chronic imipramine. Psychopharmacology, Berlin *68:* 1–5 (1980)
48   Naitoh, P.; et al.: Psychophysiological changes after prolonged deprivation of sleep. Biol. Psychiat. *3:* 309–320 (1971)
49   Palmblad, J.; Akerstedt, T.; Fröberg, J.; Melander, A.; Schenk, H. von: Thyroid and adrenomedullary reactions during sleep deprivation. Acta endor., Copenh. *90:* 233-239 (1979).
50   Papousek, M.: Chronobiologische Aspekte der Zyklothymie. Fortschr. Neurol. Psychiat. *43:* 381–440 (1975).
51   Pflug, B.; Tölle, R.: Disturbance of the 24-hour rhythm in endogenous depression and the treatment of endogenous depression by sleep deprivation. Int. Pharmacopsychiat. *6:* 187–196 (1971).
52   Pflug, B.; Tölle, R.: Therapie endogener Depressionen durch Schlafentzug – praktische und theoretische Konsequenzen. Nervenarzt *42:* 117–124 (1971).
53   Pflug, B.; Tölle, R.: Die Wirkung des Schlafentzugs auf die Symptomatik der endogenen Depressionen; in Jovanovic, The nature of sleep, pp. 177–180 (Fischer, Stuttgart 1973).
54   Pflug, B.; et al.: Depression and daily temperature. A long-term study. Acta psychiat. scand. *54:* 254–266 (1976).
55   Pflug, B.: The effect of sleep deprivation on depressed patients. Acta psychiat. scand. *53:* 148–158 (1976).
56   Philipp, M.: Depressionsverlauf nach Schlafentzug. Nervenarzt *49:* 120–123 (1978).
57   Pittendrigh, C.S.: Circadian surfaces and the diversity of possible roles of circadian organization in photoperiodic induction. Proc. natn. Acad. Sci. USA *69:* 2734–2737 (1972).
58   Pittendrigh, C.S.; Daan, S.: A functional analysis of circadian pacemakers in nocturnal rodents. V. Pacemaker structure: a clock for all seasons. J. comp. Physiol. *106:* 333–355 (1976).
59   Platman, S.K.; Fieve, R.: Sleep in depression and mania. Br. J. Psychiat. *116:* 219 (1970).
60   Post, R.M.; Kotin, J.; Goodwin, F.K.: Effects of sleep deprivation on mood and central amine metabolism in depressed patients. Archs gen. Psychiat. *33:* 627–632 (1976).

List of References of Symposium A

61 Rechtschaffen, A.; Kales, A.: A manual of standardized terminology techniques and scoring system for sleep stages of human subjects. Publ. No. 204 (Public Health Service, US Government Printing Office, Washington (1968).
62 Rudolf, G.A.E.; Tölle, R.: The course of the night with total sleep deprivation as antidepressant therapy. Waking Sleeping 2: 83–91 (1978).
63 Rudolf, G.A.E.; Schilgen, B.; Tölle, R.: Antidepressive Behandlung mittels Schlafentzug. Nervenarzt 48: 1–11 (1977).
64 Rudolf, G.A.E.; Tölle, R.: Circadian rhythm of circulatory functions in depressives and on sleep deprivation. Int. Pharmacopsychiat. 12: 174–183 (1977).
65 Rudolf, G.A.E.; Tölle, R.: Sleep deprivation and circadian rhythm in depression. Psychiat. Clin. 11: 198–212 (1978).
66 Rüther, E.; Zander, K.J.; Nedopil, N.: Telemetric investigations in psychiatric research; in Klewe, Kimmich, Biotelemetry, vol. IV, pp. 235–238 (Karger, Basel 1978).
67 Sachar, E.J.; et al.: Disrupted 24-hour patterns of cortisol secretion in psychotic depression. Archs gen. Psychiat. 28: 19–24 (1973).
68 Scheyen, J.D. van: Slaapdeprivatie bij de behandeling van unipolaire (endogene) vitale depressies. Ned. Tijdschr. Geneesk. 14: 564–568 (1977).
69 Schilgen, B., Tölle, R.: Partial sleep deprivation as therapy for depression. Archs, gen. Psychiat. 37: 267–271 (1980).
70 Schmocker, M.; Baumann, P.; Reyerd, F.; Heimann, H.: Der Schlafentzug. Eine klinische, psychophysiologische und biochemische Untersuchung. Arch. Psychiat. Nervenkrankh. 221: 111–122 (1975).
71 Schulte, W.: Kombinierte Psycho- und Pharmacotherapie bei Melancholikern; in Kranz, Petrilliwitsch, Probleme pharmacopsychiatrischer Kombination und Langzeit Behandlungen, pp. 150–169 (Karger, Basel 1966).
72 Schulte, W.: Zum Problem der Provokation und Kupierung von Melancholische Phasen: Schweizer Arch. Neurol. Neurochir. Psychiat. 109: 427–435 (1971).
73 Silberman, E.K.; Post, R.M.: Atypicality in primary depressive illness. Proc. Annu. Meet. of the Society of Biological Psychiatry, Boston 1980, abstr. No. 51, p. 82.
74 Sitaram, N.; Gillin, J.C.; Bunney, W.E., Jr.: Circadian variation in the time of 'switch' of a patient with 48-hour manic-depressive cycles. Biol. Psychiat. 13: 567–574 (1978).
75 Sitaram, N.; Gillin, J.C.; Bunney, W.E., Jr.: The switch process in manic-depressive illness. Acta psychiat. scand. 58: 267–278 (1978).
76 Snyder, F.; comment in: Weitzman, E.D.; Goldmacher, D.; Kripke, D.F.; MacGregor, P.; Kream, J.; Hellman, L.: Reversal of sleep-waking cycle: effect of sleep stage pattern and certain neuroendocrine rhythms. Trans. Am. Neurol. Ass. 93: 153–157 (1968).
77 Spitzer, R.L.; Endicott, J.; Robins, E.: Research diagnostic criteria for a selected group of functional disorders (Biometric Research, N.Y. Psych. Inst., New York 1975).
78 Spitzer, R.L.; Endicott, J.; Robins, E.: Clinical criteria for psychiatric diagnosis and DSM III. Am. J. Psychiat. 32: 1186–1192 (1975).
79 Svendsen, K.: Sleep deprivation therapy in depression. Acta psychiat. scand. 54: 184–192 (1976).
80 Vogel, G.W.; et al. REM deprivation. II. The effect on depressed patients. Archs gen. Psychiat. 18: 301–311 (1968).
81 Vogel, G.W.; Vogel, F.; McAbee, R.S.; Thurmond, A.J.: Improvement of depression by REM sleep deprivation: new findings and a theory. Archs gen. Psychiat. 37: 247–253 (1980).

82 Vosz, A.; Kind, H.: Ambulante Behandlung endogener Depression durch Schlafentzug. Schweiz. Rdsch. Med. *63:* 564–565 (1974).
83 Waldmann, H.: Die Tagesschwankung in der Depression als rhythmisches Phänomen. Fortschr. Neurol. Psychiat. *40:* 83–104 (1972).
84 Waldmann, H.: Schlafdauer und psychopathologische Tagesrhythmik bei Depressiven; in Jovanovic. Die Natur des Schlafes; pp. 184–187 (Fischer, Stuttgart 1973).
85a Webb, W.B.; Agnew, H.W., Jr.: Sleep and waking in a time-free environment. Aerospace Med. *45:* 617–622 (1974).
85b Wehr, T.A.; Wirz-Justice, A.; Goodwin, F.K.; Duncan, W.; Gillin, J.C.: Phase advance of the sleep-wake cycle as an antidepressant. Science *206:* 710–713 (1979).
86 Wehr, T.A.; Goodwin, F.K.: Biological rhythms and psychiatry; in Arieti, Brodie, American Handbook of psychiatry: 2nd ed., vol. VII (Basic Books, New York, in press, 1980).
87 Wehr, T.A.; Goodwin, F.K.; Wirz-Justice, A.; Lewy, A.J.: Uncoupling of circadian rhythms in manic-depressive illness (submitted for publication, 1980).
88 Wehr, T.A.; Wirz-Justice, A.; Goodwin, F.K.: Advanced circadian rhythms and a sleep-sensitive switch mechanism: an internal coincidence model of depression (unpublished).
89a Weitzman, E.D.; Nogeire, C.; Perlow, M.; Fukushima, D.; Sassin, J.; McGregor, P.; Gallagher, T.F.; Hellman, L.: Effects of a prolonged 3-hour sleep-wake cycle on sleep stages, plasma cortisol, growth hormone, and body temperature in man. J. clin. Endocr. Metab. *38:* 1018–1030 (1974).
89b Wever, R.A.: The circadian system of man: results of experiments under temporal isolation (Springer, New York 1979).
89c Williams, R.L.; Karacan, I.; Hursch, C.J.: EEG of human sleep: clinical applications (Wiley, New York 1974).
90 Wirz-Justice, A.; Kafka, M.S.; Naber, D.; Wehr, T.A.: Circadian rhythms in rat brain alpha- and beta-adrenergic receptors are modified by chronic imipramine. Life Sci. *27:* 341–347 (1980)
91 Wirz-Justice, A.; Wehr, T.A.; Goodwin, F.K.; Kafka, M.S.; Naber, D.; Marangos, P.J.; Campbell, I.C.: Antidepressant drugs slow circadian rhythms in behaviour and brain neurotransmitter receptors. Psychopharmacol. Bull. *16:* 45–47 (1980).
92 Wirz-Justice, A.; Wehr, T.A.: Uncoupling of circadian rhythms in hamsters and man; in Koella, Sleep 1980 (this volume).
93 Wirz-Justice, A.; Wehr, T.A.; Goodwin, F.K.; Campbell, I.C.; Murphy, D.L.: Antidepressant drugs slow and uncouple circadian rhythms (submitted for publication, 1980).
94 Wirz-Justice, A.; et al.: Sleep deprivation and clomipramine in endogenous depression. Lancet *ii:* 912 (1976).
95 Wirz-Justice, A.; Tobler, I.; Borbely, A.A.; Kafka, M.S.; Naber, D.; Marangos, P.: Sleep deprivation: effects on circadian rhythms of rat brain receptors. Psychiatry Res. (in press, 1981).
96 Zander, K.J.; Ackenheil, M.; Rüther, E.: Prolonged and repeated sleep deprivation in patients with endogenous depression. Effect on EEG, psychopathology, psychophysiological and biochemical parameters. 11th CINP abstr. Wien 1978, pp. 446.
97 Zander, K.J.; et al.: Long term neuroleptic treatment of chronic schizophrenic patients: clinical and biochemical effects of withdrawal. Psychopharmacology, Berlin (submitted).
98 Zulley, J.: Der Einfluss von Zeitgebern auf den Schlaf des Menschen (Fischer, Frankfurt 1979).

Addresses of Authors

Dr. M. Gillberg, National Defense Research Institute, Section for Psychophysiology (FOA 541), S–10450 Stockholm (Sweden)

Dr. R.H. van den Hoofdakker, Department of Biological Psychiatry, University Psychiatric Clinic, NL–9713 EZ Groningen (The Netherlands)

Dr. E. Rüther, Psychiatric Clinic, University of Münich, Nussbaumstrasse 7, D–8000 Munich 2 (FRG)

Dr. T.W. Uhde, National Institute of Mental Health, Bethesda, MD 20205 (USA)

Dr. A. Wirz-Justice, National Institute of Mental Health, Clinical Psychobiology Branch, Bethesda, MD 20205 (USA), and Psychiatrische Universitätsklinik, CH–4025 Basel (Switzerland)

B. Circadian Rest-Activity and Sleep-Wake Rhythms

Chairman: *A.A. Borbély,* Zürich
Participants: *G. Chouvet,* Lyon; *C. Eastman,* Chicago, Ill.; *G.A. Groos,* Leiden; *H. Schulz,* Munich; *A. Wirz-Justice,* Basel

Introduction[1]

A.A. Borbély, Zürich, Switzerland

For the first time at a European Sleep Research Congress a symposium has been devoted to circadian rhythms. This reflects not only the recent important advances in rhythm research, but also the growing interest and relevance of the field for sleep research.

The mechanisms generating circadian rhythms are still incompletely understood. Nevertheless, in the last few years the suprachiasmatic nucleus (SCN) has been recognized as a crucial brain structure and a candidate for the pacemaker of circadian rhythms. The discovery that SCN lesions abolish the circadian rhythms of drinking and motor activity [71], was followed in rapid succession by analogous findings for other circadian rhythms, including those of sleep and waking [38–40]. The SCN receives a direct input from the retina via the retino-hypothalamic tract, a pathway mediating the entrainment of circadian rhythms by light. *Gerard Groos* demonstrates in his contribution that the electrical stimulation of the SCN leads to prominent changes in the circadian feeding rhythm. He also documents the responsiveness of units in the SCN to the environmental luminance level, and shows in his long-term recordings that the firing level of SCN units exhibits a circadian rhythm. *Guy Chouvet, Jean-Louis Valatx* and their co-workers discuss the genetic factors determining sleep rhythms. Based on their studies in inbred mouse strains, they conclude that sleep duration and sleep rhythms may be controlled by separate factors, and that even the rhythms of the substates of sleep may be dissociated from each other.

The following two contributions deal with aspects of circadian rhythms that are common to rodents and man. *Charmane Eastman* recorded brain

[1] For references see compound list on page 79.

temperature and sleep in rats maintained under a 22-h light-dark cycle, and observed abrupt phase shifts in the rhythms. She proposes a single oscillator phase-shift model to account not only for the rat data, but also for the internal desynchronization of humans living in temporal isolation. *Anna Wirz-Justice* and *Thomas Wehr* are also concerned with the internal dissociation of circadian rhythms. On the basis of their data they propose a functional analogy between the two oscillators controlling the REM sleep rhythm and the sleep-wake rhythm in man, and those controlling the evening and morning bouts of activity in the hamster. Furthermore, they present evidence that in animals and man circadian rhythms may be affected in comparable ways by antidepressant drugs. *Hartmut Schulz* focuses his presentation on the altered REM-sleep distribution in endogenously depressed patients, and describes 3 models to account for the changes. He discusses in some detail

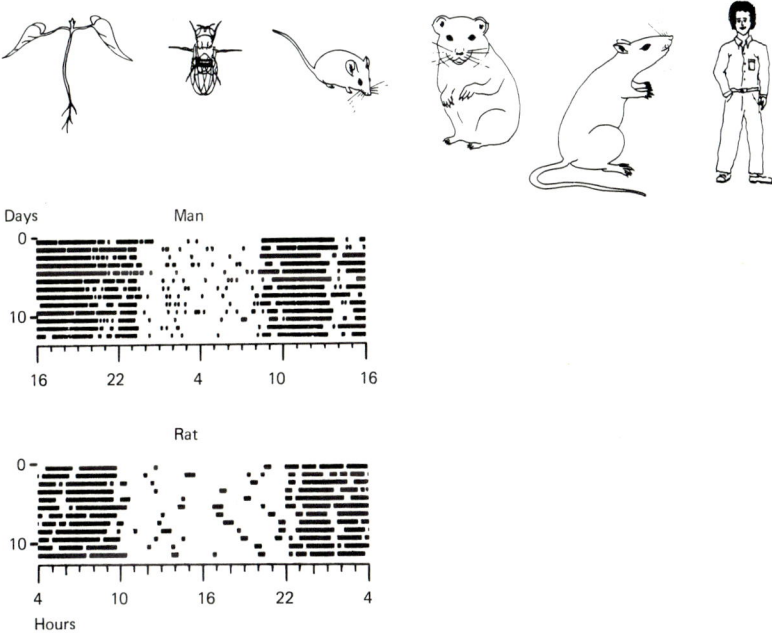

Fig. 1. Illustrating the similarity of the rest-activity pattern in man and rat. Motor activity in man was measured by a wrist-worn solid-state activity monitor, in the rat (light from 1000 to 2200 hours) by a mechano-electric transducer. The top drawings illustrate not only the mammalian species used in the experiments of the symposium, but indicate that circadian rest-activity rhythms, in contrast to sleep, are present also in invertebrates (e.g. insects) and, in a more figurative sense, even in plants (e.g. leaf movements).

the chronobiological model which may explain the occurrence of sleep onset REM episodes under seemingly disparate conditions (e.g. in babies, narcoleptics and depressed patients).

A common feature of the experiments reported in this symposium is the long duration of the recordings which usually extend over several days or weeks. Such studies necessitate special procedures for recording and data analysis which differ considerably from the techniques used in conventional, short-term sleep studies. However, as the reports of this symposium amply demonstrate, the time and effort invested in long-term experiments is justified. Such experiments lead not only to new and unexpected findings but, more importantly, they contribute to enlarging the conceptual framework in which sleep is examined. The statement that, from a circadian vantage point, sleep corresponds largely to the rest period, and waking to the activity period, may seem trivial. Nevertheless, the implications of this correspondence may be far-reaching, since circadian rest-activity rhythms are basically similar not only between man and other mammals (fig. 1), but also between vertebrates and invertebrates for whom sleep cannot be defined by the conventional criteria. The investigation of the circadian facet of sleep may reveal homologies with the ubiquitous circadian rest period, and thereby expand the realm of sleep research to hitherto unexplored phylogenetic territories.

Sleep 1980. 5th Eur. Congr. Sleep Res., Amsterdam 1980, pp. 42–51
(Karger, Basel 1981)

The Suprachiasmatic Nuclei as a Central Pacemaker Timing the Rest-Activity Cycle[2, 3]

G.A. Groos, Leiden, The Netherlands

A remarkable aspect of the circadian rest-activity cycle (and the closely related sleep-wakefulness cycle) is its precise timing with respect to the daily

---

[2] Part of the present work was supported by grants from the European Science Foundation. The author is grateful to Dr. R. Mason for his help with the preparation of the manuscript.

[3] For references see compound list on page 79.

light-dark (LD) cycle in the environment. Under normal or artificial illumination cycles mammals are active during either the dark (nocturnal species) or the light part of the LD cycle (diurnal species). This does not mean, however, that the rest-activity rhythm is a simple response to the periodic LD signal. Animals living in constant light or darkness will exhibit a free-running rest-activity and sleep-wakefulness rhythm the period of which is somewhat different from 24 h [8]. The free-running period in constant darkness is thought to reflect the intrinsic frequency of an endogenous pacemaker driving the overt rhythm. In order to adjust the intrinsic period to 24 h this pacemaker is subject to photic entrainment by daily LD cycles. The notion of an internal pacemaker (or 'biological clock') is conceptually attractive but needs to be experimentally substantiated. In adult mammals extraretinal photic entrainment does not occur [30, 32, 61] and yet the pacemaker is entrained by illumination cycles. Consequently the biological clock must be coupled to the retina. Thus, the search for the biological clock has focussed on structures related to the visual pathways [61].

Figure 2 presents an outline of the mammalian visual system. The major projection from the retina passes via the optic nerves, optic chiasm and optic tract to the lateral geniculate complex, the dorsal division of which gives rise to a prominent projection to the visual cortex. A second retinal pathway terminates in the tectal and pretectal areas. The idealized experiment illustrated in the right part of figure 2 summarizes the effects of lesions of these terminal structures of the optic pathway on steady-state entrainment of a circadian rhythm. None of the lesions affect rhythmicity as such, nor its entrainment to an LD cycle. Similarly lesions of the accessory optic system to the midbrain tegmentum leave entrained rhythms undisturbed. Yet binocular enucleation of the eyes invariably results in free-running of circadian rhythms [61]. Only the retino- hypothalamic projection to the suprachiasmatic nuclei (SCN) remains to mediate photic entrainment. Complete lesions of the SCN, however, do not simply result in disruption of entrainment, as can be seen in the lower right part of figure 2, but also cause arrhythmicity. This effect has been demonstrated for various circadian rhythms in rodent species including the sleep-wakefulness and rest-activity rhythms [61]. In primates, however, SCN lesions have more complex effects. Whereas the sleep-wakefulness rhythm of the squirrel monkey is abolished by bilateral SCN lesions the temperature rhythm has been observed to persist [26]. Nevertheless, the disruption of circadian organization after destruction of the SCN suggests that this cell group may be an important circadian pacemaker in many mammalian species.

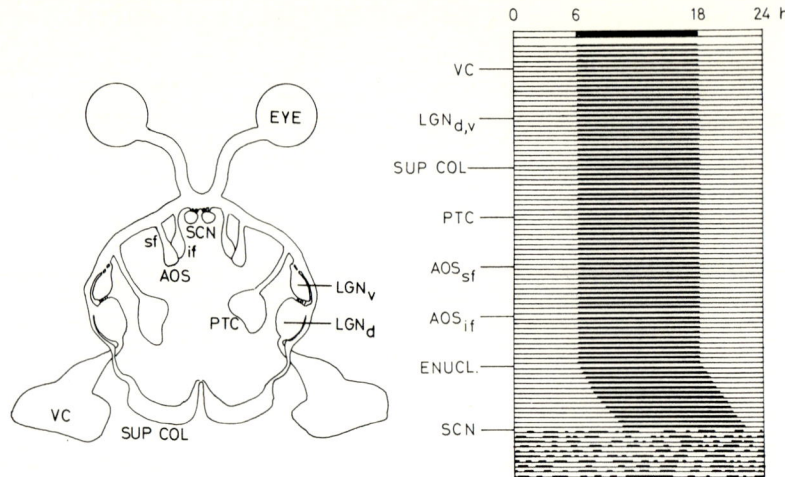

Fig. 2. The mammalian visual system and the pathway of entrainment. On the left the visual system is represented schematically. The retinal outflow projects to the suprachiasmatic nuclei (SCN), to the midbrain via the inferior and superior fasciculi of the accessory optic system (AOS_{if}, AOS_{sf}), the pretectal complex (PTC), superior colliculi (SUP COL), the ventral and dorsal lateral geniculate nuclei (LGN_v, LGN_d) and from the latter to the visual cortex (VC). The figure on the right illustrates the rest-activity rhythm of a nocturnal animal entrained to a light-dark cycle and its response to subsequent lesions of various parts of the visual system. Only after binocular enucleation (Enucl) the rhythm becomes free-running, while SCN lesions result in arrhythmicity.

The SCN appear bilaterally as two parvocellular cell groups situated medially in the hypothalamus just dorsal to the optic chiasm (fig. 3). The retino-hypothalamic projection is partly decussated and terminates in the ventral and lateral portions of the caudal SCN. These features are phylogenetically invariant as they have been observed in mammals ranging from rodents to chimpanzees [51, 61].

Interestingly some of the small diameter cells in the SCN appear to have a neurosecretory function and may be involved in the regulation of the oestrous cycle which itself is found to be dependent on the circadian timekeeping function of the SCN [61]. The retino-hypothalamic fibres make Gray type I and Gray type II synaptic contacts on the dendritic stems and spines of SCN cells [33, 61]. The relative number of symmetrical (Gray type II) and asymmetrical (Gray type I) optic synapses is modifiable under the influence of environmental lighting conditions [33]. This structural plasticity of the retino-suprachiasmatic system is of interest in the light of the well-

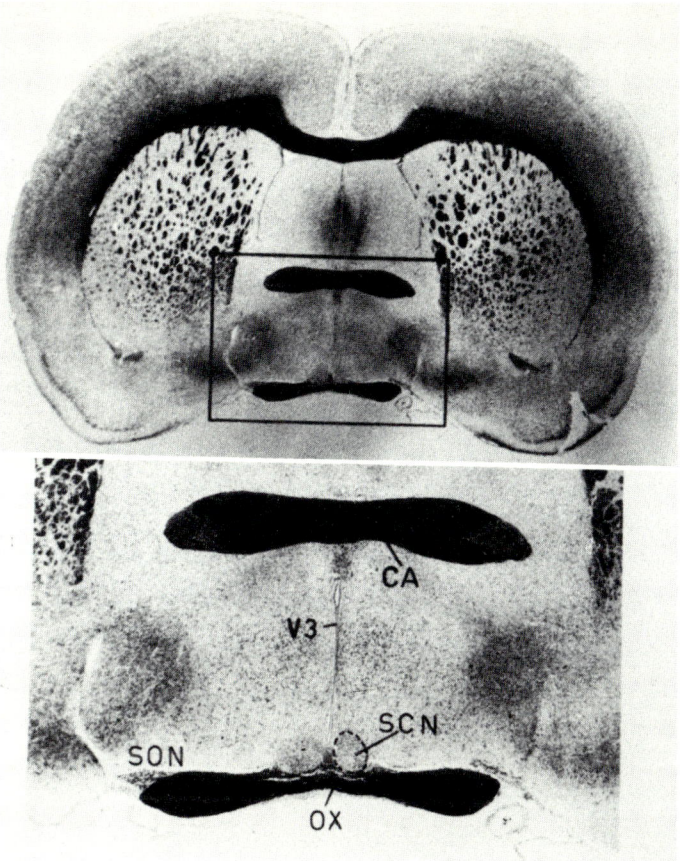

Fig. 3. Localization and light microscopic appearance of the SCN in a transverse section of the rat brain. Fibre tracts appear dark. CA = Anterior commissure; V3 = third ventricle; SON = supraoptic nucleus; SCN = suprachiasmatic nucleus; OX = optic chiasm.

known dependence of rhythm properties on the previously experienced lighting conditions [8]. The importance of the visual input to the SCN is further demonstrated by the fact that the structural and functional development of the SCN is critically dependent on the presence of its retinal innervation [61].

In addition to retinal afferents the SCN receive neuronal inputs from other hypothalamic structures including the anterior and medial preoptic areas, the ventromedial hypothalamic and arcuate nuclei. Extrahypothalamic inputs originate in the ventral part of the lateral geniculate body, the medial septal nucleus and serotonergic cells in the B7 and B8 nuclei of the raphe complex [61]. Lesions of any of these structures do not result in arrhythmicity or the disruption of normal photic entrainment, however [61]. Thus, the SCN appear to control circadian rhythms independent of their neuronal inputs. The role of the 5-HT projection from the raphe complex deserves special attention in future studies. While raphe lesions have no apparent effects on circadian rhythms [61] experimental reduction of the total brain 5-HT content results in transient arrhythmicity followed by a phase shift of the activity rhythm [36]. Moreover, recent experiments involving the administration of tricyclics and MAO inhibitors have demonstrated impressive effect on the phase control of the circadian rest-activity cycle [see the contribution of *Wirz-Justice* in this volume].

Control of the circadian system is mediated by SCN efferents as was demonstrated by *Stephan and Nunez* [70] who made partial or complete cuts around the SCN and subsequently observed arrhythmicity in sleep- wakefulness, temperature and water intake. The SCN have diverse projections terminating in the periventricular nucleus, the ventromedial and arcuate hypothalamic nuclei, the median eminence and the midbrain central gray [61]. The control of the melatonin rhythm of the pineal gland by the SCN is mediated by a descending pathway via the cervical sympathetic ganglion which, in turn, innervates the pineal [61, 87]. So, although the initial stage of SCN output to the circadian system is neuronal, the efferents may eventually control hormonal rhythms.

The evidence supporting the notion that the SCN is an important pacemaker of the circadian system is largely based on lesion experiments. Such studies have implicated the retino-hypothalamic projection to the SCN as the pathway of photic entrainment. Further they demonstrate that lesions of the SCN or its efferents result in arrhythmicity and, significantly, the latter effects can most probably not be ascribed to the control of the SCN by its afferents [61]. Non-destructive techniques have led to findings consistent with the lesion data. Injections of small amounts of local anaesthetics or neurotoxic agents (e.g. TTX) in the vicinity of the SCN result in reversible arrhythmicity [87; *Groos*, unpublished observations]. Electrical stimulation of the SCN in rats and hamsters has been observed to alter the phase and/ or period of free-running activity and food intake rhythms [*Rusak and Groos*, in preparation]. These effects are illustrated in figure 4. One most compelling piece of evidence will probably never be obtained, however. If it could be shown that transplantation of the SCN of a donor animal to an

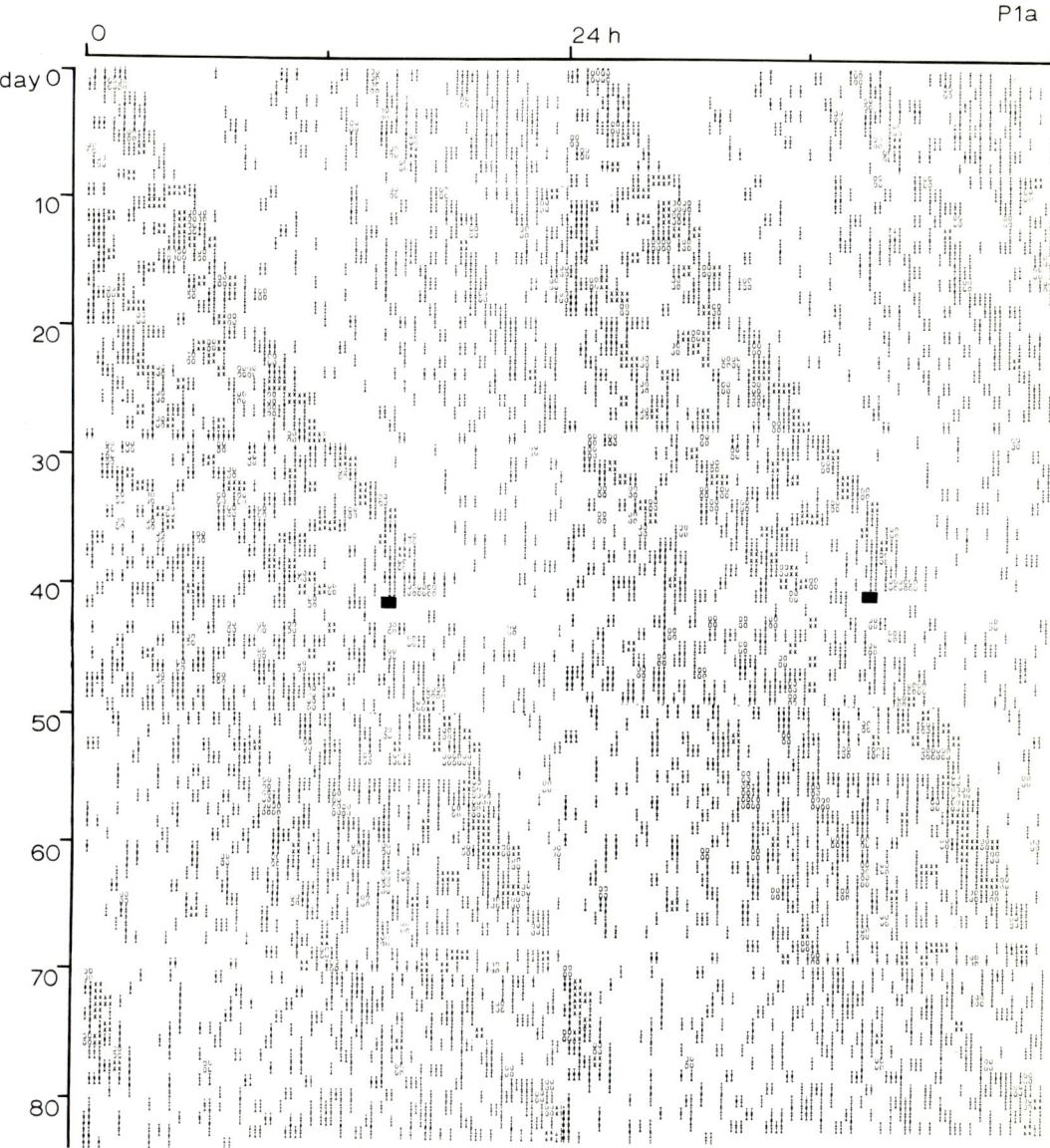

Fig. 4. Alteration of the free-running circadian food intake rhythm by electrical stimulation of the rat SCN. The record on the left shows the food intake rhythm of a blinded rat before and after electrical SCN stimulation (indicated by the dark square). The period before stimulation was 24.41 ± 0.03 h. Following stimulation the period shortened to 24.29 ± 0.05 h. In addition to a period change a phase shift was observed.

arrhythmic SCN lesioned recipient would turn the latter into a rhythmic animal this would establish the importance of the SCN beyond all doubt. Excision of the SCN, however, implies severing all its neuronal efferents. Obviously such an SCN would never be able to exert neuronal control over the circadian system of the recipient animal.

A physiological study of the SCN should be aimed at solving two basic problems. First, the behaviour of those SCN neurones receiving visual fibres will have to be studied in relation to the process of photic entrainment. Secondly, if the SCN is postulated to be responsible for the self-sustaining rhythms in neuro-endocrine and behavioural processes the mechanism of rhythm generation should be accounted for. Relating to the mechanisms of photic entrainment, it is now known that there is a class of SCN neurones which shows a clear and consistent response to illumination of the retina. After adaptation to a dark background these cells typically respond to a stepwise transition to some higher illumination level with either a sustained increase or a sustained decrease of their discharge rate [28, 31]. Thus the visual SCN cells can be divided into light- activated and light-suppressed neurones (fig. 5). Each type either increases or decreases its discharge

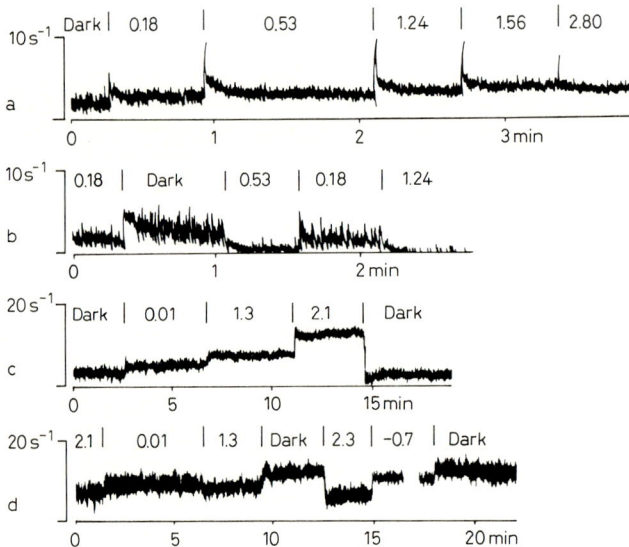

Fig. 5. Discharge rates of SCN neurones at different luminance levels. The light-activated cell type is illustrated in *a* for the cat and in *c* for the rat. Two examples of light-suppressed cells are shown in *b* and *d* for the cat and rat respectively. The numbers above each record refer to the ambient luminance in log cd · m^{-2}.

monotonically as a function of the ambient luminance. Moreover, they have large receptive fields which integrate photic energy from large areas in the visual field [31]. At present it cannot be stated whether this latter property is a result of convergence of many photoreceptors onto one ganglion cell projecting to the SCN [58] or of convergence of many retinal ganglion cells onto individual SCN neurones as was proposed by *Groos and Mason* [31]. In any case, the visual properties of SCN cells suggest that the retino-suprachiasmatic system has differentiated towards coding luminance and gross temporal luminance gradients. Such a development seems very appropriate in terms of the SCN's function to detect the daily cycle in overall environmental luminance to synchronize endogenous circadian rhythms. Much less is known about the neuropharmacology of the retinal projection to the SCN. Both types of visual SCN cells respond to micro-iontophoretic application of acetylcholine, serotonin, norepinephrine, and dopamine [54] but only in the case of acetylcholine is there some additional evidence that this transmitter is involved in photic entrainment [87].

Considering the physiological problems related to rhythm generation lesion studies carried out by *Rusak and Zucker* [61] pointed to the possibility that the SCN may contain a network of neuronal oscillators each with a period different from 24 h. Earlier *Pavlidis* [57] suggested a model for circadian rhythm generation which involves a network of coupled high frequency oscillators. For such network he showed that its macroscopic, or compound, behaviour will exhibit rhythmicity with a period which is an increasing function of the number of such oscillators and the strength of their coupling. Although the Pavlidis model was developed for biochemical oscillators it is possible to apply it to a network of coupled high frequency neuronal oscillators. Interestingly neuronal oscillators have been observed in in vitro preparations of the rat SCN. Moreover, in anaesthetized rats, evidence has been obtained that such oscillating neurones are tightly coupled. Unfortunately the number of such cells in the SCN is far too small to postulate a network of neuronal oscillators as a mechanism for generating a much slower circadian rhythm [29].

If rats are entrained to an LD cycle and subsequently anaesthetized and kept in constant dim light it is possible to record an endogenous circadian rhythm in multiple (i.e. from 3–10 neurones) unit activity from the SCN (fig. 6). Such rhythms are not uniquely found in the SCN but also in other hypothalamic nuclei and the hippocampus. *Inouye and Kawamura* [41] demonstrated that the SCN rhythm persists whereas other rhythms in the brain are abolished if the SCN is isolated from other nervous tissue by a

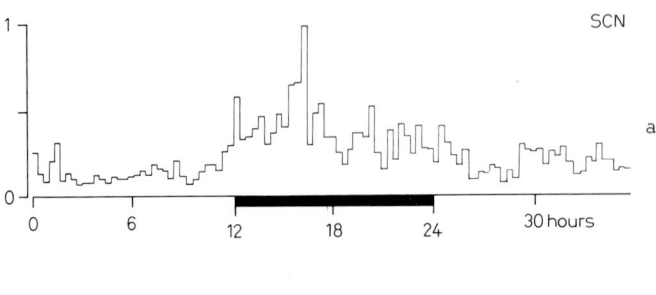
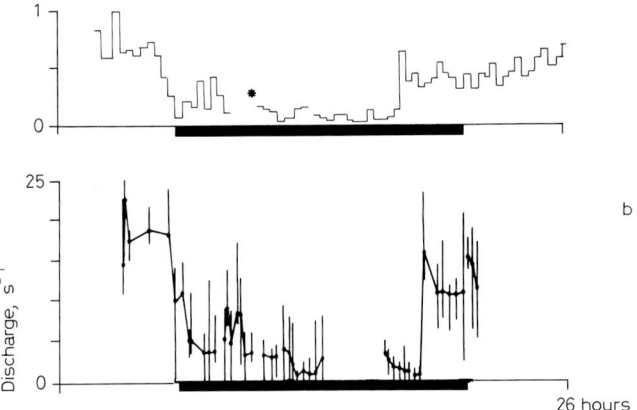

Fig. 6. Multiple unit activity rhythms recorded in the SCN (a) and preoptic area (b) of the anaesthetized rat. The dark bars indicate the extrapolated dark phase of the LD cycle to which the animals had been exposed prior to the experiment. The recordings were made in continuous dim light. The multiple unit activity was averaged for 20-min intervals and normalized to the maximum value of the rhythm. Single preoptic cell discharge rates are shown in b (lower graph). Note the close phase correspondence between the single cell and multiple unit rhythms.

circular cut. This finding suggests that the neuronal rhythm is intrinsic to the SCN and drives the other rhythms outside these nuclei. This conclusion is supported by the observation that of all brain structures only the SCN exhibits an endogenous circadian rhythm in metabolic activity [66]. It is likely that the multiple unit activity rhythms reflect circadian fluctuations in the discharge of single neurones. Figure 6 illustrates this point for the rhythm in the preoptic area. Single cells change their discharge rate in close correspondence with the two daily transitions observed in the multiple unit activity rhythm. In summary it can be tentatively suggested that the SCN is a unique structure in that it contains neurones which exhibit a circadian rhythm in metabolic and neuronal activity. It is of great importance to further substantiate this idea as it could represent the first major step in the discovery of the biological mechanism underlying the circadian rhythms, not only in sleep-wakefulness and rest-activity but also in a large number of other physiological functions.

Sleep 1980. 5th Eur. Congr. Sleep Res., Amsterdam 1980, pp. 51–59
(Karger, Basel 1981)

Circadian Rhythms of Sleep-Waking Cycle: a Genetic Approach[4, 5]

G. Chouvet, Lyon; *H. Emptoz*, Villeurbane; *B. Burtschy*, Paris; *H. L. Routhier*, Paris; *J. L. Valatx*, Lyon, France

Introduction

In animals and in man, the study of sleep has shown large interindividual variations especially for the duration of slow wave sleep (SWS) and paradoxical sleep (PS). While environmental effects upon sleep are well documented, hereditary factors which could be involved in these variations have not been extensively studied [73]. Moreover, circadian rhythms of sleep have been recorded under various experimental conditions (different light and food schedules, temperature, etc.) and after lesions of the central nervous system in order to demonstrate the existence of one or several internal oscillators [15, 53]. Only very few data deal with the genetic control of the regulation of the sleep rhythms, i.e. whether there are some hereditary factors controlling the internal oscillator and its synchronization by the environment. In mice, *Oliverio* [55] has shown the genetic determination of the

[4] This work was supported by INSERM U52, U171, CNRS LA 162 and DRET.
[5] For references see compound list on page 79.

level of locomotor activity. Moreover, in SEC and Balb/c strains, the circadian rhythms of wheel running activity disappeared in continuous light or darkness, while in C57 strain it remained quite stable. A genetic study of this sensitivity to the environment, however, has not yet been performed. In this short presentation we would like to demonstrate that there are several hereditary factors which influence the sleep rhythms and seem to be independent from those involved in the regulation of sleep duration. For this purpose, we have used the inbred strains of mice known for their high degree of homozygoty (over 150 generations of brother-sister matings).

Material and Methods

Animals. The subjects were at least 12-week-old male mice (n = 76) at the beginning of the experiments. Two highly inbred strains, C57BR/cdOrl (BR) and BALB/cOrl (C), their reciprocal F1 hybrids (C.BR F1 and BR.C F1) and backcrosses (C.BR F1 ×BR and C.BR F1× C) were used. All the mice were chronically implanted with cortical and muscular electrodes and continuously recorded for a week under the same standard conditions: L(7–19)–D(19–7) schedule, constant ambient temperature (24 ± 1 °C) food and water ad libitum and 10 days of habituation before polygraphic recordings [72].

Statistical Analysis. By visual interpretation, all the recordings were scored as W, SWS or PS by 30-s epochs and stored on a digital file. The daily (24h) night and day durations of SWS and PS were calculated. Circadian rhythms were computed by means of harmonic regression techniques [42]. The temporal variations of hourly SWS and PS durations were estimated by computing around the daily mean level the best-fitting sine wave function (amplitude and phase) of the fundamental (24-h period), and its first three harmonic components (12-, 8- and 6-h periods). As each animal may be characterized by several parameters (SWS and PS durations, amplitude and phase of their rhythms) a statistical cluster analysis was used in order to define each strain in a multidimensional space.

(a) A first way was to compute for each strain and with bivariates distributions (which have the advantage to allow planar representations) a 'dispersion ellipse' featuring the clustering of the observations. This ellipse satisfies the following equation:

$$\frac{1}{1-r^2}\left[\left(\frac{x-\bar{x}}{\sigma_x}\right)^2 - 2r\left(\frac{x-\bar{x}}{\sigma_x}\right)\left(\frac{y-\bar{y}}{\sigma_y}\right) + \left(\frac{y-\bar{y}}{\sigma_y}\right)^2\right] = \chi^2_\alpha(2), \tag{1}$$

were x, y, σ_x, σ_y represent the means and standard deviations of the two chosen variables (x, y), and r their correlation coefficient. According to the numerical value of χ^2 (2 degrees of freedom), it is possible to plot an ellipse which includes 90, 95, 99%...of the observations (α level of probability). When computed on one strain (considered as a population of several observations) this ellipse is representative of the strain and may be used for a further classification analysis. Indeed, when dealing with two (or more) strains, each one being characterized by its own dispersion ellipse, it is possible to compute the statistical distance (as estimated by the χ^2 value of formula 1) between an individual observation (from a mouse to be classified) and the barycenter (center of the ellipse) of the reference strain. The similarity (or proximity) with a strain is associated with the lowest χ^2 value.

(b) A second way was to perform on all the observations a principal components analysis which reduces by linear combinations of the variables the dimension of the multidimensional cluster and allows its projection in a reduced space. The main operation in this factorial analysis is to determine the principal axis of inertia of the cluster around its center of gravity and to represent the cluster according to these new axes, usually by projecting it on one plane determined by the two first principal axes.

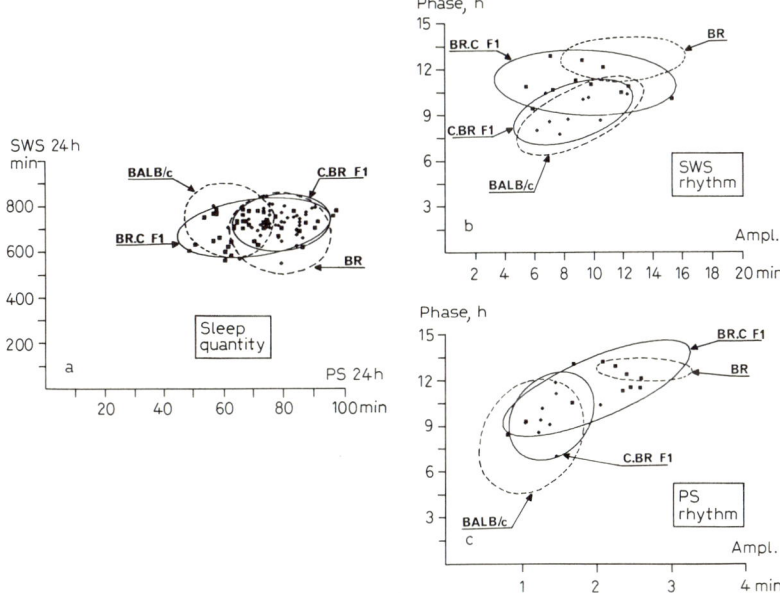

Fig. 7. Planar representation of SWS and PS daily quantities *(a)* and circadian rhythms (as estimated by amplitude and phase of the 24-h best-fitting sine wave) of SWS *(b)* and PS *(c)* hourly mean durations. Only the parental strains (BR and Balb/c) and the F1 hybrids are represented. For each representation and for illustration clarity purposes, the individual observations on parental strains are not featured and are replaced by their 'dispersion ellipse' (interrupted lines). The F1 hybrids are represented with different symbols according to the crossing (dark losanges for C.BR F1 with mother from Balb/c strain, and black squares for BR.C F1) inside their corresponding 'dispersion ellipse' (continuous line). Each individual observation features 1 day for sleep daily durations *(a)* and one mouse (5 continuous days) for SWS *(b)* and PS *(c)* rhythms, as a precise determination of the parameters of a rhythm requires several periods of the fundamental. This explains the discrepancy between the number of experimental points in *(a)* and *B, C*.

Results

Sleep Durations. Figure 7a shows that the daily SWS durations are not statistically different in the two parental strains. The daily PS durations are more discriminant, the parental ellipse being well separated with a small overlapping. The daily PS duration was higher in BR strain (80 min/24 h or 3.3 min/h) than in Balb/c strain (60 min/24 h or 2.5 min/h). The reciprocal F1 hybrids were not completely identical. In C.BR F1, the daily PS durations are similar to those of BR strain (F1 ellipse is included in the BR ellipse) and in BR.C F1 the PS durations are more variable than those of the parental strains.

Circadian Rhythms. In spite of its variability, the amplitude of the fundamental (24 h) in BR strain was 2–4 times higher than that of the 12 h component, while in C strain both components were in the same range. To

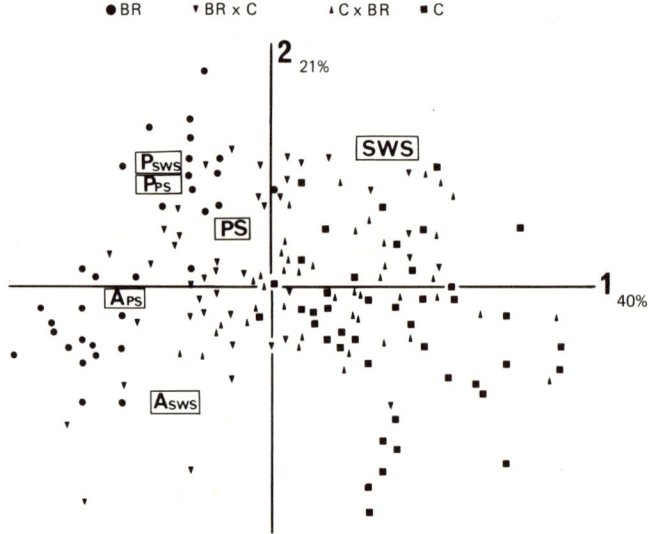

Fig. 8. A principal components analysis was performed on the 2 parental strains and their reciprocal F1 hybrids. Each individual point features a 6-dimensions observation summarizing the sleep data of one animal computed on a single 24-h continuous polygraphic recording: daily duration of SWS and PS, amplitude (A) and phase (P) of the best-fitting sine wave (24-h period) around the hourly mean level of SWS and PS. On this planar representation with the 2 first principal axes as reference (explaining respectively 40 and 21% of the total variance), the parental strains are well differentiated (BR strain clusters on the left and C strain on the right) meanwhile the C.BR F1 cluster overlaps the C one and the BR.C F1 cluster is intermediate between the parental clusters.

simplify the interpretation of the results, we have taken in account only the fundamental component (24 h).

As shown in figure 7b and c, the amplitude and phase of SWS and PS rhythms are more discriminant for separating the parental strains than the sleep durations: the ellipse of BR and C strain are non over-lapping for SWS and PS as well. *The reciprocal F1 hybrids* were different: BR.C F1 are more variable and intermediate between parental strains and C.BR F1 were similar to the C parent (its ellipse is included in that of C strain), while PS and SWS daily durations were related to BR strain (fig. 7a). These results on the relationships between parental strains and their reciprocal F1 hybrids were confirmed by a further principal components analysis (fig. 8).

For the backcrosses which represent the recombination of all the parental characters, the classification was done by calculating the global distance separating each mouse from each parental strain and classifying the animal according to the lowest χ^2 value (fig. 9). Table I summarizes the results.

Mice classified 'BR type' are nearer to BR barycenter than those of C strain and 'C type' means that mice are nearer to C than to BR (without implying an equality of the parameters). The repartition of the backcrosses indicates that BR type was dominant for SWS rhythms (70% in F1×BR

Table I. Repartition of mice from backcrosses F1×BR and F1×C for SWS and PS rhythms and according to their coat color. This table was filled from the values of χ^2 distances

		SWS rhythm	PS rhythm	SWS and PS rhythm
F1×BR				
All the mice (n=20)	BR	14	7	7
	C type	6	13	6
Brown coat (n=10)	BR type	7	4	4
	C type	3	6	3
Agouti coat (n=10)	BR type	7	3	3
	C type	3	7	3
F1×C				
All the mice (n=16)	BR type	8	3	3
	C type	8	13	8
Albino coat (n=9)	BR type	4	0	0
	C type	5	9	5
Agouti coat (n=7)	BR type	4	3	3
	C type	3	4	3

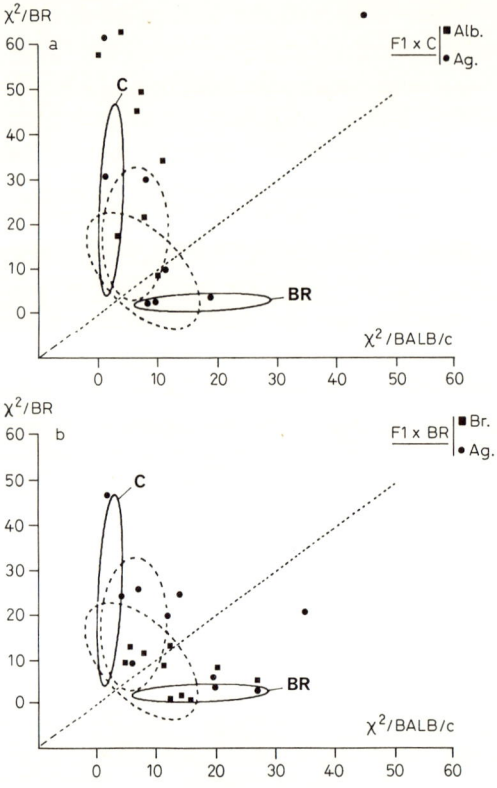

Fig. 9. Statistical classification of backcross mice. The coordinates are expressed as similarity values corresponding to the sum of the 3 statistical distances (χ^2 values) computed with formula 1 (see Material and Methods) in each of the 3 planar representations of figure 1a–c. The abcissae represent the χ^2 distance when BALB/c is used as strain reference, and ordinates when BR strain is used as reference. The parental strains (continuous ellipses) are characterized by small distances when compared with their related strain, versus large distances from the alternative strain. F1 hybrids are represented with interrupted lines ellipses (C.BR F1 near C strain, BR.C F1 overlapping C and BR strains). Each experimental point features a mouse (5 days of recordings) with a different symbol according to the crossing and the color: *a* Crossings between F1 (mother) and BALB/c. Black squares = albinos; Black circles = agouti. *b* Crossings between F1 (mother) and BR. Black squares = brown; Black circles = agouti. The first diagonal in dotted lines divides the similarity space in two subspaces associated with each parental strain.

and 50% in F1×C) without linkage with the color genes (equal repartition in the different coat colored mice). For PS rhythms, the repartition was different from that of SWS, C type seeming to be dominant with a linkage with the albino gene.

Discussion

It is noteworthy that by means of different statistical methods we reach the same results. In the parental strains, the SWS daily durations are not statistically different, so only the PS durations can separate them. Furthermore, sleep rhythms appear to be more discriminant than durations, in spite of the variability of the amplitudes and acrophases. Indeed, for the same sleep duration, it is theoretically possible to get many repartitions around the clock. In C mice, the variations of the position of the acrophasis are important; this might be due to the 12-h harmonic, the amplitude of which is often as high as the fundamental which can be shifted to early in the morning. The variability of the C strain has been observed in continuous light exposition [55] where the rhythms of activity completely disappeared. One can hypothesize the existence of some genetic factors which make the mouse more sensitive to alterations in its environment. To check this hypothesis it would be interesting to record sleep rhythms in continuous light or darkness in F1 hybrids and backcrosses.

In *C.BR F1 hybrids*, the PS duration is similar to that of BR mice, but their rhythms are close to those of C mice (fig. 7). These results indicate that durations may be independently determined from the sleep rhythms. This fact is supported by the results of backcrosses in which several recombinations are observed as shown in figure 10.

Backcrosses represent the recombination of all the genes of the parents and each mouse is different from another. For one pair of parameters it is possible to classify each mouse as BR or C type according to the values of the associated χ^2 distances. From this classification, the percentages of each parental type compared to the theoretical values of the mendelian laws may indicate if the transmission is dominant or recessive and due to the effects of a single or several genes. Our results show that the mode of inheritance of the SWS rhythm is different from that of the PS rhythm.

The transmission of the SWS rhythm of the BR type appears to be dominant over the C type (F1×BR: 70%; F1×C: 50%, and might be due to a major gene effect. For PS rhythms, the genetic control seems to be more complex. The C type is apparently dominant in all the backcrosses

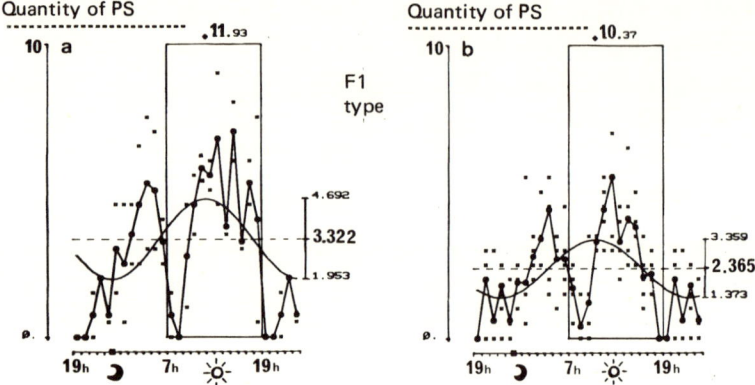
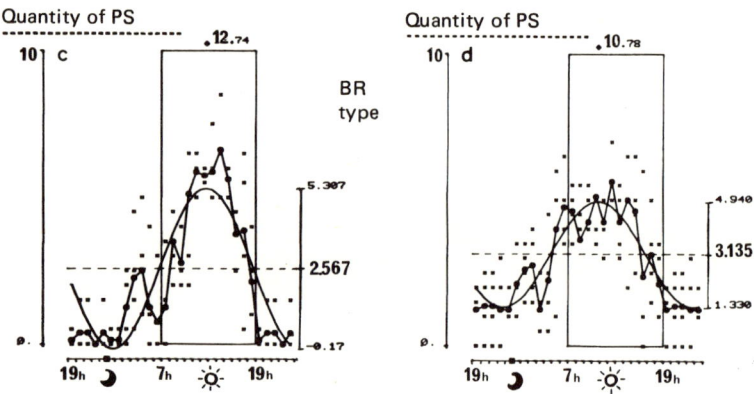

Fig. 10. Various recombinations of temporal variations of hourly PS durations in backcross mice. Each mouse is represented by the daily measurement (individual points) of hourly PS duration, its mean temporal variation over the nycthemere (continous line with black circles) and the 24-h best-fitting sine wave (continuous line). Top: Backcross mice labelled 'F1 type' because they display a PS daily mean level of BR (3.3) with a related c rhythm *(a)* or a daily mean level of c (2.36) with a F1 rhythm *(b)*. Bottom: Backcross mice labelled 'BR type' as they display a PS rhythm related to BR strain around a daily mean level of BALB/c *(c)* or around the daily mean level of BR, however, with the BALB/c amplitude and acrophasis. Ordinates: PS durations per hour in minutes.

(F1×BR: 65%; F1×C: 81%). Two hypotheses might explain this unusual repartition: (1) a maternal effect. Indeed, in all the backcrosses mothers were the C.BR F1 hybrids which have the C type of rhythms. As for PS durations, PS rhythms might be more sensitive to the environment (maternal behavior). (2) Linkage with the color genes. In F1×C, all the albino animals have the C type of PS while, in agouti animals, there is an equal repartition of the parental types. Thus, some of the genes involved in the C rhythms of PS may be located around the albino gene on chromosome 7.

This difference in the inheritance of SWS and PS rhythms implies that they can be dissociated. In fact, in backcrosses, the association of BR type of SWS with the C type of PS rhythms was often observed (30% of the mice), but the reverse (C type of SWS and BR type of PS) was never observed. If one may hypothesize some separate mechanisms for the regulation of SWS and PS rhythms, our results seem to suggest that they should have some common mechanisms.

One question remains to be elucidated: What are the mechanisms supporting this genetic regulation (biochemical or anatomical alterations of the telereceptors or/and of the central nervous system)?

Sleep 1980. 5th Eur. Congr. Sleep Res., Amsterdam 1980, pp. 59–64
(Karger, Basel 1981)

Circadian Rhythms of Rats at the Limits of Entrainment and a Phase-Shift Model of 'Spontaneous Internal Desynchronization' in Humans[6,7]

Charmane Eastman, Chicago, Ill., USA

Rats were exposed to a light-dark (LD) cycle that proved too short to produce absolute entrainment (synchronization) of the circadian rhythms.

---

[6] Supported by Grant MH-4151 from NIMH to Dr. *Allan Rechtschaffen*. A more complete account of this work can be found in *Eastman, C.I.*, Circadian Rhythms of Temperature, Waking, and Activity in the Rat: Dissociations, Desynchronizations, and Disintegrations; PhD dissertation, Chicago (1980).

[7] For references see compound list on page 79.

Complex data were produced that could be explained by either single or multiple oscillator models. One of the single oscillator models, the 'phase-shift model', was then applied to human data from other laboratories on cases of 'spontaneous internal desynchronization'.

The circadian rhythms of temperature and sleep-wake were monitored simultaneously and continuously for many weeks or months in male Sprague-Dawley rats. Temperature was recorded from thermistors implanted on the surface of the brain. Sleep-wake was scored in 30-s epochs by an automated computerized system using cortical EEG and theta activity [12, 13].

Figure 11 shows the data from a rat in a 22-h LD cycle. The 'wheel-running style' graph shows that the temperature rhythm appeared to free-run through the LD cycle with a scalloping pattern. The wake rhythm displayed

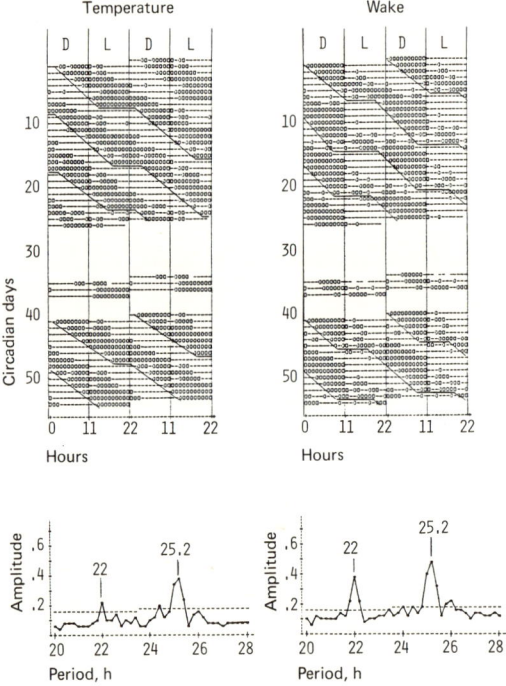

Fig. 11. Top: Double-plotted 'wheel-running style' graphs of rat C2 in LD: 11,11. Each circadian day (22-h day) is smoothed once in hourly values. Then successive circadian days are plotted one beneath the other. '0' indicates hourly values above the daily mean. Diagonal lines are drawn in to emphasize the pattern. Bottom: Periodograms [23] of the data shown in the top graphs. The dashed horizontal line represents the 95% confidence limit.

a similar pattern except that the band of concentrated waking appeared suppressed as it passed through the light portion of the LD cycle. This suppression can be attributed to masking, where the waking activity of nocturnal animals is inhibited by light [2, 14]. The periodograms show two periods in each rhythm. All the animals in the 22-h LD cycle showed similar 'wheel-running style' graphs and two periodogram peaks, one at 22 h corresponding to the LD cycle and one at about 25 h.

These data can be explained as internal desynchronization between oscillators which is forced by the LD cycle, where one oscillator remains entrained to the LD cycle while the other free-runs through the LD cycle with a period of about 25 h. To illustrate this two-oscillator theory, computer models were made by superimposing two sine-waves; one had a period of 22 h and one had a period of about 25 h. A masking component, or a lowering of values occurring during the light period, was added to the wake rhythm. These two-oscillator models matched the rat data very well.

These data can also be explained by a single oscillator which is relatively coordinated to the LD cycle [4, 79] or by a similar concept, the 'phase-shift model', in which a single oscillator free-runs and abruptly phase-shifts each time a certain phase relationship to the LD cycle is reached (see the pattern in fig. 11, top). Computer models of the single oscillator 'phase-shift model' also matched the rat data very well.

Humans in temporal isolation often change from 'normal', synchronized, free-running circadian rhythms to 'spontaneous internal desynchronization', where multiple periodogram peaks and scalloped patterns are produced. This phenomenon has been used to support a multiple oscillator model of human circadian rhythms [11, 82]. The phase-shift model used to explain the rat data was expanded and applied to human 'spontaneous internal desynchronization'. Computer models were made by the method shown in figure 12. The thick sine wave represents the single driving oscillator which controls the rhythms of both temperature and sleep-wake. Temperature follows the oscillator and a masking component is added; temperature is raised by a constant amount during waking and lowered by a constant amount during sleep. Sleep onset usually occurs at the low points of the oscillation, but occasionally a low is skipped and the subject stays up later. When sleep finally occurs, it has an unusually long duration and extends through the following low. Finally, sleep onset begins again at the low points in the oscillation and has a shorter duration.

The masking of temperature by sleep and wake and the relationship between temperature and the long and short sleeps during 'internal desyn-

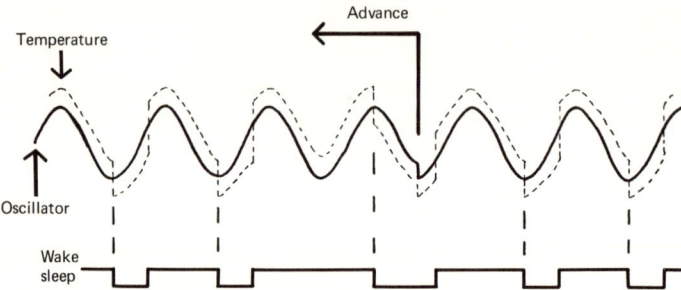

Fig. 12. Schematic illustration of the method used to produce human phase-shift models of 'spontaneous internal desynchronization'.

chronization' has been documented by others [5, 19, 90]. The human phase-shift model adds a shift (advance) of the oscillator of a few hours during the long sleeps. In other words, the model proposes feedback from the sleep-wake rhythm to the circadian oscillator.

One function of sleep may be to help entrain human circadian rhythms in the normal environment. People have, on the average, a free-running period of 25 h [11, 82] which must be advanced by 1 h each day for entrainment to the 24-h day. In normal free-runs, sleep onset begins close to the temperature low, but during entrainment, sleep onset precedes the temperature low by about 6 h [82]. Sleep which occurs much earlier than the temperature low may function to help advance the circadian oscillator by the amount necessary for entrainment. Sleep deprivation studies shows slight delay of circadian rhythms [10] which may be due to the absence of the advances produced by sleep.

Human phase-shift models made by the method described above (fig. 12) matched actual cases of 'spontaneous internal desynchronization' very well (e.g. fig. 13). The model in figure 13 was made with a single sine wave representing a single driving oscillator which free-ran with a period of 25.7 h in both sections A and B. The pattern we know as 'internal desynchronization' was produced because the driving oscillator was repeatedly shifted. When the shifts occur regularly, as in this case, they may be attributed to a second 'shift-inducing' oscillator analogous to the LD cycle in figure 11. This oscillator could induce the subject to stay up late past his temperature low. Calculations by *Wever* [81, 82] raise the possibility that this shift-inducing oscillator could be the moon or an internal oscillator entrained to the moon. However, in many cases of 'internal desynchronization'

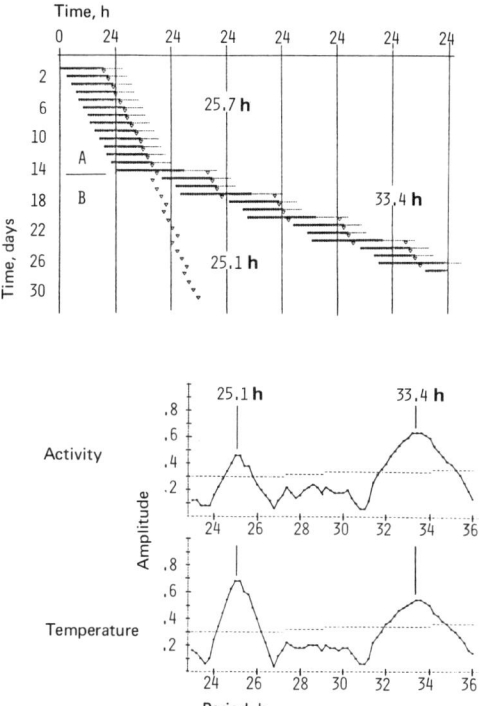

Fig. 13. Phase-shift model of 'spontaneous internal desynchronization' designed to match human data collected by *Aschoff's* group [11], figure 11; [81], figure 7a; [82], figure 27; [80], figure 3; [6], figure 3; [7], figure 20. Top: Temporal course of the rhythms of activity and temperature presented successively one period beneath the other. Thick horizontal bars represent wake, thin horizontal lines represent sleep. Triangles represent the lows of body temperature. Section A, synchronized free-running rhythms. Section B, 'spontaneous internal desynchronization'. Bottom: Periodograms produced by the model data during the 'internal desynchronization' shown in section B.

This phase-shift model was produced by the method shown in figure 12. A 25.7-h driving oscillator controlled the rhythms of temperature and sleep. Starting on day 14, sleep onset skipped every fourth low point in the oscillation. When a low point was skipped, sleep onset was delayed by 12.85 h, until the oscillation reached its peak. The ensuing long sleep advanced the oscillation 2.34 h. Normal length sleeps were 8 h; long sleeps were 14 h.

in the literature the shifts do not occur regularly. In these cases the subject may stay up late for any number of reasons, such as to finish an interesting book. In any case, in the phase-shift model 'internal desynchronization' is produced by repeated shifts of the driving oscillator which controls both the rhythms of temperature and activity.

This theory stands in contrast to the traditional interpretation of *Aschoff and Wever* [11, 82] in which 'temperature' and 'activity' oscillators uncouple and run with their own natural frequencies causing spontaneous internal desynchronization. In the traditional model it is necessary to postulate 'activity' oscillators with periods that are either very long (between 30 and 40 h) or very short (between 15 and 20 h). The existence of oscillators with periods of this length has often raised skepticism because they are not in the usual circadian range [3, 9, 49]. In the phase-shift model it is not necessary to postulate oscillators with periods outside of the usual circadian range. The human phase-shift model presented here does not in any way disprove or rule out the traditional model of spontaneous internal desynchronization. It is merely offered as a feasible alternative. We might learn more about circadian rhythms if we explore the ways in which such abnormalities might be produced while the circadian system functions as a single unit.

Sleep 1980. 5th Eur. Congr. Sleep Res., Amsterdam 1980, pp. 64–72
(Karger, Basel 1981)

Uncoupling of Circadian Rhythms in Hamsters and Man[8]

Anna Wirz-Justice[9], Basel, Switzerland; *Thomas A. Wehr*, Bethesda, Md., USA

The mammalian circadian system has been most intensively studied in two species: humans and hamsters. In each species a model which involves two coupled oscillators has been proposed to account for the formal properties of the system. In the human model, a strong oscillator that controls body temperature, rapid-eye-movement (REM) sleep propensity, and cortisol is frequency-coupled to a weak oscillator that controls sleep [82]. In the hamster model, oscillators that control an evening ('E') and a morning ('M')

[8] For references see compound list on page 79.
[9] Visiting Fellow of the Swiss Foundation for Biomedical Research.

activity bout (the well-known bimodal activity pattern in nocturnal rodents), are also frequency coupled [59]. These two oscillator models were developed to explain the fact that two circadian sub-systems in the same organism may dissociate from one another and run with different frequencies when the organism is isolated from external 24-h time cues.

This paper summarizes data obtained from long-term monitoring of the rest-activity cycle of both humans and hamsters [75, 84, 85]. The results of these observations suggests a closer relationship between the two models than previously recognized. From the phenomenological similarities, we propose that the strong oscillator driving the circadian temperature and REM sleep rhythms in humans is analogous to that driving both the temperature rhythm and the evening bout of activity in hamsters, and that the weak oscillator driving the sleep-wake cycle in humans is analogous to that driving the morning activity bout in hamsters. If, as we suggest, there are certain conditions where the temperature rhythm can uncouple from the rest-activity cycle in the hamster, this species would then provide an animal model for the more difficult studies of the human circadian system under temporal isolation.

In free-running hamsters, oscillator uncoupling (as reflected in the 'splitting' of two activity components) can be provoked by constant bright light (LL) [59] and oestrogen withdrawal [52]. As yet, there are no published studies of the temperature rhythm under split conditions. We have also observed that two components of the activity rhythm can uncouple and run at different frequencies under conditions of constant dim red light (RR) [86] and under chronic treatment with antidepressant drugs [84, 85]. The two oscillators driving the components peeling off from the end of activity (m) and those associated with the onset of activity (e) may be identical to the 'M' and 'E' oscillators, respectively, known to uncouple in LL [59]. This is the first description of uncoupling of activity components occurring at light intensities below 100–200 lx in nocturnal rodents [59], although a similar effect in RR has been found in the cockroach [83]. There are, however, important differences between the 'splitting' of activity components in LL and the uncoupling of activity components in RR. In LL, the activity bout which occurs late in the subjective night (m) most often shortens and traverses the active phase, crossing the onset activity bout (e) and establishing a new stable antiphase [59]. In RR, m usually lengthens and traverses the rest phase, with a resultant phase-advance of e. Examples of the varieties of spontaneous uncoupling in RR observed in a large population of hamsters are shown in figure 14. As a control, figure 14a shows a long-term actogram

RR

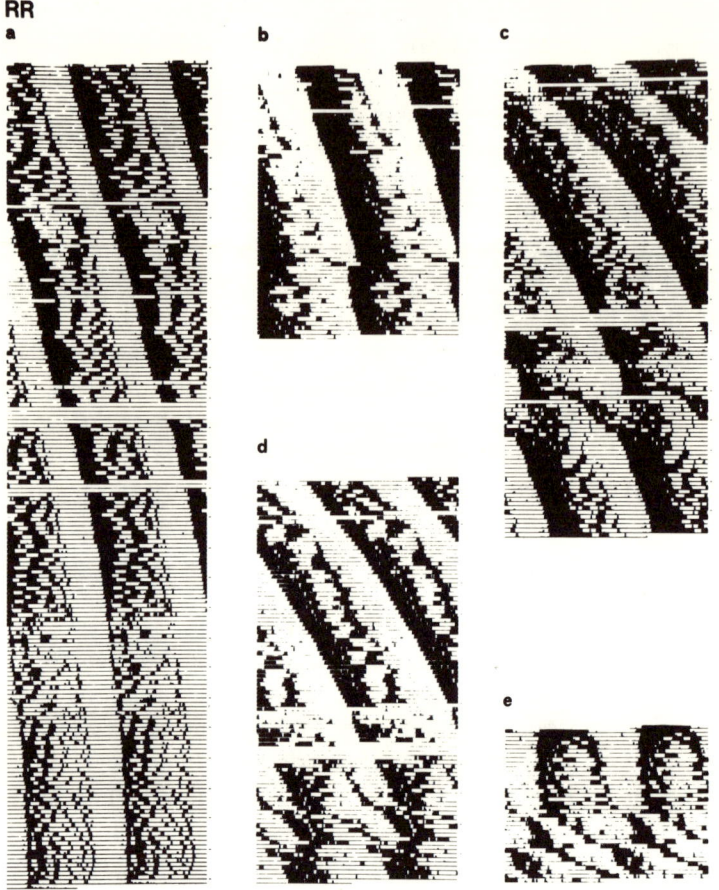

Fig. 14. Long-term actograms of hamster motor activity rhythms in constant dim red light (RR). The actograms are double plotted to facilitate inspection of the records. Details are described in the text.

of a hamster where no uncoupling occurred. There is a clearly delineated activity phase and a rest phase, with a strong *e*, weaker *m*, but noteworthy are the presence of other activity bouts within the active phase and their different frequencies over time. In figure 14b, *m* occurs unusually late in the middle of the rest phase, and is associated with a phase-advance of *e* in the next cycle. In figure 14c, the process is more gradual, *m* lengthening and traversing the rest phase, with a phase-advance of *e* (a distinction between this interpretation, or whether *m* coalesces with *e*, cannot be made in these two examples). In figure 14d, *m* lengthens, traverses the rest phase with a phase-advance of *e*, but then continues to traverse the activity phase with its own long free-running period, beating in and out of phase with *e*. The example in figure 14e was rarely seen, where an apparent 'split' condition with a stable 180° antiphase of *e* and *m* occurred. And underlying rhythmic component of shorter frequency is also visible in this actogram. In all cases, *e* is more strongly defined than *m* in onset precision, intensity, and duration of the activity bout, and has a shorter free-running circadian period.

The *Pittendrigh and Daan* [59] model postulates that the intrinsic period of the oscillator driving *e* is a positive and that of *m* a negative function of light intensity. Our observations support this model, *e* being shorter in RR than in LL, and *m* being longer in RR than in LL. However, computer simulations have demonstrated that the presumed dependency of the periods of *e* and *m* on light intensity is not a necessary condition for splitting [21]. Analogous behaviours arise when the strength of coupling between the oscillators is assumed to be a function of light intensity.

This pattern of lengthening of *m* and uncoupling of *m* and *e* is accelerated by chronic antidepressant drug administration. Two drugs from different classes of antidepressants, a monoamine oxidase inhibitor, clorgyline, and a tricyclic, imipramine, have been found to lengthen the free-running period of the circadian rest-activity cycle, as well as in some animals apparently preferentially lengthening the period of *m* [75, 84]. Some of the varieties of drug-induced lengthening and uncoupling are exemplified by actograms in figure 15. The numbers of animals showing these effects are summarized in table II.

It can be seen that clorgyline, but not imipramine administration induced sedation for a few days. In figure 15a and e lengthening of the entire coupled circadian rest-activity cycle by the drugs is shown. In figure 15b

Table II. Effect of chronic antidepressant drugs in circadian rhythm parameters in female hamsters

Treatment	N	Lengthening of the free-running period (τ)	Prepost-treatment $\Delta\tau$, h	Uncoupling and relative coordination
Controls	20	4/20	$+0.04 \pm 0.02$	5/20
Clorgyline 2 mg/kg/day	18	17/18	-0.61 ± 0.21***	9/18
Imipramine 20 mg/kg/day	19	12/19	-0.36 ± 0.11**	8/19

Paired t-test: ***$p<0.001$; **$p<0.02$.

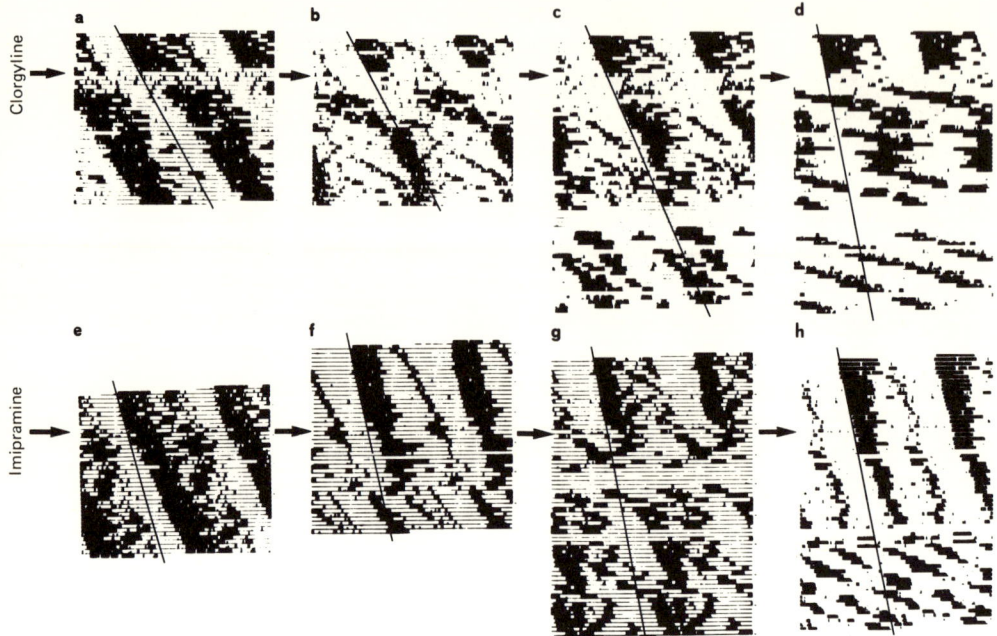

Fig. 15. Actograms of hamster motor activity rhythms in RR before and after (arrow) implantation of osmotic minipumps containing clorgyline (released at the rate of 2 mg/kg/day for 2 weeks) (a–d) or imipramine (20 mg/kg/day for 2 weeks) (e–h). The line drawn through activity onset represents the free-running circadian period for each hamster during the last 10 days before drug treatment. Details for each actogram are described in the text.

and c, f and g, lengthening with uncoupling of *m* and *e* occurs. In figure 15d the rest-activity cycle is markedly lengthened and appears to show relative coordination with an underlying shorter circadian component. In figure 15h, the drug first induces lengthening of the entire coupled rest-activity cycle, followed by a splitting of *m* and *e,* and similar to figure 14e, with an underlying rhythmic component of shorter frequency.

The effects of the antidepressant drugs are therefore complex, varying in the time before onset and in persistence. They were administered in osmotic minipumps (Alzet®) which released drug at a constant rate for 2 weeks. In most cases, the drug effects lasted much longer. Reversal of an effect on withdrawal of the agent is an important criterion in drug studies. However, it is known that agents directly affecting the circadian system (e.g. oestradiol, [52]) often show long-lasting after-effects, and this appears to be the case with antidepressant drugs.

In free-running human subjects the rest-activity cycle does not show the two activity components found in nocturnal rodents, but can uncouple as a unit from the circadian temperature rhythm, showing a relative coordination between the two oscillations [82]. The temperature rhythm remains in the circadian range with a free-running period of ca 25 h, whereas the rest-activity cycle can lengthen to 30–50 h. Both the fast and the slow frequency can be seen within a single variable, the rest-activity cycle (fig. 16a). Comparison of such an actogram in man with an actogram in a drug-treated hamster (fig. 16b) suggest that the rest-activity cycle broke away from the underlying shorter circadian temperature rhythm in both cases. Concomitant measurement of the circadian temperature and rest-activity rhythms

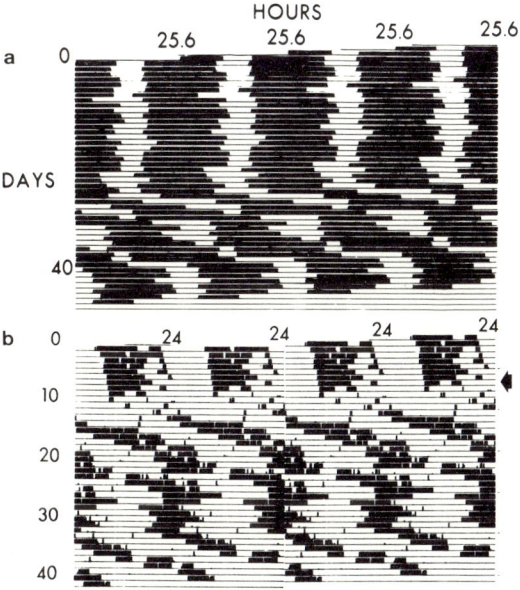

Fig. 16. a Circadian rest-activity cycle for a free-running normal human subject redrawn from *Wever* [82]. The rest-activity cycle is drawn on a 25.6-h axis (the calculated free-running period for this subject before uncoupling of the rest-activity and temperature rhythms). The actogram is plotted four times to facilitate inspection of the records. On day 27 the rest-activity cycle lengthens and shows relative coordination with the underlying circadian temperature rhythm. *b* Circadian rest-activity cycle for a free-running hamster before and after (arrow) treatment with the antidepressant drug clorgyline. The actogram is plotted four times on a 24-h axis. After drug treatment the rest-activity cycle lengthens to 27.45 h and between days 20 and 40 appears to show relative coordination with an underlying shorter circadian component.

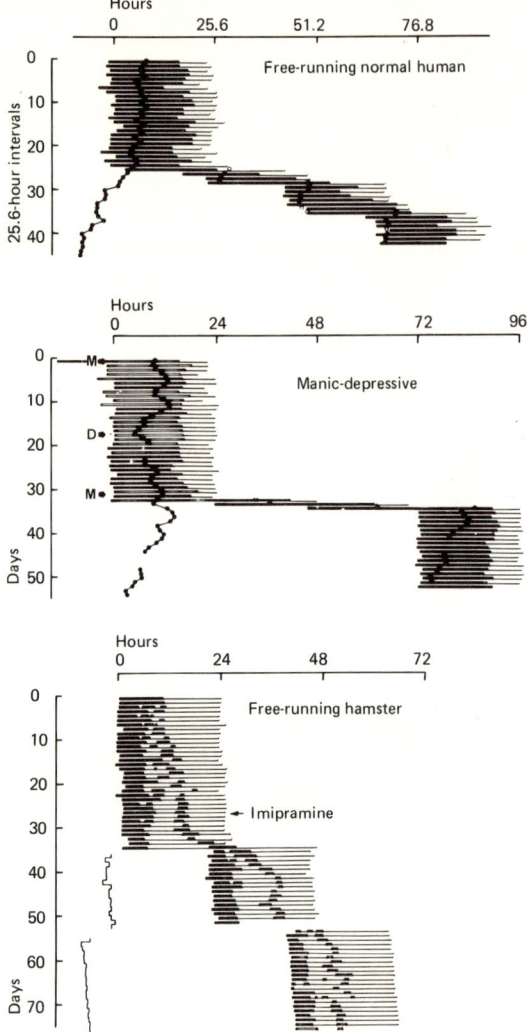

Fig. 17. Rest-activity cycle of a free-running normal human subject, redrawn from *Wever* [82], a manic-depressive patient, and a free-running hamster treated with imipramine. All are single-plotted: where the rest-activity cycle lengthens it traverses multiple plot of the axis re-presenting the circadian day. The maximum of the circadian temperature rhythm is plotted as closed circles on the appropriate time of the activity cycle; at the break-away point where the rest-activity cycle lengthens there is a parallel phase-advance of the temperature rhythm. In the free-running subject this occurs spontaneously after about three weeks under isolation from time cues; in the depressed patient this occurs near the switch out of depression into mania (clinical state indicated by D = depression, M = mania). In the imipramine-treated

Table III. Similarities between humans and hamsters in circadian rhythm characteristics

1 Free-running circadian period (τ_{DD}) > 24.0 h
2 Under constant conditions uncoupling of 2 circadian components can occur
3 Uncoupling can occur in either direction (in hamsters RR and LL have opposite signs on the uncoupling)
4 There can be relative coordination between these two components
5 In hamsters, $\tau_{DD} < \tau_{LL}$: it is not clear whether man obeys Aschoff's rule, as a diurnal mammal showing a shorter circadian period in LL than in DD. However, findings in bunker experiments with blind subjects [45b] and normal subjects in complete darkness [6, pp. 88–89] indicate that $\tau_{DD/blind} < \tau_{LL/sighted}$ suggestive of man in this respect also being similar to a hamster

---

under conditions where uncoupling of circadian rhythms occur is required to confirm or disprove this apparent parallel between the species.

A further analogy is suggested in figure 17. Here both temperature and activity rhythms have been measured in a free-running normal subject and in a manic-depressive patient hospitalized on a psychiatric ward. The switch out of depression into mania is accompanied by an uncoupling of the rest-activity cycle and the temperature rhythm, even though the subject is entrained (described in detail in ref. [75]). The switch out of depression into mania is promoted by antidepressant drugs. Thus, the analogy with a hamster treated with imipramine, where the activity component *m* breaks away from *e*, may be clinically relevant and related to the mode of action of these drugs. Even though in the hamster temperature was not measured, the phase-advance of *e* occurring when *m* lengthens is remarkably similar to the phase-advance of the temperature rhythm in the human subjects at the break-away point.

The actograms in figure 15 where *m* lenghtened and broke away from *e* can also be considered in this light: that an oscillator 'E' driving the *e* activity bout also drives the circadian temperature rhythm, and that these two are tightly coupled and remain in the circadian range, whereas an oscillator 'M' driving the *m* activity bout has a much larger range of entrainment. Antidepressant drugs, apart from their intrinsic interest as agents that can

---

hamster, where temperature was not measured, an outline of the activity onset of the *e* component is shown undergoing a phase-advance each time the *m* activity component lengthens and traverses the rest phase, analogous to the phase-advance of the temperature rhythm in the other two examples.

modulate frequency of biological oscillations, therefore provide a new tool for investigating the circadian system. They can induce rapid uncoupling of circadian rhythms, a phenomenon which usually requires at least 60 days in LL [52, 59].

Table III summarizes the similarities between human and hamster circadian rhythm characteristics, and opens up consideration of the potential usefulness of this animal model.

Sleep 1980. 5th Eur. Congr. Sleep Res., Amsterdam 1980, pp. 72–79
(Karger, Basel 1981)

Sleep Onset REM Episodes in Depression[10, 11]

Hartmut Schulz[12], München, FRG

Introduction

Most depressive patients show an abnormal sleep pattern with frequent sleep interruptions and a characteristic redistribution of the sleep stages. The amount of delta sleep is reduced and the amount of REM sleep is increased in the first half of sleep [48]. The first REM episode occurs prematurely [44] and its eye movement density is considerably higher than in normals [25, 64].

While REM sleep latencies in normals are distributed unimodally with a mean value of 60–70 min, those of depressed patients are distributed bimodally with one peak in the first 20 min after sleep onset (sleep onset REM episodes, SOREMs) and the other at about 60 min after sleep onset [63]. In the following the REM latencies of a larger group of depressive patients will be presented and the results will be discussed within the framework of chronobiology.

---

[10] Dedicated to Prof. *D. Ploog* on his 60th birthday.

[11] For references see compound list on page 79.

[12] The author is member of a research group studying chronobiological aspects of depression at the Max-Planck-Institute for Psychiatry, München.

Methods

18 endogenously depressed patients were studied for 14–21 days. 9 of the patients were further studied during a 14-day follow-up after full remission. All patients were free of medication during the period of study (exceptions are documented elsewhere [63], table IV). Bed rest was between 2200 and 0700 hours. Sleep onset was defined as the first epoch stage 2 sleep. REM latency was defined as the time interval between sleep onset and the first epoch of REM sleep. The results from 6 of the depressed and 4 of the remitted patients were already dealt with in the above-mentioned report.

Results

The distribution of the REM latencies for the depressed and the remitted patients are shown in figure 18. For the depressed patients this distribution is clearly bimodal with peaks between 0 and 5 min and between 55

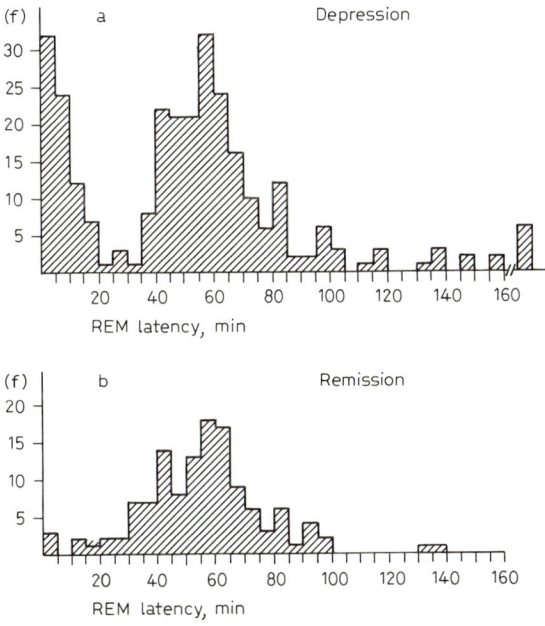

Fig. 18. a Frequency (f) distribution of the REM latencies of 18 endogenously depressed patients (n = 281 sleep episodes). *b* Distribution of the REM latencies of n = 127 sleep episodes of 9 of these patients (patients No. 2, 5, 6, 7, 10, 11, 12, 13, 14 according to table IV) after full remission. The x-axis gives the latency in minutes, the y-axis the absolute frequency.

and 60 min and a trough between 20 and 35 min after sleep onset. The distribution of the remitted patients is unimodal with the median in the interval between 55 and 60 min.

For each patient the frequency of REM latencies for the interval 0–20, 21–40 and more than 40 min are given in table IV. The percentage of sleep episodes with a SOREM (REM latency ≤ 20 min) varies between 87 and 0% for this group of patients.

While 11 patients had at least 1 SOREM during depression, 7 patients had none. In most patients, who had SOREMs, the frequency of REM latencies in the interval 21–40 min is smaller than in the adjacent intervals. This indicates that the bimodal distribution is not simply the result of the combination of two subgroups of patients, namely those with SOREMs and

Table IV. Subject identification and summary of REM sleep measures for 18 endogenously depressed patients

Patient[1]	Sex	Age	Number of analyzed bed rests	REM latency, min			SOREM %
				≤ 20	21–40	> 41	
1	m	47	21	18	1	2	84.2
2	f	56	12	8	3	1	66.7
3	f	53	12	7	1	4	58.3
4	f	67	14	7	1	6	50.0
5	m	46	21	9	2	10	42.9
6	m	33	3	1	0	2	33.3
7	f	46	20	6	0	14	30.0
8	m	43	18	5	2	11	27.8
9	f	52	12	3	0	9	25.0
10	m	33	21	5	0	16	23.8
11	m	31	21	3	0	18	14.3
12	f	69	14	0	3	11	0.0
13	f	57	14	0	0	14	0.0
14	m	30	20	0	0	20	0.0
15	f	58	14	0	0	14	0.0
16	m	45	10	0	0	10	0.0
17	m	48	19	0	0	19	0.0
18	f	57	15	0	0	15	0.0

[1] The data of 6 of the present patients have already been published [63]. The patient numbers 1–6 of the earlier study correspond to the patient numbers 5, 10, 6, 11, 2 and 9 in the present table. In the previous publication the age of patient No. 11 was given erroneously as 30 instead of 31 years.

those without, but rather that this distribution is typical for the majority of patients studied.

Discussion

SOREMs in Depressive Patients. SOREMs have been described as a typical feature of the sleep of depressed patients since the earlier electrophysiological sleep studies in these patients [69]. A reanalysis of REM latencies of 269 sleep episodes of 14 depressed patients, which were extracted from the literature, confirms the assumption that a bimodal distribution of REM latencies is typical for many of these patients (fig. 19).

A comparison of the different studies is hampered by the fact that the definition of the REM latency is not standardized. REM latency may be

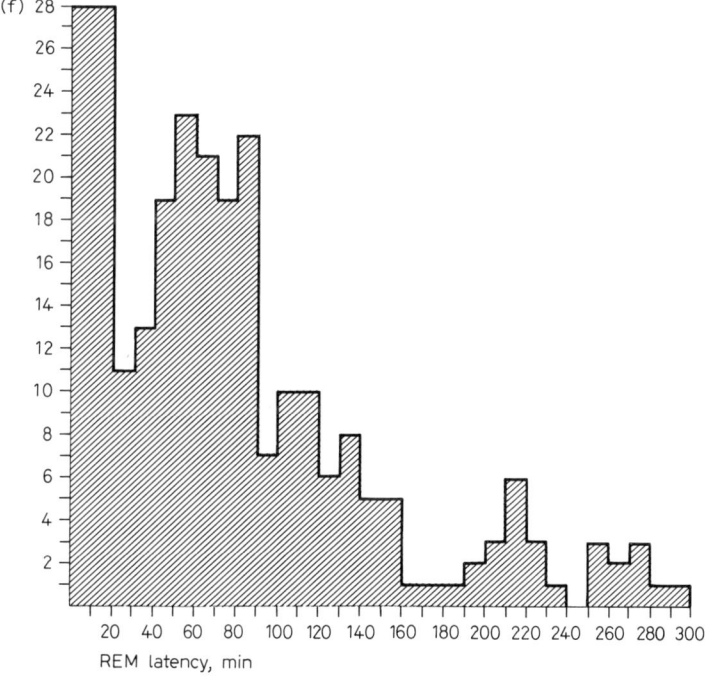

Fig. 19. REM latency distribution of 14 depressed patients (n = 269 sleep episodes). The data were extracted from the following publications: [45a] (N = 6 patients, n = 76 sleep episodes); [47] (N = 5, n = 118); [27] (N = 1, n = 10); [34] (N = 1, n = 32); [44] (N = 1, n = 33).

either the time interval between sleep onset and the first REM episode or the time asleep within this time interval. Furthermore, sleep onset may be defined as the first epoch stage 1 or stage 2 sleep after lights off. Nevertheless, the overall distribution is bimodal with a trough between 20 and 40 min after sleep onset. A bimodal form of distribution may also be inferred from the data of *Snyder* ([68], table 3). He reports intraindividually short mean REM latencies (mostly under 20 min) for psychotic depressed patients and longer mean REM latencies (mostly over 40 min) for non-psychotic patients. There is only a small overlap of the ranges in the critical interval of 20–40 min. In this, as in other studies, SOREMs are presumably not the result of a pharmacologically induced REM rebound [48], because the patients have either had no antidepressive medication before the investigation or the medication has been withdrawn for a longer period of time ([68], table 2).

Coble et al. [18] who reanalyzed 737 sleep episodes of 22 drug-free depressed inpatients also found a bimodal REM latency distribution. About 18% of the nights show REM latencies ≤ 10 min. The second peak of the distribution is located between 40 and 60 min.

SOREMs in Other Patients and Healthy Subjects. During ontogenesis SOREMs occur in the first months of life. More than half of the sleep episodes during the first 5 months begin with a REM episode [56]. As can be seen from data of a study of *Salzarulo et al.* [62] on continuously fed babies, the REM latency distribution is bimodal (fig. 20b). In adults SOREMs are not only seen in depressed patients but also in patients with a diagnosis of narcolepsy-cataplexy [60]. *Montplaisir et al.* [50] have additionally shown that the distribution of the REM latencies in such patients is bimodal (fig. 20d).

Healthy persons exhibit SOREMs only under special conditions, such as ultradian sleep-wake cycles (e.g. '90-min days' [16], '180-min days' [78]), rapid time zone shifts [24], and occasionally in isolation from external *zeitgebers* which normally entrain the sleep-wake cycle to 24 h [19, 88]. One example of SOREMs in free-running sleep-wake cycles is given in figure 20c, (single case data, adapted from [17]). Finally, SOREMs have also been observed in a subject living on a 48-h day [46].

The Sleep Onset Gate. The common characteristics of these studies and some models which could explain the occurrence of SOREMs shall now be discussed. The models are: (a) the chronobiological model; (b) the REM pressure model, and (c) the neurochemical model. Only the chronobiological

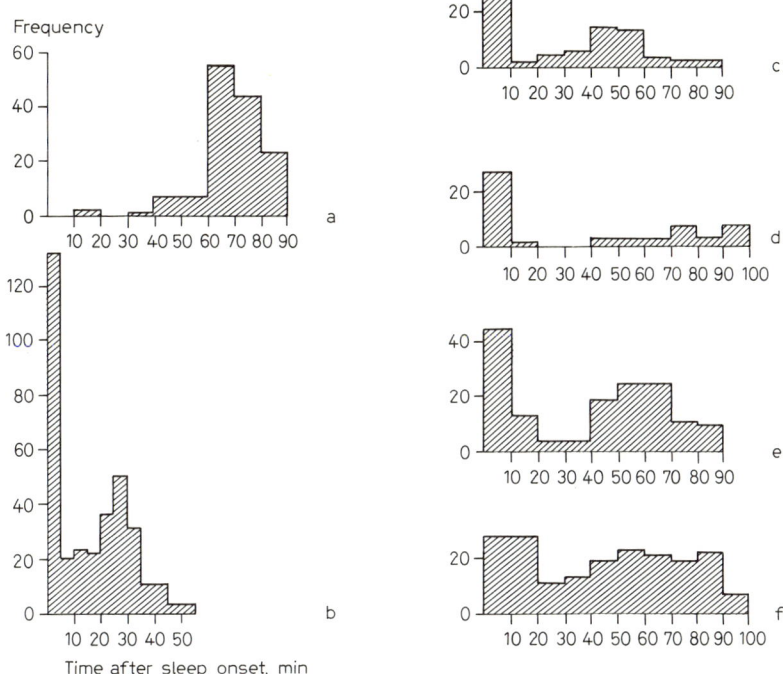

Fig. 20. REM latency distributions from different experiments. *a* shows the REM latency distribution of N = 6 young healthy subjects, n = 140 sleep onsets [unpublished data]. *b* N = 15 babies, n = 339 sleep onsets [unpublished data from Dr. *P. Salzarulo*, for further information see [62]. *c* N = 1 subject, isolated from external *zeitgebers*, n = 74 sleep onsets [17]. *d* N = 20 narcoleptic patients, n = 54 sleep onsets [50]. *e* N = 11 endogenously depressed patients, n = 183 sleep onsets (part of the present data, see table IV). *f* N = 14 depressed patients, n = 269 sleep onsets (same data as in fig. 19). A typical feature of the distributions *b–f* is the bimodality.

model will be discussed in some detail because it is the basis of our reasoning on the results on REM latency.

The results which were presented above support the hypothesis that sleep onset has a gating function for REM sleep. REM sleep may be released in the proximity of sleep onset or about 60 min later. While under normal conditions the gate is closed at sleep onset and open at the end of the first NREM episode, it can be open at sleep onset under the various conditions described earlier in this paper. A similar gating mechanism was observed in depressed patients after forced awakenings in the middle of the night sleep episode. The distribution of the REM latencies after the post-awakening sleep onset was bimodal with peaks at 5 and 50 min [65].

Discontinuous REM latency distributions, which may be caused by a gating mechanism, have been known in sleep research for a long time. As early as 1957, *Dement and Kleitman* [22] described the phenomenon of skipped or missed first REM episodes which are followed by a NREM episode of normal duration.

The probability that the gate for REM sleep is open at sleep onset may depend on a threshold for triggering of REM sleep. Different factors which influence this REM threshold have been proposed in the literature. Those relevant for the explanation of shortened REM latencies in depressed patients will be discussed here.

A *chronobiological hypothesis* was developed in the last few years when it became evident that REM sleep propensity varies with the time of day [37, 43, 74, 78]. Recently it could be shown that REM propensity and the distribution of REM sleep within total sleep depends on the phase relationship between bed rest and body core temperature [20, 89, 91]. The peak of the circadian REM sleep rhythm coincides with the trough of the circadian body temperature cycle. In the light of this interaction hypothesis between body temperature and REM sleep the seemingly disparate conditions where SOREMs occur show some striking similarity. In all these conditions a monophasic 24-h rest-activity cycle is either absent (babies; ultradian sleep-wake schedules) or attenuated (narcolepsy). Thus, the triggering of REM sleep at sleep onset may be facilitated either by a reduced amplitude of the circadian rest-activity or arousal cycle [63] or by a phase advance of the body temperature cycle relative to sleep as in free-running sleep-wake cycles [20, 46, 89].

Phase advance of different circadian rhythms has been reported as occurring in manic-depressive patients [77]. These authors proposed that a phase advance may possibly induce a depressive phase. They also demonstrated that an experimentally induced phase shift of the bed rest may have a recuperative effect in these patients [76].

Snyder [69] considered *REM pressure* as a critical factor for the triggering of REM sleep. According to this hypothesis, there is a deficit of REM sleep early in depression. A pressure for REM sleep builds up at this time. Later in the course of the illness this deficit will be compensated by a high REM sleep propensity. The assumed REM pressure could explain the premature occurrence of REM sleep at sleep onset.

A *neurochemical hypothesis* was proposed by *Sitaram et al.* [67]. They could show that an increased cholinergic activity, which was induced by the application of physostigmine or arecholine, reduced the REM latency. REM induction by cholinergic mechanisms was not only shown in humans but also in the cat [1] and the rat [35].

At the moment it seems premature to decide on the appropriateness of the different models or to envisage their possible combination. The aim of the present study was to clarify the phenomenon of shortened REM latencies in depressed patients and to propose possible explanations of this phenomenon.

Acknowledgements

I thank Mrs. *Renate Stillger* and Mrs. *Christa Eversmeyer* for their careful technical assistance.

List of References of Symposium B

1 Amatruda, T.T., III; Black, D. A.; McKenna, T.M.; McCarley, R.W.; Hobson, J.A.: Sleep cycle control and cholinergic mechanisms: differential effects of carbachol injections at pontine brain stem sites. Brain Res. *98:* 501–515 (1975).
2 Aschoff, J.: Exogenous and endogenous components in circadian rhythms. Cold Spring Harb. Symp. quant. Biol. *25:* 11–28 (1960).
3 Aschoff, J.: Circadian rhythms in man. Science *148:* 1427–1432 (1965).
4 Aschoff, J.: Response curves in circadian periodicity; in Aschoff, Circadian clocks, pp. 95–111 (North-Holland, Amsterdam 1965).
5 Aschoff, J.: Circadian rhythm of activity and of body temperature; in Hardy, Gagge, Stolwijk, Physiological and behavioral temperature regulation, pp. 905–919 (Thomas, Springfield 1970).
6 Aschoff, J.: Features of circadian rhythms relevant for the design of shift schedules. Ergonomics *21:* 739–754 (1978).
7 Aschoff, J.: Circadian rhythms: general features and endocrinological aspects; in Krieger, Endocrine rhythms, pp. 1–61 (Raven Press, New York 1979).
8 Aschoff, J.: Circadian rhythms: influences of internal and external factors on the period measured in constant conditions. Z. Tierpsychol. *49:* 225–249 (1979).
9 Aschoff, J.; Gerecke, U.; Wever, R.: Desynchronization of human circadian rhythms. Jap. J. Physiol. *17:* 450–457 (1967).
10 Aschoff, J.; Hoffman, K.; Pohl, H.; Wever, R.: Re-entrainment of circadian rhythms after phase shifts of the Zeitgeber. Chronobiologia *2:* 23–78 (1975).

11   Aschoff, J.; Wever, R.: Human circadian rhythms: a multi-oscillatory system. Fed. Proc. *35:* 2326–2332 (1976).
12   Bergmann, B.: A comparison of rat sleep-wake behavior to measured electrophysiological categories defined by a statistical model; unpublished doct. diss., Chicago (1980).
13   Bergmann, B.; Rosenberg, R.; Eastman, C.; Metz, J.; Rechtschaffen, A.: A parametric system for scoring sleep and arousal in the rat. Sleep Res. *6:* 206 (1977).
14   Borbély, A.: Effects of light on sleep and activity rhythms. Prog. Neurobiol. *10:* 1–31 (1978).
15   Borbély, A.A.: Effects of light and circadian rhythms on the occurrence of REM sleep in the rat. Sleep *2:* 289–298 (1980).
16   Carskadon, M. A.; Dement, W.C.: Sleep studies on a 90-minute day. Electroenceph. clin. Neurophysiol. *39:* 145–155 (1975).
17   Chouvet, G.; Mouret, J.; Coindet, J.; Siffre, M.; Jouvet, M.: Périodicité bicircadienne du cycle veille-sommeil dans des conditions hors du temps. Etude polygraphique. Electroenceph. clin. Neurophysiol. *37:* 367–380 (1974).
18   Coble, P. A.; Kupfer, D.J.; Spiker, D.G.; Neil, J.F.: Distribution of REM latency in depression. I. Proc. 20th Annu. Meet. Ass. for the Psychophysiological Study of Sleep, Mexico 1980, p. 158.
19   Czeisler, C.A.: Human circadian physiology: internal organization of temperature, sleep-wake and neuroendocrine rhythms monitored in an environment free of time cues; PhD diss., Stanford (1978).
20   Czeisler, C.A.; Zimmerman, J.C.; Ronda, J.M.; Moore-Ede, M.C.; Weitzman, E.D.: Timing of REM sleep is coupled to the circadian rhythm of body temperature in man. Sleep *2:* 329–346 (1980).
21   Daan, S.; Berde, C.: Two coupled oscillators: simulations of the circadian pacemaker in mammalian activity rhythms. J. theor. Biol. *70:* 297–313 (1978).
22   Dement, W.; Kleitman, N.: Cyclic variations in EEG during sleep and their relation to eye movements, body motility, and dreaming. Electroenceph. clin. Neurophysiol. *9:* 673–690 (1957).
23   Dörrscheidt, G.; Beck, L.: Advanced methods for evaluating characteristic parameters (τ, α, ρ) of circadian rhythms. J. Math. Biol. *2:* 107–121 (1975).
24   Endo, S.; Yamamoto, T.; Sasaki, M.: Effects of time zone changes on sleep: West-east flight and east-west flight. Jikeikai med. J. *25:* 249–268 (1978).
25   Foster, F.G.; Kupfer, D.J.; Coble, P.; McPartland, R.J.: Rapid eye movement sleep density. Archs gen. Psychiat. *33:* 1119–1123 (1976).
26   Fuller, C.A.; Sulzman, F.M.; Moore-Ede, M.C.: The effect of suprachiasmatic lesions on circadian rhythms in the squirrel monkey *(Saimiri sciureus)*. Soc. Neurosci. Abstr. *3:* 162 (1977).
27   Green, W.J.; Stajduhar, P.P.: The effect of ECT on the sleep-dream cycle in a psychotic depression. J. nerv. ment. Dis. *143:* 123–134 (1966).
28   Groos, G.A.; Mason, R.: Maintained discharge of rat suprachiasmatic neurons at different adaptation levels. Neurosci. Lett. *8:* 59–64 (1978).
29   Groos, G.A.; Hendriks, J.: Regularly firing neurones in the rat suprachiasmatic nucleus. Experientia *35:* 1597–1599 (1979).
30   Groos, G.A.: Electrophysiological evidence for the absence of photosensitive neurones in the rat suprachiasmatic nucleus. IRCS med. Sci. *7:* 342–343 (1979).

31 Groos, G.A.; Mason, R.: Visual properties of rat and cat suprachiasmatic neurones. J. comp. Physiol. *135:* 349–356 (1980).
32 Groos, G.A.; Kooy van der, D.: Functional absence of brain photoreceptors mediating entrainment of circadian rhythms in the adult rat. Experientia (in press).
33 Güldner, F.H.; Ingham, C.A.: Plasticity in synaptic appositions of optic nerve afferents under different lighting conditions. Neurosci. Lett. *14:* 235–240 (1979).
34 Hawkins, D.R.; Mendels, J.; Scott, J.; Bensch, G.; Teachy, W.: The psychophysiology of sleep in psychotic depression: a longitudinal study. Psychosomat. Med. *29:* 329–344 (1967).
35 Hill, S.Y.; Reyes, R.B.; Kupfer, D.J.: Physostigmine induction of REM sleep in imipramine treated rats. Commun. Psychopharmacol. *3:* 261–266 (1979).
36 Honma, K.; Watanabe, K.; Hiroshige, T.: Effects of parachlorophenylalanine and 5,6-dihydroxytryptamine on the free-running rhythms of locomotor activity and plasma corticosterone in the rat exposed to continuous light. Brain Res. *169:* 531–544 (1979).
37 Hume, K.I.; Mills, J.N.: Rhythms of REM and slow-wave sleep in subjects living on abnormal time schedules. Waking Sleeping *1:* 291–296 (1977).
38 Ibuka, N.; Kawamura, H.: Loss of circadian rhythm in sleep-wakefulness cycle in the rat by suprachiasmatic nucleus lesions. Brain Res. *96:* 76–81 (1975).
39 Ibuka, N.; Inouyé, S.T.; Kawamura, H.: Analysis of sleep-wakefulness rhythms in male rats after suprachiasmatic nucleus lesions and ocular enucleation. Brain Res. *122:* 33–47 (1977).
40 Ibuka, N.: Sleep-wakefulness rhythms in mice after suprachiasmatic nucleus lesions. Waking Sleeping *4:* 167–173 (1980).
41 Inouye, S.T.; Kawamura, H.: Persistence of circadian rhythmicity in a mammalian hypothalamic island containing the suprachiasmatic nucleus. Proc. natn. Acad. Sci. USA *76:* 5962–5966 (1979).
42 Kan, J.P.; Chouvet, G.; Hery, F.; Debilly, G.; Mermet, A.; Glowinski, J.; Pujol, J.F.: Daily variations of various parameters of serotonin metabolism in the rat brain. I. Circadian variations of tryptophane-5-hydroxylase in the raphé nuclei and the striatum. Brain Res. *123:* 125–136 (1977).
43 Karacan, I.; Finley, W.W.; Williams, R.L.; Hursch, C.J.: Changes in stage 1-REM and stage 4 sleep during naps. Biol. Psychiat. *2:* 261–265 (1970).
44 Kupfer, D.J.: REM latency: a psychobiologic marker for primary depressive disease. Biol. Psychiat. *11:* 159–174 (1976).
45a Lowy, F.H.; Cleghorn, J.M.; McClure, D.J.: Sleep patterns in depression. J. nerv. ment. Dis. *153:* 10–26 (1971).
45b Lund, R.: Circadiane Periodik physiologischer und psychologischer Variablen bei 7 blinden Versuchspersonen mit und ohne Zeitgeber; Diss., München (1974).
46 Meddis, R.: Human circadian rhythms and the 48 hour day. Nature, Lond. *218:* 964–965 (1968).
47 Mendels, J.; Hawkins, D.R.: Sleep and depression. IV. Longitudinal studies. J. nerv. ment. Dis. *153:* 251–272 (1971).
48 Mendelson, W.B.; Gillin, J.C.; Wyatt, R.J.: Human sleep and its disorders (Plenum Press, New York 1977).
49 Mills, J.: Transmission processes between clock and manifestations; in Mills, Biological aspects of circadian rhythms, pp. 27–83 (Plenum Press, New York 1973).

50 Montplaisir, J.; Billiard, M.; Takahashi, S.; Bell, I.R.; Guilleminault, C.; Dement, W.C.: Twenty-four-hour recording in REM-narcoleptics with special reference to nocturnal sleep disruption. Biol. Psychiat. *13:* 73–89 (1978).

51 Moore, R.Y.: Retino-hypothalamic projections in mammals: a comparative study. Brain Res. *49:* 403–409 (1973).

52 Morin, L.P.: Effect of ovarian hormones on synchrony of hamster circadian rhythms. Physiol. Behav. *24:* 741–749 (1980).

53 Mouret, J.; Coindet, J.; Chouvet, G.: Effet de la pinéalectomie sur les états et les rythmes de sommeil du rat mâle. Brain Res. *81:* 97–105 (1974).

54 Nishino, H.; Kolzumi, K.: Responses of neurons in the suprachiasmatic nuclei of the hypothalamus to putative transmitters. Brain Res. *120:* 167–172 (1977).

55 Oliverio, A.: Sleep and activity rhythms in mice. Waking Sleeping *4:* 155–166 (1980).

56 Paul, K.; Dittrichova, J.: The process of falling asleep in infancy. Activitas nerv. sup. *19:* 272–273 (1977).

57 Pavlidis, T.: Qualitative similarities between the behavior of coupled oscillators and circadian rhythms. Bull. math. Biol. *40:* 675–692 (1978).

58 Pickard, G.E.: Morphological characteristics of retinal ganglion cells projecting to the suprachiasmatic nucleus: a horseradish peroxidase study. Brain Res. *183:* 458–465 (1980).

59 Pittendrigh, C.S.; Daan, S.: A functional analysis of circadian pacemakers in nocturnal rodents. V. Pacemaker structure: a clock for all seasons. J. comp. Physiol. *106:* 333–355 (1976).

60 Rechtschaffen, A.; Wolpert, E.A.; Dement, W.C.; Michtell, S.A.; Fisher, C.: Nocturnal sleep of narcoleptics. Electroenceph. clin. Neurophysiol. *15:* 599–609 (1963).

61 Rusak, B.; Zucker, I.: Neural regulation of circadian rhythms. Physiol. Rev. *59:* 449–526 (1979).

62 Salzarulo, P.; Fagioli, I.; Salomon, F.; Ricour, C.; Raimbault, G.; Ambrosi, S.; Cicchi, O.; Duhamel, J.F.; Rigoard, M.T.: Sleep patterns in infants under continuous feeding from birth. Electroenceph. clin. Neurophysiol. *49:* 330–336 (1980).

63 Schulz, H.; Lund, R.; Cording, C.; Dirlich, G.: Bimodal distribution of REM sleep latencies in depression. Biol. Psychiat. *14:* 595–600 (1979).

64 Schulz, H.; Trojan, B.: A comparison of eye movement density in normal subjects and in depressed patients before and after remission. Sleep Res. *8:* 49 (1979).

65 Schulz, H.; Zulley, J.: Consistency and persistency of the REM-NREM sleep rhythm (in press).

66 Schwartz, W.J.; Davidsen, L.C.; Smith, C.B.: In vivo metabolic activity of a putative circadian oscillator, the rat suprachiasmatic nucleus. J. comp. Neurol. *189:* 157–167 (1980).

67 Sitaram, N.; Mendelson, W.B.; Wyatt, R.J.; Gillin, J.C.: The time-dependent induction of REM sleep and arousal by physostigmine infusion during normal human sleep. Brain Res. *122:* 562–567 (1977).

68 Snyder, F.: Electrographic studies of sleep in depression.; in Kline, Laska, Computers and electronic devices in psychiatry, pp. 272–303 (Grune & Stratton, New York 1968).

69 Snyder, F.: NIH studies of EEG sleep in affective illness; in Williams, Katz, Shields, Recent advances in the psychobiology of the depressive illnesses, pp. 171–192 (US Government Printing Office, Washington 1972).

70 Stephan, F.K.; Nunez, A.A.: Elimination of circadian rhythms in drinking, activity, sleep and temperature by isolation of the suprachiasmatic nuclei. Behav. Biol. *20:* 1–16 (1977).

71 Stephan, F.K.; Zucker, I.: Circadian rhythms in drinking and motor activity of rats are eliminated by hypothalamic lesions. Proc. natn. Acad. Sci. USA *69:* 1583–1586 (1972).

72 Valatx, J.L.; Bugat, R.: Facteurs génétiques dans le déterminism du cycle veille-sommeil chez la souris. Brain Res. *69:* 315–330 (1974).

73 Valatx, J.L.; Cespuglio, R.; Paut, L.; Bailey, D.W.: Etude génétique du sommeil paradoxal chez la souris. I. Liaison possible avec les gènes de coloration. Waking Sleeping *4:* 175–184 (1980).

74 Webb, W.B.; Agnew, H.W., Jr.; Williams, K.L.: Effect on sleep of a sleep period time displacement. Aerospace Med. *42:* 152–155 (1971).

75 Wehr, T.A., Goodwin, F.K.; Wirz-Justice, A.; Lewy, A.J.: Uncoupling of circadian oscillators in manic-depressive illness (submitted, 1980).

76 Wehr, T.A.; Wirz-Justice, A.; Goodwin, F.K.; Duncan, W.; Gillin, J.C.: Phase advance of the circadian sleep-wake cycle as an antidepressant. Science *206:* 710–713 (1979).

77 Wehr, T.A.; Muscettola, G.; Goodwin, F.K.: Urinary 3-methoxy-4-hydroxyphenylglycol circadian rhythm. Archs gen. Psychiat. *37:* 257–263 (1980).

78 Weitzman, E.D.; Nogeire, C.; Perlow, M.; Fukushima, D.; Sassin, G.; McGregor, P.; Gallagher, T.F.; Hellman, L.: Effects of a prolonged 3-hour sleep-wake cycle on sleep stages, plasma cortisol, growth hormone, and body temperature in man. J. clin. Endocr. Metab. *38:* 1018–1030 (1974).

79 Wever, R.: Virtual synchronization towards the limits of the range of entrainment. J. theor. Biol. *36:* 119–132 (1972).

80 Wever, R.: ELF-effects on human circadian rhythms; in Persinger, ELF and VLF electromagnetic field effects, pp. 101–144 (Plenum Press, New York 1974).

81 Wever, R.: The circadian multi-oscillator system of man. Int. J. Chronobiol. *3:* 19–55 (1975).

82 Wever, R.A.: The circadian system of man: results of experiments under temporal isolation (Springer, New York 1979).

83 Wiedenmann, G.: Two activity peaks in the circadian rhythm of the cockroach *Leucophaea maderae.* J. interdiscipl. Cycle Res. *8:* 378–383 (1977).

84 Wirz-Justice, A.; Wehr, T.A.; Goodwin, F.K.; Kafka, M.S.; Naber, D.; Marangos, P.J.; Campbell, I.C.: Antidepressant drugs slow circadian rhythms in behavior and brain neurotransmitter receptors. Psychopharmacol. Bull. *16:* 45–47 (1980).

85 Wirz-Justice, A.; Wehr, T.A.; Goodwin, F.K.; Campbell, I.C.; Murphy, D.L.: Antidepressant drugs slow and uncouple circadian rhythms (submitted, 1980).

86 Wirz-Justice, A.; Wehr, T.A.; Goodwin, F.K.: Spontaneous uncoupling of hamster circadian rhythms (unpublished).

87 Zatz, M.; Brownstein, M.J.: Intraventricular carbachol mimics the effects of light on the circadian rhythm in the rat pineal gland. Science *203:* 358–360 (1979).

88 Zulley, J.: Der Einfluss von Zeitgebern auf den Schlaf des Menschen (Fischer, Frankfurt 1979).

89 Zulley, J.: Distribution of REM sleep in entrained 24-hour and free-running sleep-wake cycles. Sleep *2:* 377–389 (1980).

90 Zulley, J.: Comparison of long and short sleep durations in free-running sleep-wake cycles (this volume, Workshop D).

91 Zulley, J.; Schulz, H.: Sleep and body temperature in free-running sleep-wake cycles; in Popoviciu, Asgian, Badiu, Sleep 1978. 4th Eur. Congr. on Sleep Res., Tirgu-Mures 1978, pp. 341–344 (Karger, Basel 1980).

Addresses of Authors

Prof. A. A. Borbély, Pharmakologisches Institut der Universität Zürich, Gloriastrasse 32, CH-8006 Zürich (Switzerland)

Dr. G. Chouvet, INSERM U171, Département de Médecine Expérimentale, Université Claude Bernard, 8 Avenue Rockefeller, F-69008 Lyon (France)

Dr. Ch. Eastman, Sleep Laboratory, University of Chicago, 5741 South Drexel Avenue, Chicago, IL 60637 (USA)

Dr. G. A. Groos, Department of Physiology and Physiological Physics, University of Leiden, Leiden (The Netherlands)

Dr. H. Schulz, Max-Planck-Institute for Psychiatry, München (FRG)

Dr. A. Wirz-Justice, National Institute of Mental Health, Clinical Psychobiology Branch, Bethesda, MD 20205 (USA), and Psychiatrische Universitätsklinik, CH-4025 Basel (Switzerland)

C. Sleep and Emotional Stress

Chairmen: *D. Schneider-Helmert*, Königsfelden; *P. Visser*, Amsterdam
Participants: *T. Åkerstedt*, Stockholm; *J. Bastiaans*, Leiden; *R. Cluydts*, Amsterdam; *J. A. Horne*, Loughborough; *J. De Koninck*, Ottawa

Introduction

D. Schneider-Helmert, Königsfelden, Switzerland;
P. Visser, Amsterdam, The Netherlands

The present symposium evoked gratitude for the presentation of thorough and strict results from investigations of important parts of very complicated problems of emotional stress in daily life. Yet a gap remains between the results of experimental laboratory and field studies and the complaints of so many people about their everyday sleep problems and their dissatisfaction with their sleep. As a model is impossible which both describes all complicated aspects of emotional stress and sleep, and gives at the same time sufficient opportunities for the simplification necessary for research, one has to accept various limitations and specific approaches to parts only of the subject.

Bastiaans showed psychodynamic factors, personality traits and problem-solving strategies to be involved in sleep disturbances. The fact that shortening of sleep duration is not a stressor per se and not an important factor in coping with stressors during day and working, as both *Åkerstedt, Gillberg* and *Horne* convincingly have demonstrated, underlines the relevance of biological factors as rhythmicity and psychological factors as attribution. One of the important determinants of the subjective evaluation of sleep is the mood state before and after the sleep period and of the dream, as was investigated in detail by *Cluydts and Visser*. *De Koninck* indicated that the adaptive potency of dreaming in stressful situations depends on a subject's coping style, which might be influenced by manipulation of dreaming. *Schneider-Helmert* showed the usefulness of stress concepts for different clinical insomnia problems. With the analysis of psychosomatic insomniacs the importance of social dimensions was clearly demonstrated.

Though the desired final understanding of emotional stress and sleep has not yet been reached, important building blocks for this very complicated matter can be found in the present papers.

Sleep 1980. 5th Eur. Congr. Sleep Res., Amsterdam 1980, pp. 86–91
(Karger, Basel 1981)

The Psychosomatic Approach of Sleep Disturbances[1]

J. Bastiaans, Leiden, The Netherlands

In psychosomatic medicine psychosomatic syndromes usually are regarded to be the outcome of intensive neurotic efforts of the patients to escape from painful experiences caused by psycho-social stress situations. Generally these situations are marked by unsolvable conflicts and/or traumatic events. Many psychiatric and psychomatic disorders are marked by insomnia. *Kales* [47] describes that over 85% of insomniacs suffer from psychological disturbances. *Leigh and Reiser* [56] state that psychological distress is the most common cause of insomnia. The distress may be related to interpersonal or intrapsychic conflicts or to anxiety associated with a medical disease. Anxiety and depression are the most common psychiatric syndromes causing insomnia. Anxiety is more often associated with difficulty in falling asleep, while depression is often associated with early-morning awakening.

The symptomatology of reactions to unsolvable conflicts or psychic traumatization is summarized in table I.

The most severe reaction is given when in states of mental or bodily shock. Initial psychic shock can be of shorter or longer duration. When consciousness is not lost, the first symptoms of pain and alarm manifest themselves in feeling states of powerlessness, helplessness and hopelessness. The more the individual is incapable to cope with the stress situation the more these feeling states may persist. If the stressors activate the alarm and ar-

[1] For references see compound list on page 114.

Table I. The model of symptom formation due to traumatizing stress

Phases of adaptation syndrome	Emotional	Intellectual
Shock	fainting, no emotions	paralysis of intellectual functioning in dream or coma
Alarm	alarm emotions and feelings; affects, anxiety, pain terror, panic, extreme insecurity and nervousness, hyperesthesia, agitation, hesitation, doubt	hyperactivity in perception and thinking
Adaptation with accent on fight	overactivity, destruction vs construction, ranging from hostile action to protest; sadism and overproduction	intellectual overactivity
Adaptation with accent on fight and flight	depression, masochism, disappointment; manifest or masked mourning	insufficient investment of intellectual capacities in action, production, communications; everything lacks sense
Adaptation with accent on flight	indifference, dislike or disgust of action, work, contact; feeling lonely; emotional autism and apathy	failing functioning in orientation, perception, memory, conceptualization, thinking
Exhaustion	feelings of weakness, asthenia, sleep paralysis, feeling of impotence	chronic asthenia of intellectual functions; narrowing of mental horizon

Increased need for: security, protection, contact, oral satisfaction, compensation.

ousal reactions sufficiently, the organism is prepared to find coping and adaptation solutions with the help of the biological defence mechanisms of fight and flight. A most painful experience is rigid fixation in alarm symptomatology. Most individuals try to escape from such a fixation. Psychosomatic syndromes usually are the outcome of preference for coping styles related to controlled fight behaviour, whereas depressive syndromes are pre-

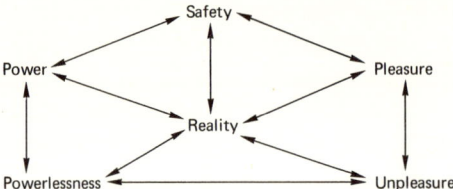

Fig. 1. The safety model.

dominantly the result of dominance of flight behaviour or of inefficient use of both fight and flight behaviour.

A disharmonious or inefficient use of psychological and biological defence mechanisms may also result in sleep disturbances. Some sleep disturbances may be regarded as the outcome of persistance of a state of alarm marked by pain, anxiety, guilt or shame. Other sleep disturbances may be the outcome of an inefficient effort to escape from such a state. Seen from the outside certain coping styles seem to be successful in the wake state whereas they are unsuccessful or inefficient in the various stages of sleep. In these stages a threatful experience of powerlessness can be activated subconsciously in such an intensity that the patient cannot sleep anymore. At least such an experience may activate the defence mechanisms to such a degree that normal sleep relaxation becomes impossible.

The essence of these views is based on what has been described by *Sandler* [61] as the safety concept. *Sandler* states that the feeling state of safety is the best indicator for successful coping with the environment. *Bastiaans* [9] adds that under normal circumstances an optimal feeling state of safety is obtained in positive and pleasurable communication and interaction (fig. 1).

But frequently human communication is blocked so much that the only source left for getting safety and security is reinforcing one's own power position. This strategy usually implies isolation of a person or of a group in psychic 'armours', 'forts' or 'castles'. Such an isolation may lead to increased aggression and hostility, also to self centredness, narcissism and/or other defences against being fixated in a state of powerlessness. If passivity and relaxation are equated subjectively with powerlessness, normal relaxation of the individual becomes impossible, also in the sleep.

To what extent insomnia or hypersomnia may be the outcome of an incapability to guarantee for oneself the needed or wished-for safety state obtained from pleasurable communication has not been explored systematically for all psychosomatic and psychiatric disorders. Much is known about sleep disturbances in anxiety states and in depressive disorders. Much is known about sleeping disturbances of victims of severe traumatization, e.g. ex-prisoners of war, ex-concentration camp inmates, ex-hostages, victims of traffic or other accidents. However, it has not been explored systematically to what extent all the different psychosomatic disorders are marked by sleep disturbances. *Stonehill and Crisp* [70] describe the relationship between nutrition, sleep and mood in patients suffering from anorexia nervosa. Well known is the fact that patients suffering from coronary artery disease (CAD) are unable to sleep more than a few hours at night. Due to their personality type A, as described by *Friedman* and *Rosenman* [32], they may suffer from their over-responsibility, over-impulsivity, over-activity and over-sociability, which are in fact personality traits which press the patients to continue these activity patterns even during the sleep. In many of these patients over-activity is a defence against fear for passivity and powerlessness. As soon as the patients awake they are preoccupied with all the tasks they have to perform in the subsequent hours and days.

It is well-known how asthmatics may suffer from nightmares and/or night terrors which are accompanied by or result in asthmatic attacks. However, it is not yet known for all psychosomatic disorders to what extent sleep disturbances are an important aspect of the psychosomatic syndrome. The fact that nearly all psychosomatic patients have problems with passivity and relaxation may suggest that sleep disturbances frequently must occur as a result of the inner tension states. What really matters is that many sleep disturbances and many psychosomatic disorders are the result of psychosocial stress and of insufficient or blocked communication.

Recently *Cohen-Matthijsen* [20] investigated sleep disturbances among young children, especially children of 3–5 years of age. Very clearly she could prove how the mother-child interaction is the main determinant for disturbed sleep behaviour. Of 600 interviewed mothers 19% reported a sleep disturbance in their child. More than 16% of the mothers felt these sleep disturbances were a burden to her. They felt dissatisfied with the manner in which they dealt with the problems of the child. In these cases one could speak of a conflict between mother and child about the sleep of the child. The investigation made clear that mothers with children, who both slept late and woke up during the night, showed neurotic problems with a higher frequency than mothers of children who slept well. A successful brief psychotherapy of the mother resulted in the disappearance of the sleep disturbances of the child.

Against this background it could be hypothesized that also for adults and adolescents a comparable investigation might reveal how sleep disturbances in these groups may be determined by communication problems, e.g. between marital partners or at least between family members living together in housing facilities which may be far from optimal with respect to the possibility of creating a climate of safety and security which is a conditio sine qua non for optimal sleep.

It has not yet been investigated systematically how sleep and wake behaviour of one member of the family may give rise to sleep disturbances of one or more of the other members of the same family. Usually these disturbances are treated with the aid of sleeping pills instead of with the help of an adequate technique of psychotherapy which may diminish the intensity of the actual stress situation. Insufficient communication and relaxation behaviour of partners or other family members may be one of the major causes of chronic sleep disturbances among family members. In the same way insufficient possibilities to talk together about the daily stresses before falling asleep may have the same result.

In this respect relaxation training, in whatever form, may enhance the possibility of coping with sleep disturbances efficiently.

Against the background of table I it may be clear that unsuccessful adaptation to multiple stress factors may give rise to the development of syndromes of well-known psychopathology. When there is no solution for coping and adaptation, or when pathological preference for insufficient coping adaptation styles last too long, the final outcome is a state of exhaustion. Such a state may be marked by an increased need for sleep. At the same time such a state may give rise to sleep paralysis, which may be regarded to be the expression of the inner feeling state which implies also a complete inability to sleep.

In psychosomatic medicine it has been described how preventive measures have to be directed to the elimination of those factors that cause tension states. Though psychopharmaco-therapy usually is the method of choice, this kind of therapy hardly eliminates the causative factors, especially those inherent to disturbed communication.

In his book *The Psychology of Sleep*, Foulkes [27] states that sleep is marked by decreased receptivity for sensory stimulation and decreased aim-directed motor activity. Usually psychosomatic patients remain occupied with what they have to do, even during the sleep. In this respect they belong to the group of poor sleepers of which *Johns* et al. [44] state that knowledge of sleep habits may be useful in determining the efficiency of a person's psychological defences in dealing with the stresses to which he is subjected. The defences of psychosomatic patients do not allow these patients to switch off completely stimulation from the outside, especially when this stimulation has already activated the processes of controlled and suppressed vivid imagination and fantasy.

Bastiaans [8] concludes that psychotherapy in psychosomatic medicine has to be directed to a behavioural change which results in a shift from forced adaptation into flexible and smooth adaptation. Such a shift may be promoted by the use of different techniques, for example: (1) relaxation therapies (hypnosis, autogenic training, yoga, etc.); (2) techniques which alleviate the pressure of an all too strict superego; (3) techniques which strengthen the ego capacities for adaptation to such an extent that adaptation to reality no longer implies a strain; (4) techniques which focus on the decrease of situational or environmental stresses and demands; (5) techniques which focus on improvement of adaptation by the use of psychopharmaca.

All these techniques may lead to a disappearance of the psychosomatic symptoms and also to an improvement of sleep.

Finally one may conclude that sleep is a psychosomatic process in the broader sense of the word. Disturbed sleep may increase psychosomatic tension states, but at the same time these tension states may lead to sleep disturbances. Therefore, the psychosomatic approach of sleep disorders is as important as a complementary somato-psychic approach.

Sleep 1980. 5th Eur. Congr. Sleep Res., Amsterdam 1980, pp. 91–95
(Karger, Basel 1981)

Dream and Mood Regulation in Stressful Situations[2]

J. De Koninck, Ottawa, Canada

While it is generally recognized that a good night of sleep is an excellent remedy for emotional stress, it is not known whether the dreaming activity

[2] For references see compound list on page 114.

which goes on during sleep has any contribution. At first sight, it is difficult to understand why one would expect from dreams relief of real life emotional stress. It has been well established by normative studies that dreams are generally unpleasant for the dreamer, that the most common emotion observed in dreams is apprehension, and that the most common interaction is aggression [39]. Yet, ancient as well as modern specialists of dreams have developed rationales proposing that such unpleasant dreams serve as a relief of accumulated tension and thus have a beneficial effect [15]. The discovery by *Aserinsky and Kleitman* [6] that dream recall could be enhanced drastically by awakening subjects during rapid-eye-movement (REM) sleep provided modern researchers with a new tool to examine the various aspects of dreaming activity surrounding stressful events, and to attempt to answer this intriguing question of the potential link between dreams and adaptation.

The first step has been to determine what happens to dreams following an emotionally stressful experience. The most common approach has been to induce presleep stress experimentally by putting subjects in prearranged situations such as the presentation of stressful films [25]. A second approach has been to study dreams following real life stressful experiences such as surgical operations [12]. It should be noted that the sleep laboratory situation in itself is very stressful for the subject (electrodes, strange surrounding) and sometimes even more than the manipulations of the researchers. Dreams obtained from the first nights in the laboratory have thus also been studied for this purpose.

The results have generally shown that elements of presleep stressful experiences do find their way into dreams either in a direct or an indirect fashion. Furthermore, real life stress, or stress which has more personal relevance to the subject appears to have more impact on dreams [12]. However, increases in negative dream affect do not necessarily follow presleep stress. For example, *Visser et al.* [71] have observed more pleasant dreams following the presentation of a stressful film.

These observations lead us to a series of questions. The central one is: Are these changes which are observed in dreams following stressful experiences linked to an adaptive mechanism? If so, does this mechanism require that dreams be remembered? And if we have to remember them, is interpretation or work on their content required? Are some dreams more useful than others? Following a car accident, is it better to experience dreams during which the accident is revived in a direct or a symbolic fashion as the mastery notion calls for, or is it better to have pleasant dreams as a compensation process would suggest [24]. One way to try to answer these questions has been to not only measure levels of stress experienced before sleep, but also to measure stress levels again in the morning and to look for relationships between stress reduction and the characteristics of dreams collected during the night. Although this approach still presents a number of limitations, interesting observations arise from studies conducted up to now. First, it appears that forcing subjects to remember dreams by awakenings can have a detrimental effect on morning mood and potentially short-term adaptation to stress. A good example of this comes from a recent study by *Koulack* [52]. In this case, subjects slept in the laboratory for 4 consecutive nights, the first 2 serving as adaptation. On the third and fourth nights, subjects were asked prior to sleep to complete intelligence tests. Half of the subjects were given difficult versions of the tests which they could not complete (ego-threatening condition), while the other half were given easy versions. On night 3, subjects were allowed to sleep without interruption while on night 4, they were awakened in REM sleep for dream collection. The results showed that on night 3, when subjects were left to sleep uninterrupted, their mood anxiety decreased from presleep to postsleep. However, on night 4, when they were forced to remember their dreams, their anxiety tended to increase from presleep to postsleep. This effect was statistically significant for the subject in the difficult condition who actually experienced more anxious dreams than the subjects in the easy condition. It is of course difficult to determine to what extent this effect is due to the awakenings per se, to the REM deprivation which comes with REM awakenings, or to the fact that the subjects were made aware of their dreams. It does suggest at any rate that inducing REM dream recall may not have a short-term beneficial effect. It is interesting to note that this observation is congruent with Freud's notion that dreams that fulfill their function are those which are not interrupted by awakening and do not find their way into consciousness [30].

A second observation pertains to the usefulness of incorporations of stressful elements into dreams. Here the experimental evidence is for the time being contradictory. *Cohen and Cox* [19] have found that subjects who incorporated into dreams elements from an ego-threatening presleep experience similar to the one just described felt better about this experience in the morning than those who had not had such incorporations. An apparently contradictory finding has been obtained by *De Koninck and Koulack* [25]. In this case, subjects were presented with a stressful film prior to sleep and a second time upon awakening in the morning. During the night, half of the subjects were presented with the soundtrack of the film below waking threshold during REM periods in order to induce them to incorporate elements of the film into their dreams. Contrary to expectations, those subjects who experienced anxiety in the morning while viewing the film again were those who had incorporated elements of the films into their dreams. These findings contradict those of *Cohen and Cox* [19] and suggest that incorporations of stressful elements can have detrimental effects. However, in both studies, several methodological problems limit the comparability and generalization of the findings. Clearly more research is required.

In the study by *De Koninck and Koulack* [25], an interesting observation was made. Subjective mood checklist measures were obtained prior to sleep, in the dreams, and in the morning. It was observed that the correlations between postsleep mood and dream mood were statistically significant on a number of dimensions and higher than between presleep mood and dream mood. A similar finding was observed in the study by *Koulack et al.* [52] mentioned earlier, where in the difficult tests condition group, the correlation between dream anxiety and morning mood anxiety was positive and significant but not that between presleep anxiety and dream anxiety. These suggest that in stressful situations dream mood and dream affect can have an impact on morning mood.

The results of a recent study [69] further suggest that, in cases of stressful experiences, the subject's conscious perception of its impact may be an important factor in determining the effects of remembered dreams on morning mood. Here a group of women kept a home dream diary over two menstrual cycles. Each day, prior to sleep, they answered a menstrual distress questionnaire which allowed to measure the presleep stress experienced throughout the phases. It was observed that, regardless of the level of presleep stress, the subjects experienced more anxiety and hostility in their dreams during menstrual phases as compared to the intermediate phases. It was thus interesting to see what impact these dreams would have on the morning mood of these subjects. Interestingly, it was observed that during the menstrual phases, subjects who reported experiencing menstrual stress before sleep (stressed subjects) tended to experience an improvement in mood from night to morning (that is, positive emotions increasing and negative ones decreasing) whereas those who did not experience presleep stress (nonstressed subjects) tended to have the opposite change from presleep to postsleep. A simple explanation for this finding could be that the sleep and the dreams have carried a mood regulatory function by bringing back the subjects to an intermediate level of mood in the morning. It is however interesting to speculate that in the nonstressed group, the anxious dreams may have forced into consciousness preoccupations which were not clearly acknowledged before sleep. Although such a process may appear to be detrimental in a short-term basis it may prove beneficial in the long-term. In a series of very interesting studies, *Fiss and Litchman* [26] and *Cartwright et al.* [16] have examined the long-term impact of training subjects to remember their dreams and have observed positive effects probably due to the expansion of the individual's awareness of his or her own preoccupations and coping mechanisms.

Generally, it can be seen that it is too early to claim that dreams have an adaptive function in stressful situations. With respect to their short-term value at least, the results are very inconclusive and in some cases contradictory. If, however, further studies demonstrate that certain types of dreams have some adaptive value, an interesting avenue can be pursued. Since it has been demonstrated that presleep suggestion can be very successful in influencing dream content [72], it may be possible to use it to induce these types of dreams. In a recent study [14], we were able with presleep suggestion to induce a group of snake phobic subjects to have more pleasant dreams even though they had been exposed to a snake prior to sleep. Another group of snake phobic subjects were successfully induced to have more unpleasant dreams. Due to the fact, however, that all subjects adapted very rapidly to the sight of the snakes, we were not able to test the usefulness of these dream manipulations. Further research may demonstrate that learning to control dream content and manipulate their affect level could have positive effects on morning mood and general adaptation.

Sleep 1980. 5th Eur. Congr. Sleep Res., Amsterdam 1980, pp. 95–97
(Karger, Basel 1981)

Sleep Deprivation, Stress and Sleep Function[3]

J. A. Horne, Loughborough, England

Biochemical and physiological studies of total sleep deprivation (TSD) in man have been oriented towards the stress perspective, believing that

---

[3] For references see compound list on page 114.

TSD is a putative stressor. But perhaps surprisingly, little significant effect has been found [42]:

(1) Out of nine studies assessing corticosteroid activity, over 24–205 h TSD, none reported significant change. (2) Out of five studies measuring one or more of: blood haematocrit, sedimentation, glucose, free fatty acids, none reported significant change. (3) Of ten studies monitoring urinary catecholamines over 24–205 h TSD, only one reported any change (increase). (4) Heart rate has increased in three studies, decreased in five and shown no change in nine. (5) Systolic blood pressure has decreased in three studies and remained unchanged in nine.

However, if uncertainty or other means of reducing subject confidence are introduced, then corticosteroid and catecholamine output will be increased [28].

Sleep is thought to have body restorative functions, but because of the stress-oriented measurements, there is little data relating to this putative sleep function. From what data are available, there is no sign of adverse reactions [42]:

(1) Two studies [13, 40] have assessed physical work capacity (physiological ability to do physical work), sampling daily over 120 h TSD; both found no change. (2) Although the two studies measuring blood adenylate activities [57, 58] reported 'emergency responses' with ATP levels over 123–220 h TSD, these interpretations were based on one measurement approximately half way through TSD; at the end of TSD levels were back to normal parameters. (3) Of seven studies reporting on creatinine excretion, only one [40] found a significant change, but the level was biologically normal. (4) Of three studies measuring urinary nitrogen (a crude guide to protein metabolism) two found no change and one [67] reported a decrease on day 1 of TSD and an increase on day 2, with balance overall.

It should be remembered, though, that there still remain several body systems where the response to TSD is not well known. For example, recently limited studies [59, 60] have begun on the immune response during TSD. Some changes were evident, but were difficult to interpret.

There are methodological problems often affecting TSD findings:

(1) Because most TSD studies are laboratory bound, subjects usually lead a relaxed existence which may offset possibly significant changes. (2) Baseline data may include an anticipatory effect which declines over TSD, to be balanced by other responses, leading to no apparent change. (3) Data tend to be averaged, and subtle individual effects may be lost in 'noise' increasing with TSD.

Whilst this review's findings may not be applicable to chronic partial sleep deprivation, where there is an absence of this biological data, it does seem that man can lose at least 1–2 h sleep per day, permanently, without any loss of well being [31, 45].

Although within the limitations described it seems that short-term TSD produces little or no physical ill effects, TSD soon brings psychological performance detriments, behavioural irritability, disorientation, some enhancement of paroxysmal EEG activity, but few other significant neurological changes [42]. Whilst these changes indicate a CNS impairment and a need for the brain to sleep rather than to rest, the performance tasks most vulnerable to sleep loss are not those which might be considered as 'cerebrally demanding' such as IQ tests and complex decision-making, but simpler, low interest long duration tasks [74]. Seemingly, the former tasks induce compensatory effort, offsetting detriments well up to 64 h TSD, and indicating a cerebral capacity to function normally under limited TSD. Usual performance levels at the simpler tasks can be maintained during limited TSD if sufficient extrinsic reward is given. Therefore, if much of human sleep were for cerebral restitution, then the extent of cerebral capacity to perform at near optimal levels during limited TSD would be less than is found. The main initial effect of TSD seems to be with motivation rather than capacity, suggesting that much of initial tiredness is due to a non-restorative sleep drive.

In theory at least, the human body and cerebrum could function adequately with less sleep. Circadian rhythms, on the whole, are not disrupted by TSD [1] and so less sleep per se should not present problems from this quarter. Less sleep per se (for example, about 2 h less per day) should not lead to stress effects, unless a subject perceives such a loss as harmful. Other stresses, of course, lead to sleep disturbance, but this is unlikely [43] to lead to less than 6 h sleep per night, and/or more than 30 min interim wakefulness, and/or more than 30 min sleep onset. From the restitutional perspective, at least, such disturbances should not present difficulties. But daytime tiredness may be a complication, related perhaps mostly to a non-restorative sleep drive. However, such tiredness is generally not the main problem with insomniacs [43]. The irony of treating so-called insomnia with hypnotics is that daytime tiredness is likely to be greater owing to hangover effects. Therefore, perhaps rather than continue searching for new hypnotics to promote sleep, we ought to be looking for compounds to reduce the more facultative form of sleep and the non-restorative form of tiredness, as well as educating people that it may not be so necessary to obtain a 'good night's sleep'.

Sleep, Stress and Recuperation[4]

T. Åkerstedt, M. Gillberg, Stockholm, Sweden

Our experience of sleep and stress has mainly concerned the effects of sleep loss on psychological processes during sleep deprivation (SD), as well as during recovery sleep after SD. In the preceding paper Dr. *Horne* made an elegant general review of SD literature. In the present paper we shall concentrate on a few special points from our own research which bear on the issue.

Sleep Loss as a Stressor

Part of our work has been directed towards trying to establish whether sleep loss causes stress in the classical sense [68] i.e. a non-specific arousal response of the organism to any demands made on it. Since stress indices, by definition, will be sensitive to arousing influences, an important point is to control such influences, particularly the activity of the subject. The ad libitum activity conditions which are found in most studies of sleep loss may, for example, confuse the interpretation of changes in stress indices seen during SD. We have, therefore, conducted several studies of SD in which particular efforts have been made to keep unwanted influences under control. This means, for example, no time cues (from clocks, daylight, media, etc.), evenly distributed intake of small and identical meals, and importantly, evenly distributed and controlled activity.

Using adrenaline excretion as an indicator of stress (among others) we have found *no* indication of the arousal response in any of the five studies conducted [2, 3, 33–35], irrespective of age or sex (fig. 2). The circadian rhythmicity, however, has been very pronounced in all studies. Taken together with the lack of effects on many other stress indicators [reviewed by Dr. *Horne* in the preceding paper] it seems safe to conclude that sleep deprivation at least does not function as a stressor per se.

---

[4] For references see compound list on page 114.

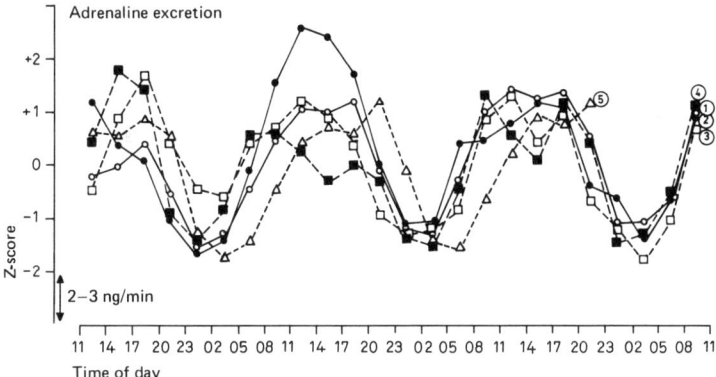

Fig. 2. Mean urinary adrenaline excretion in 3-h intervals from studies representing a total of 100 subjects. All 5 studies show significant (anova) circadian rhythms. 1 = *Fröberg et al.* [34]; 2 = *Fröberg et al.* [35]; 3 = *Fröberg* [33]; 4 = *Åkerstedt and Fröberg* [2]; 5 = *Åkerstedt and Fröberg* [3].

However, even if sleep deprivation, in itself, does not constitute a stressor, it might conceivably potentiate the reactions to an acute stressor. Overreacting and irritation often seem to accompany sleep deficits. Possibly, this could be brought about indirectly through reduction of coping capacity. In another study a performance stressor, the film version of the Stroop colour conflict test [29] was introduced after 28, 40 and 53 h of waking as well as for a non-sleep-deprived group. The adrenaline response for all exposures was quite normal, as was the cortisol response and performance [3]. Similar results have been obtained by *Bergström et al.* [11] who exposed subjects to electric shocks after 80 h of SD but only found the normal increase of heart rate. Also, *Kollar et al.* [51] administered ACTH to their subjects during sleep deprivation but found only the usual cortisol response. From these studies it appears that SD does not potentiate the response to a stressor.

In retrospect, it does not seem illogical that short, temporally well-defined, non-ego-threatening, artificial, and trivial laboratory tasks may be quite within the coping capacity not causing stress of the sleep-deprived subject, particularly if the consequences of poor performance are of minor importance. On the other hand, it also seems logical that the many demands by everyday life on integrated functioning may be much more difficult to cope with. Such demands often occur frequently, are out of the subject's control, will remain until dealt with, may involve interpersonal and maybe ego-threatening components, may cause repeated frustration of the drive to sleep, etc. This constitutes a psychosocial stressor quite different from the trivial laboratory task. Thus, it seems necessary to start performing field studies if further knowledge of sleep loss and stress shall be gained. It is important, though, that such studies should have as much experimental character as possible. Such natural experiments occur all the time but will, of course, require more effort than the laboratory experiment.

Recuperation from Sleep Loss

Our second approach to the effects of sleep loss is to use recovery sleep as a dependent variable. We have reasoned that, if sleep loss is important, it should be given priority over other needs. On the whole, studies of sleep deprivation show the well known extension of total sleep length and SWS [4, 10, 38, 49, 75]. This seems to attest to the importance of the loss of sleep although, in absolute time, far less is recovered than what is lost. However, not only is the increase of amounts of sleep out of proportion with the sleep lost, but there are also instances when extended wakefulness may be connected with *decreased* amounts of recovery sleep.

In a recent study [5] with 6 subjects, we postponed bedtime from the conventional 2300 to 0300, 0700, 1100, 1500, 1900 and 2300 hours (24 h later), respectively. This means that prior wakefulness varied between 16 and 40 h, in a balanced order, and with a week in between sessions. The subjects were urged to sleep until completely satisfied and as in the previous experiments, special attention was paid to controlling environmental factors and activity. In particular, subjects did not know how long they had been asleep when they decided to rise.

The results show (fig. 3) that total sleep length exhibits a very pronounced circadian rhythm, reaching a peak (8–10 h in length) after evening bedtimes and a trough (4–5 h) after morning/noon bedtimes. Recently, *Czeisler* [23] and *Zulley* [76] have obtained similar results in connection with experiments on circadian desynchronization. The circadian 'interference' means that when the amount of wakefulness is extended from 16 towards 28 h there is, paradoxically, a strong *negative* correlation with subsequent sleep length! Apparently, circadian rhythmicity overrules the need to compensate for moderate sleep loss.

Sleep deprivation did have some effects, however. Firstly, sleep latency was reduced to almost zero, already for the first postponement of bedtime.

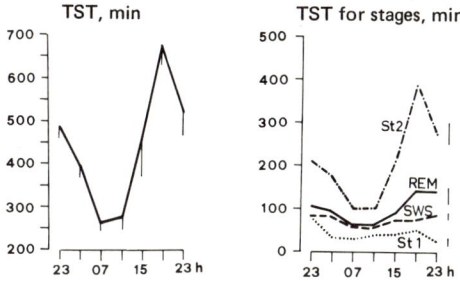

Fig. 3. Total sleep length and its distribution on stages of sleep for 7 bedtimes; mean and SE for 6 subjects (SE for sleep stages drawn as maximum SE to the right of each curve) [5].

This, very likely, helped the subjects to start sleep at times of day which otherwise would have presented difficulties [73]. Also, without SD we suspect that we would have obtained hardly more than short naps for day sleep and that the long sleep at 1900 hours was partly due to sleep deprivation 'helping' to retain sleep before the circadian rhythmicity could 'take over'.

Logically, the next question would be the recuperative value of the displaced sleep. Thus far, we have only studied the subjective ratings of sleepiness after rising. The results show that when length of wakefulness increases and length of subsequent sleep decreases, then sleepiness actually *decreases*. This paradoxical effect is apparently due to the pronounced circadian rhythmicity of psychological arousal. The circadian interference with recovery sleep suggests that moderate amounts of (acute) sleep loss are of minor importance – why would the sleep mechanism otherwise 'allow' the circadian mechanism to dominate?

Conclusions

To conclude, sleep loss clearly is no stressor in the conventional sense. Neither is it likely to contribute to the effects of trivial laboratory stressors. Furthermore, it exerts a comparatively weak pressure for recovery, at least at moderate levels of sleep loss. This does not, of course, mean that sleep is unimportant and that we sleep too much. It merely means that acute *moderate* sleep loss may be of minor short-term importance and that there is a strong need for investigating sleep loss with new types of behavioral and biological variables.

Emotional Stress, Mood State and Sleep[5]

R.J.G. Cluydts[6], Brussels, Belgium; *P. Visser*, Amsterdam, The Netherlands

Introduction

It is striking how few publications are found in the literature about emotional stress and sleep. It is difficult to investigate the relation between 'naturally' occurring stressful situations in daily life and sleep as a dependent measure in a laboratory setting.

In psychosomatic syndromes we have to deal with 'accumulating' effects of unsuccessful coping with difficult and/or unsolvable situations, traumatic events, etc. In psychiatric medicine clear evidence exists concerning the impact of psychiatric disorders on sleep. However, these rather dramatic, 'pathological' conditions are somewhat further away from the real life stresses that we want to focus on.

Mood states can be seen as a psychological variable that informs about the successfulness of the coping behaviour of an individual in a particular situation. *Crisp and Stonehill* [22] showed that isolated mood states have a selective impact on sleep: anxiety is characterized by initial insomnia, broken sleep and delayed awakening in the morning, sadness by early awakening, and anger by sleep disturbances throughout the night. According to *Kramer et al.* [53] sleep has a mood regulating function.

In experimental research on the relations between emotional stress and sleep, very often pre-sleep stimuli are used that bring about changes in mood states. As stimuli are used: aversive films, traffic noise, loud music, relaxation, psychotherapy, etc. The menstrual cycle can be an operationalization of a non-experimental, physiological stimulus. Using experimental pre-sleep stimuli we always have to deal with the problem concerning the extrapolation of these laboratory stressors to real life situations. In the study on emotional stress and sleep most of the interest is dedicated to the sleep

[5] For references see compound list on page 114.
[6] Supported by a FGWO grant No. 3.0029.77.

variables and it is striking how little importance is given to a fine, precise description of the subjects' subjective state before retiring for bed.

In many investigations the impact of the stressful situation, the stressor on the subjects is taken for granted and the different behavioural reactions of the subjects to the stressors are always underestimated.

Results

In the Brussels studies on sleep and emotional stress the effects of the menstrual cycle and of an aversive film on sleep were studied [18]. From these studies it could be concluded that the changes in a psychological variable, the mood states as measured by the profile of mood states (POMS), are conditional to changes in the physiological sleep parameters as EEG and EOG. In the study with the menstrual cycle in one subject a clear, menstrual cycle-related change in mood was detected, and in this subject a significant change in sleep variables, more precisely in the non-REM variables, was found. Of 10 subjects who saw a surgical colour movie, 5 reacted rather neutral as measured by the POMS and interviews, whereas the other 5 subjects showed changes in their mood scores, more precisely in the so-called 'negative mood' scales, especially in the factor 'tension'. Statistical analysis, see correlation coefficients in table II, showed some significant relations between mood and sleep.

The tension scores have a concordant relation with the time spent awake during the first 3 h of the night and the percentage stage 2 sleep, a discordant relation with the number of stage shifts and the amount of slow wave sleep. With the anger scale there is a very significant positive relation with the time spent awake for as well the first 3 h as for the whole night. It is remarkable how these results, obtained with changes in mood states in normal subjects, closely resemble the results by *Crisp and Stonehill* [22] in a population of psychiatric patients, especially their 'lightly disturbed' mood group, in whom the impact of mood on sleep is in the first part of the night.

The Amsterdam group studied in about the same way different disturbing endogenous influences and measured the effects in a systems approach to the sleep quality, mood states and performance during the next day [71]. The still incomplete psychophysiological theory which forms the basis of these experiments comprises a descriptive, quantitative psychological part, and a physiological part based on a neurophysiological process theory of motivation [50]. In the psychological approach the sleep quality was

Table II. Correlation coefficients between mood scores and sleep variables, 10 subjects

	Tension	Depression	Anger	Vigour	Fatigue
Number of stage shifts	−0.70*	−0.05	−0.21	−0.22	0.11
Number of awakenings	0.58	0.60	0.36	0.04	0.22
Number of REMS	−0.22	0.01	−0.41	0.08	0.33
Sleep latency	−0.13	0.02	0.35	0.31	−0.54
REM latency	0.17	−0.18	0.02	0.51	−0.11
Stage 0 (180 min)	0.72*	0.17	0.50	−0.02	−0.09
Stage 1 (180 min)	−0.15	0.00	0.24	0.34	−0.43
Stage 2 (180 min)	0.38	−0.23	−0.25	−0.48	0.53
Stage SWS (180 min)	−0.34	0.12	−0.07	0.31	−0.45
Stage REM (180 min)	−0.24	0.11	0.14	−0.39	0.45
Stage 0	0.13	0.67*	0.67*	0.12	−0.25
Stage 1	−0.57	−0.02	0.02	0.26	−0.30
Stage 2	0.80**	0.09	0.18	−0.36	0.18
Stage SWS	−0.38	0.20	0.24	0.54	−0.19
Stage REM	−0.31	0.04	−0.04	0.28	0.32

Levels of significance: ** 0.77 $p \leq 0.01$; * 0.63 $p \leq 0.05$; 0.55 $p \leq 0.10$.

measured during the night after waking up the sleeper and after awakening in the morning. The sleeper attributes his ratings on the items of a sleep quality scale and on analogue rating scales according to many different and apparently causative factors related with emotionally important events and experiences of his life history. These attributions are compared with the physiological and other psychological data.

In the physiological theory a lower level of neuronal structures as hypothalamus and mesencephalic reticular formation controls the homeostatic state of the 'milieu intérieur' or the 'drive' aspect. Deviations from the stationary state of this 'milieu intérieur' induce changes, that aim at the restoration of the steady state. The higher level in the motivational state is controlled by limbic structures as amygdala, hippocampus, infero-temporal cortex. These structures process the historical dimensions of life history, and of individual reactivity and performance of the subjects behaviour. Essential for the theory is that the restorative aspects of sleep should be studied comparable with studies of eating, drinking, sex and thermoregulation, in all of which restoration of the steady state goes with experiences of pleasure. In a still speculative interpretation of the sleep quality measured during and after the sleep period the pleasure will be indicative for the restorative soundness of the sleep, which is related with the deep-sleep stages 3 and 4, primarily reached in the first 3 h of the sleep period. Other aspects of sleep are related with the 'habit strength' which indicates in behaviour theory the mode of learning or the incentive, and with the cognitive activities. These are present primarily in the stages REM or paradoxical sleep; these stages increase in importance and in duration the longer the duration of sleep has been.

An experimental study was done in young, male subjects who were each studied during three consecutive nights and to whom just before falling asleep were shown respectively a black and white movie on the adaptation night, a travelogue colour movie on the first experimental night, and the surgical colour movie on the second experimental night, the same movie as used in the Brussels experiments as pre-sleep stressful stimulation. Table III gives some results of different physiological and psychological parameters based on the EEG recording, the ratings and the interview during and after the night sleep.

After the aversive movie sleep quality was better in the first hours of the night and the dream mood indicated after being woken up was less tense and more pleasurable. After the whole night, however, the sleep quality was lower after the aversive movie. Of some of the subjects a dream content analysis was performed of the tape recorded reports of the interview. The content material of in total 38 tapes was studied according to a simplified *Hall and Van de Castle* [39] method. Emotions and moods from this content analysis were compared with the subjects' own ratings mentioned above. Table III gives some of the calculated Spearman correlations. These results

Table III. Influence of aversive vs neutral movie on physiological and psychological variables: 9 male subjects

EEG	Aversive	Neutral	Test
Set of sleep polynome parameters	better sleep	worse sleep	Hotelling T^2;
Stage variability	12.5	18.7	$t = 2.06; p \leq 0.05$
Sigma spindle density	0.18 min^{-1}	0.36 min^{-1}	$t = 1.66; p \leq 0.05$
Sleep onset latency	22.6 min	22.6 min	n.s.
Sleep depth	4.15	3.85	n.s.
REM latency	62.4 min	70.0 min	$t = 2.78; p \leq 0.05$
Dream content			
Pleasant vs unpleasant	Spearman coeff. 0.47		n.s.
Relaxed vs tense	Spearman coeff. 0.66		$p \leq 0.01$

give support to the still speculative hypothesis that pleasurable restorative aspects of sleep are to be expected with a higher probability in the deep-sleep stages early in the night sleep, whereas the unpleasant cognitive elaborations of the aversive pre-sleep experiences are to be expected during stages REM, which are longer and more pronounced in the later sleep.

Discussion

When the Brussels and the Amsterdam results are compared they are found consistent, except for the shorter REM latency after the aversive movie. It is known that in mono-cyclic depressives the REM latency is shortened, but as in our subjects the change in mood state was most prominent in the 'tension', this appears not to be relevant. Otherwise both electrophysiological and psychological parameters show comparable trends under the influence of the exogenous and endogenous stimulation. This fact lends support to the validity and the reproducibility of the results.

This experimental setup opened possibilities to unravel the interconnected relations of the three successive phases of a stress model pre-sleep, sleep and post-sleep conditions and the different states which influence the performance on the next day. Physiological and psychological data should be seen as two different languages of the same living hard-ware, and it should be stressed not to hypothesize cause-effect relations between physiological and psychological variables. Thus, the effects on the morning after awakening can be due to the two different physiologically defined states deep sleep and REM sleep, which result in one homogeneous state, the waking state. The reported sleep quality and the performance achieved on a given task have to be accounted to a whole gamut of physiological and psychological states, each of which cannot be traced separately in the measured resulting variable. Longitudinal, 24-h studies continued over a series of days can be a more promising approach and can help to fill the gaps in the stress model asked for in sleep research [41]. Another difficult experimental problem may be noted here: sleep quality measurements are based on attributions by the subject, which is both actor and observer.

Clinical and Conceptual Aspects of Sleep and Emotional Stress[7]

D. Schneider-Helmert, Königsfelden, Switzerland

In this paper, some concepts of emotional stress and psychophysiological response will be discussed with respect to their relevance for sleep and sleep disorders by considering correspondances in the clinical field.

System Approach

The stress concept is especially favored for *pathogenetic models*, stress being regarded as a precursor of disease, if not its actual cause. This applies especially to explanations of complex phenomena such as psychosomatic disorders. If emotional stress is considered in its relation to sleep, one should be aware of the continuation of psychic functions during sleep, their stream being sporadically revealed in the form of dreams and by overnight changes of mood. Apart from this, both sleep and stress can be considered as independent or dependent variables. (These aspects are discussed in detail by the other contributions to this symposium.) A complexity like that suggests to develop a cybernetic system.

A fundamental scheme was presented by *Kagan and Levi* [46]: Psychosocial stimuli and a psychobiological program determine the stress reaction which may develop to precursors of disease or disease, depending on interacting variables; all states influence the preceding states, but not in a specified manner. If looking for a correspondance in the clinical field at this stage of concept formation in rather vague and general terms, one should concentrate on correlations on the level of frequency distribution rather than to try to separate and operationalize particular components of the system. In a survey of 524 patients upon their admission to our psychiatric clinic [37] we found the following dependencies (fig. 4):

---

[7] For references see compound list on page 114.

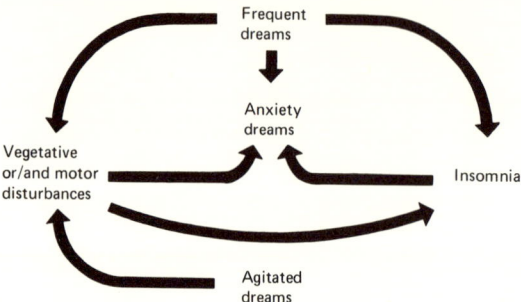

Fig. 4. Correlations of symptoms in a population of 524 psychiatric patients.

Patients with frequent dreams are more often subject to anxiety dreams and insomnia than patients without dreams; nocturnal vegetative and/or motor disturbances are more likely in patients with frequent and agitated dreams, and they are a factor for insomnia and anxiety dreams, the latter being more frequent in insomniacs than in good sleepers; finally, the daytime mood and performance was worse in insomniac than in good sleeping patients.

This clinical constellation certainly fits the concept of a psychobiological program which is determined by genetic factors and past experience, and the combination of psychiatric illness with various disturbances of sleep has to be recognized as a stress condition with psychosocial stimuli involved. As a psychiatrist I am tempted to speculate about casual priorities among the various phenomena, but this would exceed the scope of our findings that represent a system of closed loops without definite origin and end. However, some clarification of the complicated interconnections can be achieved by tracing them back to the four basic dimensions: stress, sleep, performance, satisfaction. Upon wide experience and numerous laboratory studies their mutual relations can be described in a two-dimensional system according to the following rules (fig. 5):

(1) The better the performance, the more a subject is satisfied. (2) Inverse relation between stress and sleep, i.e. less sleep under high stress conditions. (3) Performance and satisfaction are related to stress in a parabolic mode: A certain amount of stress is necessary to attain optimal performance as it improves motivation; the more stress increases, the more anxiety is induced and performance and satisfaction are diminished. (4) The same holds true for the relation of performance and satisfaction to sleep: They are low with poor sleep as well as with excessive sleep, and a given subject has his optimal range of sleep. (5) Taking all these aspects together, different conditions can be defined along this parabolic curve: The optimal range represents the condition of health; the right lower corner represents the condition of insomnia; the left lower corner may represent certain types of hypersomnia.

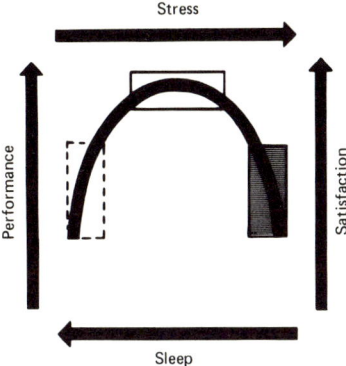

Fig. 5. Schematic representation of basic relations.

How are specific factors of the pathogenetic model applicable to sleep, especially stressor and transition from stress to disease?

The Stressor Problem

A variety of laboratory experiments were designed for testing the effects of acute stress on sleep, usually in the form of applying standardized stimuli or activities to normal volunteers before bedtime. The results of these studies were either weak or contradictory and rise several questions concerning the experimental approach per se [63]: The high stressor standardization may lead to divergent individual stress conditions and to a wide scale of responses, depending on a subject's momentary psychophysiological state, past experience and personality factors, thus levelling the results; or an artificially elicited stress may be too weak for overloading the homeostatic control of the waking-sleep regulation. As a possibility to overcome these difficulties I propose that we consider *severe real life stress situations* that sometimes occur by chance as intervening variables during ongoing sleep laboratory experiments. In our material I found four events of unequivocal stress during experimental periods free of other independent variables, so that we could evaluate the nights of stress in comparison with the preceding and following nights individually.

The subjects and stress conditions were: (1) Female, aged 36 years: Got at the same day a call from police saying that the corps of her missing father was possibly found, and a demand of a friend in divorce to live with her temporarely. (2) Male, aged 34 years: Unexpected death of a collaborator. (3) Female, aged 49 years: Upsetting discussions with her friend who suddenly made opposition against sleep experiment. (4) Male, aged 49 years: Acute induction of anxiety by experimentator error prior to sleep recording. Subjects 1 and 2 were insomniacs, subjects 3 and 4 normal sleepers.

The main significant results ($p < 0.05$) for the deviation of sleep in stress conditions from the values of the 2 adjacent nights were: Sleep latency 104% above baseline mean, sleep efficiency 19% below baseline mean, stage 1 proportion of sleep time 52% above baseline mean.

This tentative analysis suggests that acute severe life stress indeed influences sleep considerably into the direction of acute insomnia.

Transition from Stress Response to Disease

Christie and McBrearty [17] pointed to the fact that a large somatic response is not necessarily indicative of a stressed system. On the contrary it may be avantageous to mobilize one's metabolic resources to respond to a challenge. Whether or not homeostatic *recovery* is attained may depend on the adequacy of homeostatic control mechanisms and the speed with which the response ceases. *Lader* [55] refered to a similar model but elaborated more detailed the interactions between external stimuli and intrapersonal factors such as information processing on a conscious or subconscious level, emotion, CNS arousal and peripheral physiological changes. He stressed the significance of *CNS arousals* that reduce the subject's ability to adapt to a continuing or repetitive stimulus. 'Once a high level of arousal has been attained the condition tends to become self-perpetuating. In time, the function of the bodily system showing the largest responses becomes impaired' (p. 306). Obviously, these concepts are of specific interest for relations between emotional stress and sleep disorders.

Which clinical evidence can be found for the transition of an insomniac stress reaction into an insomniac state, i.e. from a psychophysiological response into a chronic disorder? *Bastiaans* [this symposium] has treated the problem from a general psychosomatic view and has shown that distinct attitudes and conflict-solving strategies are responsible for the evolution of such disorders. There is no reason why insomnia should not develop along the same lines in a rather specific way. We first proposed the concept of the *psychosomatic insomnia*, originally based upon 3 case studies [65]. The further elaboration of this concept [64, 66] led to the following description of this type of insomnia: Onset with severe emotional stress; multiple func-

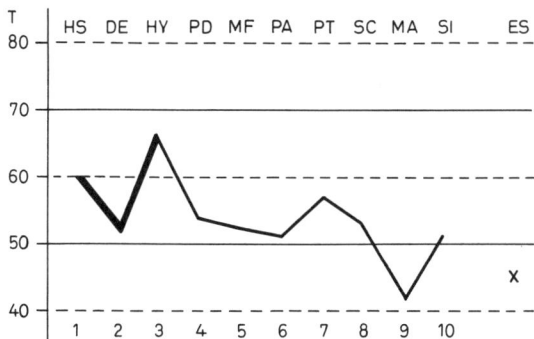

Fig. 6. Mean MMPI scores of 10 psychosomatic insomniacs.

tional disturbances, but in time concentration on insomnia which becomes chronic and is resistant against usual therapeutic measures like hypnotics and pharmacological or psychological relaxation; the personality of psychosomatic insomniacs is characterized by rigid attitude with precariousness, compensative self-control, denial of emotional problems, increased anxiety and depression, attempt to attain and guarantee interpersonal relations by high performance though insomnia impaires continuous activity.

This can be confirmed with data of 10 insomniacs who meet these criteria and who were examined in our sleep laboratory. They are 5 females, aged 26–54, mean age 43.2 years and 5 males, aged 31–60, mean age 38.2 years.

The mean MMPI profile (fig. 6) shows indeed the typical personality traits of psychosomatic patients [62], i.e. V-shape of the scales number 1–3, the highest scales being hysteria (HY), hypochondriasis (HS) and psychasthenia (PT), and low ego strength (ES). It should be noticed that depression (DE) is only in the lower part of all scales. Data characterizing the nocturnal behavior (table IV) give a clear indication of insomnia. The most striking fact is the wide interindividual variability as indicated by the differences between the minima and maxima of individual means. The more detailed consideration of the mean sleep data confirmes the severity of insomnia by frequent arousals (7.7/h sleep), too high stage 1 (10%) and too low REM sleep (18%) proportions. The mean Stanford Sleepiness Scale ratings (2.2–3.0) show a considerably reduced alertness during the whole daytime. In the psychodynamic view, the insomnia of these patients can be interpreted as a psychophysiological regression. The sleep disorder is the price for an adaptation to acute emotional stress and continued social adjustment with the

Table IV. Sleep behavior of 10 psychosomatic insomniacs, derived from 36 polysomnograms (excluding first laboratory nights)

Duration (min) of:	Total mean	Individual means	
		min.	max.
Time in bed	450	296	487
'Real' sleep (S2 – 4 + SR)	251	84	364
Sleep latency	52	10	155
Wake in sleep period	65	15	141
Early morning wake	39	3	206

installation of a neurotic balance. There is no question that all aforementioned personality traits and these neurotic mechanisms cause a permanent CNS stimulation.

To sum up the points treated so far, clinical findings justify to apply stress concepts to sleep disturbances. This aimes at a basic feedback control system, at the stressor concept insofar as real life stress situations are considered, and at the role of CNS arousal for the transition of acute stress response into chronic disorder, for which psychosomatic insomnia is a good example.

Sense of the Crisis

I see two possible answers to the question, what could be the sense of such a development from stress response to disease for the organism. (1) A subjective and primary gain results from the insomniac reaction in the way that restricted sleep enables one to extend conscious control over the situation, reduces fears of dreams, and decreases tension by transforming an emotional problem with its demands on responsability into a distant bodily problem. On the other side the desire for more sleep may be an attempt to avoid conscious confrontation with a stressful situation and related problems. (2) The experience of insomnia and the complaints about sleep disturbances change the psychosocial framework. Insomnia may induce attention, excuse, sympathy, consideration, or impatience and rejection, depending upon the personality traits of the patient as well as his interactions with life partners. Positive reactions may stimulate the symptoms that promise even more positive feedback (secondary gain). Negative reactions may as well provoke augmentation of symptoms as the patient tries to change the interactional set by an increased appeal to the environment. When resignation takes place, the insomnia is definitively chronified.

Social Factors

In face of these interrelations and vicious circles, the significance of social factors becomes evident. I wish to outline a few particular aspects: The patient with a breakdown of his sleep-wake-rhythm is confronted with the conformity pressure of his social environment in that he should sleep when he feels alert, inquiet, nervous but must work when he is sleepy. *Costell and Leiderman* [21] demonstrated by experiments that psychophysiological arousal occurs specifically in persons who maintain individual behavior against conformity pressure. This could contribute to sleep problems. Higher arousality was also found in introverted as compared to extraverted personalities [36], and traits of introversion like internalization and depression were predominantly found among short/poor sleepers [48]. That introverts are superior on tasks involving vigilance [54] can of course be attributed to their higher arousal. But considering insomniacs as highly aroused and thus better performing persons seems not compatible with their claims that sleep loss impaires their daytime functioning. The clues for interpreting these apparently paradoxical results may be found in the psychosocial field: Insomniacs show a greater need to appear normal or acceptable [7], and short sleepers often get a special motivation for good performance from their attempt to compensate with work a lack of social integration [64, 66]. At the same time, their self-perception of efficiency is impaired by scepticism which reduces the correspondent satisfaction, increases tension during leisure hours and thus furthers sleep problems.

This specifically human dimension of interactional and social factors calls for an extension of the original stress model and indicates a field of sleep research to be further developed in the future.

List of References of Symposium C

1. Åkerstedt, T.: Altered sleep-wake patterns and circadian rhythms. Acta physiol. scand. *469:* suppl., pp. 1–48 (1979).
2. Åkerstedt, T.; Fröberg, J. E.: Psychophysiological circadian rhythms in women during 72 h of sleep deprivation. Waking Sleeping *1:* 387–394 (1977).
3. Åkerstedt, T.; Fröberg J. E.: Sleep, stress, and sleep deprivation in relation to circadian rhythms in catecholamine excretion. Biol. Psychol. *8:* 69–80 (1979).
4. Åkerstedt, T.; Gillberg, M.: Effects of sleep deprivation on memory and sleep latencies in connection with repeated awakenings from sleep. Psychophysiology *16:* 49–55 (1979).
5. Åkerstedt, T.; Gillberg, M.: The circadian rhythm of spontaneously terminated sleep (in preparation).
6. Aserinsky, E.; Kleitman, N.: Regularly occuring periods of eye motility and concomitant phenomena during sleep. Sciene *118:* 273–274 (1953).
7. Baekeland, F.; Hartmann, E.: Sleep requirements and the characteristics of some sleepers; in Hartmann, Sleep and dreaming, Int. Psychiatric Clin. 7, No. 2, pp. 33–43 (Little, Brown, Boston 1970).
8. Bastiaans, J.: Psychotherapy in psychosomatic medicine; in Antonelli, Therapy in psychosomatic medicine. 3rd World Congr. of the ICPM, Rome (Pozzi, Rome 1975).
9. Bastiaans, J.: The implications of the specifity concept for the treatment of psychosomatic patients. Psychother. Psychosom. *28:* 285–293 (1977).
10. Berger, R.J.; Oswald, I.: Effects of sleep deprivation on behavior, subsequent sleep, and dreaming. J. ment. Sci. *108:* 457–465 (1962).
11. Bergström, B.; Gillberg, M.; Arnberg, P.: Effects of sleep loss and stress upon radar watching. J. appl. Psychol. *58:* 158–162 (1973).
12. Breger, L.; Hunter, J.; Lane, R.W.: The effects of stress on dreams. Psychol. Issues *7:* 1–210 (1971).
13. Brodan, V.; Vostěchovský, M.; Kuhn, M.; Cepelák, J.: Changes of mental and physical performance in sleep deprived healthy volunteers. Activitas nerv. sup. *11:* 175–181 (1969).
14. Brunette, R.; De Koninck, J.: The effects of presleep suggestion related to a phobic object on dream content. Sleep Res. *6:* 120 (1977).
15. Callois, R.; Grunebaum, G.F. von: Le rêve et les sociétés humaines (Gallimard, Paris 1967).
16. Cartwright, R.; Lloyd, S.; Tipton, L.W.; Wicklund, J.; Brown, J.: Effects of lab training in dream recall on psychotherapy. Sleep Res. *6:* 125 (1977).
17. Christie, M.J.; McBrearty, E.M.: Stress – response and recovery; in Mackay, Cox, Response to stress, pp. 150–155 (IPC Science and Technology Press, Guildford 1979).
18. Cluydts, R.; Visser, P.: Mood and sleep. I. Effects of the menstrual cycle. II. Effects of aversive pre-sleep stimulation. Waking Sleeping *4:* 193–203 (1980).
19. Cohen, D; Cox, C.: Neuroticism in the sleep laboratory: implications for representational and adaptive properties of dreaming. J. abnorm. Psychol. *84:* 91–198 (1975).
20. Cohen-Matthijsen, T.: Slaapstoornissen bij jonge kinderen; thesis, Amsterdam (1980).
21. Costell, R.M.; Leiderman, P.H.: Psychophysiological concomittants of social stress: the effects of conformity pressure. Psychosom. Med. *30:* 298–310 (1968).

22 Crisp, A.H.; Stonehill, E.: Sleep, nutrition and mood (Wiley&Sons, London 1976).
23 Czeisler, C.A.: Human circadian physiology; dissertation, Stanford (1979).
24 Dallet, J.: Theories of dream function. Psychol. Bull. *6:* 408–416 (1973).
25 De Koninck, J.; Koulack, D.: Dreams and adaptation to a stressful situation. J. abnorm. Psychol. *84:* 250–260 (1975).
26 Fiss, H.; Litchman, J.: Dream enhancement: an experimental approach to the adaptive function of dreaming. Sleep Res. *5:* 116 (1976).
27 Foulkes, D.: The psychology of sleep (Scribner's Sons, New York 1966).
28 Francesconi, R.P.; Stokes, J.W.; Banderet, L.E.; Kowal, D.M.: Sustained operations and sleep deprivation: effects on indices of stress. Aviat. Space environ. Med. *49:* 1271–1274 (1978).
29 Frankenhäuser, M.; Johansson, G.: Task demand as reflected in catecholamine excretion and heart rate. J. hum. Stress *2:* 15–23 (1976).
30 Freud, S.: L'interprétation des rêves (Presses Universitaires de France, Paris 1971).
31 Friedman, J.; Globus, G.; Huntley, A.; Mullaney, D.; Naitoh, P.; Johnson, L.: Performance and mood during and after gradual sleep reduction. Psychophysiology *14:* 245–250 (1977).
32 Friedman, J.; Rosenman, R.H.: Overt behavior pattern in coronary disease. J. Am. med. Ass. *173:* 1320–1325 (1960).
33 Fröberg, J.E.: Twenty-four-hour patterns in human performance, subjective, and physiological variables and differences between morning and evening active subjects. Biol. Psychol. *5:* 119–134 (1977).
34 Fröberg, J.E.; Karlsson, C.-G.; Levi, L.; Lidberg, L.: Circadian variations of catecholamine excretion, shooting range performance and self-ratings of fatigue during sleep deprivation. Biol. Psychol. *2:* 175–188 (1975).
35 Fröberg, J.E.; Karlsson, C.-G.; Levi, L.; Lidberg, L.: Psychobiological circadian rhythms during a 72 hour vigil. Försvarsmedicin. *11:* 192–201 (1975).
36 Gange, J.J.; Green, R.G.; Harkins, S.G.: Autonomic differences between extraverts and introverts during vigilance. Psychophysiology *16:* 392–397 (1979).
37 Gnirss, F.; Schneider-Helmert, D.; Schenker, J.; Winkler, V.: Schlafstörungen bei psychisch Kranken. Nervenarzt *49:* 394–401 (1978).
38 Gulevich, G.; Dement, W.; Johnson, L.: Psychiatric and EEG observations on a case of prolonged (264 hours) wakefulness. Archs gen. Psychiat. *15:* 29–36 (1966).
39 Hall, C.; Van de Castle, R.: The content analysis of dreams (Appleton Century Crofts, New York 1966).
40 Harris, W.; O'Hanlon, J.F.: A study of recovery functions in man. (US Army Technical Memorandum, 10.12, Aberdeen Research and Development Center, Maryland 1972).
41 Hofman, W.F.; Bakker, H.J.; Poelstra, P.A.M.; Visser, P.; Kumar, A.; Diest, R. van: A longitudinal study of the psychological aspects of sleep-waking cycles based on a 42 days period; in Popoviciu, Asgian, Badiu, Sleep 1978 4th Eur. Congr. on Sleep Res., Tirgu Mures 1978, pp. 649–651 (Karger, Basel 1980).
42 Horne, J.A.: A review of the biological effects of total sleep deprivation in man. Biol. Psychol. *7:* 55–102 (1978).
43 Institute of Medicine: Sleeping pills, insomnia and medical practice. Publ. IOM-79-04 (1979).

44 Johns, M.W.; Gay, T.J.A.; Masterton, J.P.; Bruce, D.W.: Relationship between sleep habits, adrenocortical activity and personality. Psychosom. Med. *33:* 499–508 (1971).
45 Johnson, L.C.; MacLeod, W.L.: Sleep and awake behavior during gradual sleep reduction. Percept. Mot. Skills *36:* 87–97 (1973).
46 Kagan, A.R.; Levi, I.: Health and environment – psychosocial stimuli. Rep. Lab. clin. Stress Res. *27* (1971).
47 Kales, A.: Treating sleep disorders. Am. Fam. Physn *8:* 158–168 (1973).
48 Kales, A.; Caldwell, A.B.; Preston, T.A.; Healey, S.; Kales, J.D.: Personality patterns in insomnia. Archs gen. Psychiat. *33:* 1128–1134 (1976).
49 Kales, A.; Tan, T.-L.; Kollar, E.J.; Naitoh, P.; Preston, T.; Malmstroem, E.J.: Sleep patterns following 205 hours of sleep deprivation. Psychosom. Med. *32:* 189–200 (1970).
50 Karli, P.: Les bases neurophysiologiques des processions de motivation. J. Physiol., Paris *72:* 503–516 (1976).
51 Kollar, E.J.; Slater, G.R.; Palmer, J.O.; Docter, R.F.; Mandell, A.J.: Stress in subjects undergoing sleep deprivation. Psychosom. Med. *28:* 101–113 (1966).
52 Koulack, D.; Prevost, F.; De Koninck, J.: Effects of stressful intellectual activity on sleep and dream content. Proc. Annu. Meet. of the Canadian Psychological Ass. Calgary 1980.
53 Kramer, M.; Roehrs, T.; Roth, T.: Mood change and the physiology of sleep. Compreh. Psychiat. *17:* 161–165 (1976).
54 Krupski, A.; Raskin, D.C.; Bakan, P.: Physiological and personality correlates of commission errors in an auditory vigilance task. Psychophysiology *8:* 304–311 (1971).
55 Lader, M.: Psychophysiological research and psychosomatic medicine; in Physiology, emotion and psychosomatic illness; Ciba Fdn Symp., 8, pp. 297–311 (Elsevier; Amsterdam 1972).
56 Leigh, H.; Reiser, M.F.: The patient, chap. 8 (Plenum Publishing, New York 1980).
57 Luby, E.D.; Frohman, C.E.; Grisell, J.L.; Lenzo, J.E.; Gottlieb, J.S.: Sleep deprivation: effects upon behaviour, thinking, motor performance and biological energy transfer systems. Psychosom. Med. *22:* 182–192 (1960).
58 Luby, K.D.; Grisell, J.L.; Frohman, C.E.; Lees, H.; Cohen, B.D.; Gottlieb, J.S.: Biochemical, psychological and behavioural responses to sleep deprivation. Ann. N.Y. Acad. Sci. *96:* 71–79 (1962).
59 Palmblad, J.; Cantell, K,; Strander, H.; Froberg, J.; Karlsson, C.-G.; Levi, L.; Granstrom, M.; Unger, P.: Stressor exposure and immunological response in man: interferon-producing capacity and phagocytosis. J. psychosom. Res. *20:* 193–199 (1976).
60 Palmblad, J.; Petrini, B.; Wasserman, J.; Åkerstedt, T.: Lymphocyte and granulocyte reactions during sleep deprivation. Psychosom. Med. *41:* 273–278 (1979).
61 Sandler, J.: The background of safety. Int. J. Psychol. *41:* 352 (1960).
62 Schneider, D.: Klinische und testpsychologische Untersuchung zur Entstehung psychosomatischer Störungen. Schweiz. Z. Psychol. *31:* 318–340 (1972).
63 Schneider, D.: Stress und Schlaf. Schweizer Arch. Neurol. Neurochir. Psychiat. *121:* 47–54 (1977).
64 Schneider, D.: Extreme hyposomnia: is it normal or pathological? 6th World Congr. Psychiat., Honolulu 1977, abstr. No. 960, p. 178 Abstractbook (1977).
65 Schneider, D.; Gnirss, F.: Three cases of extreme idiopathic hyposomnia; in Levin, Koella, Sleep 1974, pp. 460–463 (Karger, Basel 1975).

66 Schneider, D.; Gnirss, F.: Psychosomatische Insomnie. Schweizer Arch. Neurol. Neurochir. Psychiat. *120:* 131–151 (1977).
67 Scrimshaw, N.S.; Habicht, J.P.; Pellet, P.; Piche, M.L.; Cholakos, B.: Effects of sleep deprivation and reversal of diurnal activity on protein metabolism in young men. Am. J. clin. Nutr. *19:* 313–318 (1966).
68 Selye, H.: Selye's guide to stress research (Van Nostrand, New York 1980).
69 Sirois-Berliss, M.; De Koninck, J.: Dream content in relation to stress and mood fluctuations experienced during the menstrual cycle. Sleep Res. *9:* 159 (1980).
70 Stonehill, E.; Crisp, A.H.: Aspects of the relationship between sleep, weight and mood. Psychother. Psychosom. *22:* 148–158 (1973).
71 Visser, P.; Hofman, W.F.; Kumar, A.; Cluydts, R.; De Diana, I.P.F.; Marchant, P.; Bakker, H.J.; Diest, R. van; Poelstra, P.A.M.: Sleep and mood: measuring the sleep quality; in Priest, Pletscher, Ward, Sleep Research, pp. 135–145 (MTP Press; Lancaster 1979).
72 Walker, P.C.; Johnson, R.F.Q.: The influence of presleep suggestion on dream content: evidence and methodological problems. Psychol. Bull. *81:* 362–370 (1974).
73 Webb, W.B.; Agnew, H.W., Jr.: Sleep efficiency for sleep-wake cycles of varied length. Psychophysiology *12:* 637–641 (1975).
74 Wilkinson, R.T.: Sleep deprivation; in Edholm, Bacharach, Physiology of survival, pp. 399–430 (Academic Press, London 1965).
75 Williams, H.L.; Hammack, J.T.; Daly, R.L.; Dement, W.C.; Lubin, A.: Responses to auditory stimulation, sleep loss and the EEG stages of sleep. Electroenceph. clin. Neurophysiol. *16:* 269–279 (1964).
76 Zulley, J.: Der Einfluss von Zeitgebern auf den Schlaf des Menschen (Fischer, Frankfurt am Main 1979).

Addresses of Authors

Dr. T. Åkerstedt, Laboratory for Clinical Stress Research, Karolinska Institute, S-104 01 Stockholm (Sweden)

Dr. J. Bastiaans, Psychiatrische Universiteitskliniek, Rhijngeesterstraatweg 13, Jelgersmakliniek, NL-2340 BG Oegstgeest (The Netherlands)

Dr. R. Cluydts, Physiological Laboratory, Free University Brussels, Laarbeekboslaan 103, B-1050 Brussels (Belgium)

Dr. J. De Koninck, School of Psychology, University of Ottawa, Ottawa, Ont. (Canada)

Dr. J.A. Horne, Department of Human Sciences, Loughborough University, Loughborough, Leicestershire, LE11 3TU (England)

Dr. D. Schneider-Helmert, Research Department, Psychiatric Clinic, CH-5200 Königsfelden (Switzerland)

Dr. P. Visser, Psychophysiology Laboratory, Jan Swammerdam Institute, Eerste Constantyn Huygenstraat 20, NL-1054 BW Amsterdam (The Netherlands)

Part II. Workshops

A. Periodic Hypersomnia

Chairmen: *M. Billiard,* Montpellier; *B. Roth,* Prague
Participants: *Y. Hishikawa,* Osaka; *J. Montplaisir,* Montreal; *S. Nevšímalová,* Prague

Introduction

M. Billiard, Montpellier, France

Periodic hypersomnia is an infrequent though not exceptional disorder of excessive sleepiness. The Kleine-Levin syndrome is its most conspicuous paradigm. The condition still needs clarification and it was the goal of this workshop to shed some light on several points at issue.

What sort of classification can be established among the various kinds of periodic hypersomnia [*Roth and Nevšímalová*]? Is the typical Kleine-Levin syndrome limited to adolescent boys? Is there an alternative explanation to the usually hypothesized intermittent organic dysfunction in limbic or hypothalamic structures [*Billiard*]? Does the primary disorder concern sleep or wakefulness? Does periodic hypersomnia go with a disturbed GH sleep-related secretion [*Hishikawa et al.*]? Is there any relationship between periodic hypersomnia and affective symptoms [*Montplaisir*]?

The Clinical Picture of Periodic Hypersomnia

A Study of 38 Personally Observed Cases[1]

B. Roth, S. Nevšímalová, Prague, Czechoslovakia

The characteristic features of periodic hypersomnia (PH) are periodic attacks of pathological somnolence or sleep lasting 1 day to several weeks,

[1] For references see compound list on page 138.

exceptionally even months, at a rate of one attack a month down to one or two attacks in a year. In between the attacks the patients usually have a normal rhythm of sleep and wakefulness.

Recently, we summed up our material of narcolepsies and hypersomnias examined in the course of 31 years of systematic study, a total of 786 patients, 38 of whom, i.e. 4.8%, were diagnosed as having PH.

Clinical Picture and Forms of PH

Periodic hypersomnias can be classified as monosymptomatic, in which attacks of hypersomnia are the only sign of the affection, and polysymptomatic, in which periodic attacks of hypersomnia are accompanied by some other symptoms. Table I gives an idea of the number of monosymptomatic and polysymptomatic cases in our material.

Polysymptomatic cases can be divided further into the following categories: (1) *Typical Kleine-Levin syndrome* (KLS), i.e. periodic attacks of hypersomnia with bulimia and marked mental disorders, regardless of age and sex. (2) *Atypical cases of KLS and other variants of polysymptomatic PH* with some of the three main symptoms of the syndrome missing, or with the presence of symptoms not belonging in the KLS picture, or with atypical course. (3) *PH in periodic depression and in manic-depressive psychosis.* In-

Table I. Classification of periodic hypersomnias into a monosymptomatic and a polysymptomatic form

Form	Number of cases
Monosymptomatic	17
Polysymptomatic	21
Total	38

Table II. Classification of polysymptomatic cases of periodic hypersomnia

Form of affection	Number of cases
Typical Kleine-Levin syndrome	3
Atypical forms of Kleine-Levin syndrome	13
Periodic hypersomnia accompanying depressive attacks	3
Periodic hypersomnia during menses	2
Total	21

cluded in this category were only these patients who did not suffer from hypersomnia in off-depression periods. (4) *Menstruation PH*. This category was made up of female patients with PH appearing solely at the time of menses or a few days before. Table II gives an outline of our patients with the polysymptomatic form of PH.

Discussion

Monosymptomatic PH

Cases of the monosymptomatic form of PH are rather rare. Patients with this affection were described by *Andreiev* [2], *Takahashi* [42], *Bonkalo* [8], *Lobzin et al.* [30], *Vein* [44], *Davidova* [13] and others.

Material. Monosymptomatic form of PH was seen in 17 out of our 38 patients, i.e. 44.73%. The case of 1 female patient with typical PH is particularly worth mentioning. Her mother and mother's father had suffered from fully expressed idiopathic narcolepsy-cataplexy.

Polysymptomatic PH

Typical Kleine-Levin Syndrome. Among our 38 cases of PH there are only 3 patients who could be classified as typical KLS. An outline of the symptomatology of the 3 patients can be seen in table III.

Atypical Form of KLS and Other Variants of Polysymptomatic PH. An outline of the symptomatology of our 13 patients is given in table IV.

As follows from table IV, mental disturbances were found in 11 out of 13 patients, most of them quite prominent ones. 5 patients had depression, 2 had schizophrenia-like changes, 2 were confused and irritable, and 2 more displayed marked apathy.

The category of 'other symptoms' included 3 patients, mostly women, who suffered from episodes of fainting. In 1 of the women (J.V.) the attacks started with loss of consciousness. During the attack, lasting 3–7 days, the patient cannot be woken up until after 2 or 3 days the unconscious state changes into sleep. Another patient reported his mother to be suffering from difficulties very much like his own.

Hypersomnia during Depression. Over the past 15 years, the incidence of pronounced PH during depression has been repeatedly described. Most

Table III. Survey of patients with typical Kleine-Levin syndrome

Name	Sex	Age	Age of onset	Hypersomnia	Bulimia	Polydipsia	Psychic disturbances	Subfebrile temperatures
J.M.	F	37	28	+++	+++	+++	+++	++
E.P.	F	51	46	+++	+++	+++	+++	+++
V.S.	M	24	15	+++	+++	+++	+++	---

Table IV. Survey of patients with atypical form of Kleine-Levin syndrome

Name	Sex	Age	Age of onset	Hypersomnia	Bulimia	Polydipsia	Psych. disturb.	Subfebr. temp.	Atyp. course	Other symptoms
P.V.	M	30	21	+++	---	---	++	++	---	---
M.L.	F	28	16	+++	---	---	+++	---	---	---
P.K.	F	34	18	+++	+++	---	+++	---	+++	---
B.B.	F	48	33	+++	---	+++	+++	---	---	---
P.B.	F	27	15	+++	+++	+++	++	+	---	+++
E.D.	F	28	15	---	+++	---	+++	---	---	---
B.P.	F	24	15	+++	+++	+++	+	+++	+++	+++
T.S.	M	32	15	+++	+++	+++	+++	+	+++	---
J.V.	F	36	32	+++	---	---	---	---	---	+++
M.D.	F	21	18	+++	+++	+++	---	---	+++	---
J.K.	M	25	17	++	+	---	+++	---	+++	---
M.K.	F	31	16	+++	++	++	++	---	+++	---
K.V.	M	26	15	+++	+++	+++	+++	---	+++	---

Table V. Survey of patients with hypersomnia accompanying depressive attacks

Name	Sex	Age	Age of onset	Hypersomnia	Bulimia	Polydipsia	Psychic. disturb.	Subfebr. temp.	Other symptoms
J.P.	M	32	30	+++	+	---	+++	---	---
J.S.	M	49	28	+++	---	---	+++	---	---
M.K.	M	28	18	+++	---	---	+++	---	---

Table VI. Periodic hypersomnia during menses

Name	Sex	Age	Age of onset	Hypersomnia	Bulimia	Polydipsia	Psychic disturb.	Subfebr. temp.	Other symptoms
L.B.	F	30	17	+++	+	---	---	---	+++
Z.M.	F	25	16	+++	---	---	+++	---	---

authors, including us, believe this to be an independent nosological entity. The 3 cases we have been able to observe thus far are summed up in table V. As table V shows, all the 3 patients were males.

Menstruation PH. PH in young women and girls during menses or a few days before, often accompanied by bulimia, was described by several authors. 2 such cases in our own charge are presented in table VI.

In patient L.B. the attack starts with a sudden loss of consciousness, during which she tends to fall down and often even hurt herself. She has to be taken to the hospital, cannot be woken up, and has to be artificially fed and catheterized. The attack lasts 2–3 days.

In patient Z.N. the attacks of hypersomnia last 7–14 days and are accompanied by mental changes: psychiatrists describe her condition as schizophrenic psychosis.

Sleep 1980. 5th Eur. Congr. Sleep Res., Amsterdam 1980, pp. 124–127
(Karger, Basel 1981)

The Kleine-Levin Syndrome[2]

M. Billiard, Montpellier, France

After the first report by *Anfimoff* [3] on an adolescent boy who developped recurring attacks of drowsiness, overeating and anxiety, *Kleine* [25], *Lewis* [29] and *Levin* [28] reported similar cases and *Critchley and Hoffman* [12] coined the eponymous term Kleine-Levin syndrome. 20 years later *Critchley* [11] reported 11 such cases seen by him together with 15 cases of the literature and gave an outstanding description of this syndrome consisting of 'recurring episodes of undue sleepiness lasting some days, associated with an inordinate intake of food and often with abnormal behaviour'. Moreover, he emphasized 4 other clinical features, 'sex incidence whereby males are preponderantly if not wholly affected, onset in adolescence, spontaneous eventual disappearance of the syndrome and the possibility that the megaphagia is in the nature of a compulsive eating rather than a bulimia'.

[2] For references see compound list on page 138.

Since that time numerous case reports have been published, some of them including the 3 basic symptoms, typical cases of the Kleine-Levin syndrome (polysymptomatic form of periodic hypersomnia, group I, in Roth's classification), some others with one of the main symptoms missing.

This paper focuses on the typical Kleine-Levin syndrome, of which we were able to gather 72 male cases and 24 female cases, these latter having all been published after *Critchley's* description. According to the listed cases the ratio male : female would be 4 : 3 to 4 : 1. Moreover, if the predominance of onsets in adolescence is still obvious, peak age in males 17, peak age in females 14, a later onset is not rare especially in females: 6 male reports (8.3%) are from subjects between 29 and 67 and 12 female reports (50.0%) are from subjects between 26 and 66 (fig. 1).

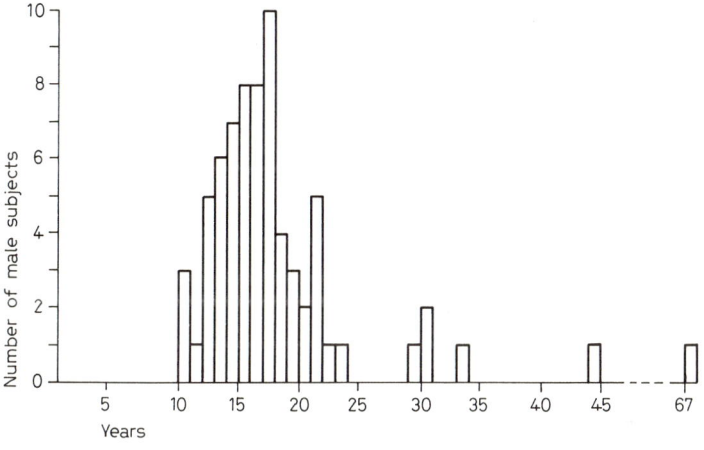

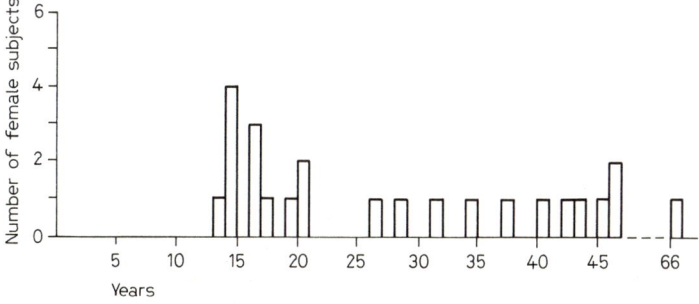

Fig. 1. Age at the time of the first hypersomnolent episode in 72 male subjects (upper diagram) and in 24 female subjects (lower diagram). Note the significant ratio of subjects, especially among females, with the first episode occurring after the age of 25.

The syndrome may appear with no obvious reason. However, an upper airway infection or a flu-like syndrome takes place a week or two prior to the first attack in 32 cases (33.3%) and various factors such as drunkenness, seasickness, enduring extremely hot or cold weather, seem to act as precipitant factors in several cases.

Excessive somnolence is of abrupt or gradual development. Sleep is both of long duration and unstable. Patients are always rousable. Vivid dreams are frequent. Incontinence does not occur. Overeating is striking and most reports support *Critchley's* contention that it is 'in the nature of a compulsive eating, rather than a bulimia'. Patients eat up all foodstuffs within sight even when of poor quality [11 (No. 3)]. Rapid weight increase during the attacks is frequent. Abnormal behaviour varies in type and frequency. Irritability, feeling of unreality – as if the persons and things were abnormally distant – are almost systematic. Confusions, delusions, visual and auditory hallucinations, disinhibited sexual behaviour are less frequent, the latter symptom having the same prevalence in males (23%) and in females (20%).

After a variable number of days, 4–7 on an average, 1–30 at most, the symptoms subside. But a short reactionary period with either depression [11 (No. 2), 18 (Nos. 1, 2), 31, 37, 38 (Nos. 1, 2), 45] sometimes with suicidal ideation or with elation, euphoria ± sleeplessness [4, 6, 8 (No. 1), 11 (Nos. 3, 5, 7), 15, 19 (No. 2), 23, 30, 36, 40] may follow.

Intervals between somnolent episodes are variable in duration but in a few cases. Therefore, the term recurring episodes is more relevant than periodic episodes.

Clinical examination of the patient during and after the attacks may show a few anomalies limited to the hypersomnolent phase. Congestion of the face, heavy sweating, dysarthria, excessive or depressed deep tendon reflexes are unexceptional. More puzzling are the findings of an extensor plantar response on one side [8 (No. 1), 11 (No. 5)] or of a horizontal nystagmus [9 (No. 1), 21]. Mental condition between attacks is not always normal. The most frequent personality pattern displayed by the Rorschach test, is one of introversion and insecurity. Furthermore, family cases of affective disorders have been reported [8 (No. 1, 2), 11 (No. 10), 14, 23, 39 (No. 1)].

Up to now no consistent biochemical anomaly has ever been demonstrated. X-Ray studies including pneumoencephalography and computerized brain scan have been performed in a significant number of patients: 12 air studies were normal whereas 9 showed either a relative enlargement of one lateral ventricle (7 cases) or a hydrocephalus (2 cases); 1 scan showed a slight ventricular dilatation. EEG studies have been numerous and some reports [22, 46] have specifically focused on this aspect. Anomalies of a non-specific character, such as paroxysmal bursts of bisynchronous generalized high voltage 5–7 cps waves, are frequent and day-time sleep onset REM periods are mentioned in 3 reports [27, 33, 46]. All-night polygraphic recordings have been performed either in symptomatic phases [16, 35] or in asymptomatic intervals [33], or in both phases [4, 27 (Nos. 1, 2), 37]. In the 4 latter reports comparison of recordings performed in symptomatic and asymptomatic phases shows the increase of total sleep time (first 3 cases), which may however be due in part to a longer total recording time, the increase of intermittent wakefulness (last 3 cases), the increase of the number of sleep stage shifts (4 cases) and the decrease of delta sleep (4 cases). In addition REM latency is decreased in the last 2 cases. Finally GH secretion studies and CSF monoamine levels may bring interesting data but are far too limited in patients with typical Kleine-Levin syndromes to draw any conclusion.

According to *Critchley,* the attacks gradually lessen in frequency and eventually cease. However, this statement needs some qualifications. Among the 96 reports under consideration a spontaneous disappearance was verified in 13 male subjects (mean duration of condition 4.8 ± 0.9 years, range 1–13 years) and in 6 female subjects (mean duration of condition 2.1 ± 0.6 years, range 1–5 years). Besides there was no disappearance of the attacks within the span of follow-up in 43 male subjects (mean duration of follow-up 5.6 ± 0.7) and in 13 female subjects (mean duration of follow-up 9.1 ± 2.2). Data were unclear in 21 subjects.

Up to a recent date treatments of the Kleine-Levin syndrome using either amphetamines (25 reports), psychotherapy (6 reports) or neuroleptics and/or ECT (4 reports) have been unsuccessful in most cases. The use of lithium may now be the answer. The results to date mainly concern atypical cases of the syndrome [1, 23, 34]. However, the relationship of these cases with the Kleine-Levin syndrome is encouraging.

The combination of abnormal sleep, excessive intake of food, sexual symptoms and may be of GH secretion anomalies may well implicate, as already widely hypothesized, a functional abnormality of the hypothalamo-limbic system. Now because of the frequent mood swing at the end of the attacks, the family records of affective disorders, the sleep polygraphic data resembling to a certain extent those found in primary depression [37] and the positive effect of lithium, one may speculate about biologic relationships between the Kleine-Levin syndrome and affective disorders.

Polysomnographic Findings and Growth Hormone Secretion in Patients with Periodic Hypersomnia[3]

Y. Hishikawa, S. Iijima, T. Tashiro, Y. Sugita, Y. Teshima, R. Matsuo, H. Kaneda, Osaka, Japan

Periodic hypersomnia is a relatively rare condition characterized by recurrent periods of abnormally prolonged sleep. There are only a small number of reports revealing on polysomnographic findings of nocturnal sleep in the patients with periodic hypersomnia [4, 7, 17, 27, 31, 32, 35, 41, 42]. As to the etiology of this disease, hypothalamic dysfunction has long been suggested by many investigators, but this suggestion has not yet been well tested with the use of recently developed endocrinological technique [20, 24, 43].

In the last 9 years, we performed polysomnographic examination in 11 patients with periodic hypersomnia including 3 patients with Kleine-Levin syndrome (6 males and 5 females, aged from 13 to 23 years) not only during the hypersomniac episode but also in the asymptomatic interval in a dimly lighted, sound-attenuated and air-conditioned room after 1 adaptation night. In 7 of them, the examination was performed during nocturnal sleep, and the remaining 4 patients and 10 normal young adults as controls underwent a 24-h examination. They were allowed to sleep undisturbed or encouraged to sleep as much as they could in bed.

In 4 patients and in 2 normal subjects, blood samples for duplicated radioimmunoassay of serum growth hormone (GH) and cortisol were taken every 30 min during the monitored night.

Polysomnographic Findings

Sleep diagrams for a 24-h period of a male patient are shown in figure 2. Total sleep time was apparently increased during the hypersomniac episode. And this was due to increased sleep in the daytime. But his nocturnal

[3] For references see compound list on page 138.

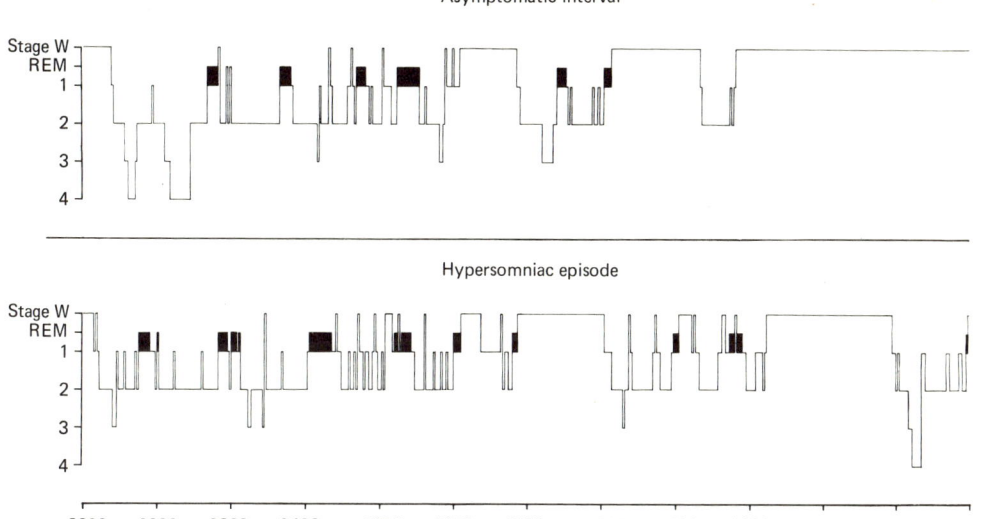

Fig. 2. Sleep diagrams of a patient with periodic hypersomnia (male, 23 years old). In the hypersomniac episode, the total sleep time was apparently longer than in the asymptomatic interval. But sleep in the hypersomniac episode was unstable with frequent interruptions by awakening and decreased amounts of deep NREM sleep (stages 3 and 4).

sleep during the hypersomniac episode was unstable with frequent changes in the sleep stages and decreased amounts of deep NREM sleep. The REM-NREM sleep cycle was rather regular in these two recordings. Similar characteristics of nocturnal sleep were found also in the other 3 patients. In a 24-h period, the 4 patients slept for 13.2 h on average in the hypersomniac episode. This total sleep time was longer by about 2 h than their total sleep time in the asymptomatic interval, but was significantly shorter than that of the normal young adults ($p<0.05$). However, nocturnal sleep of the patients was lighter and more unstable during the hypersomniac episode than in their asymptomatic interval and in the normal subjects, as described below.

The findings of nocturnal sleep in the 11 patients and 10 normal subjects are shown in table VII, where the findings are presented with respect to their recordings of 8 h after the sleep onset at night. An unexpected finding in the patients was that their total sleep time within 8 h after the sleep onset in the hypersomniac episode was significantly shorter than that in the asymptomatic interval ($p<0.05$), and it was also shorter than that in the normals ($p<0.05$). The amount of stage 1 and its percentage of the total

Table VII. Findings of nocturnal sleep during 8 h after the sleep onset in 11 patients with periodic hypersomnia and in 10 normal young adults

	Mean age years	TST min	Duration and percentage to TST of different sleep stages				Sleep latency min	REM latency[1] min	Number of awakenings
			stage 1 min (%)	stage 2 min (%)	stage 3+4 min (%)	stage REM min (%)			
Patients with periodic hypersomnia (n = 11)	18.2								
Hypersomniac episode		419.4**++	65.4++ (16.0)++	189.8***++ (45.4)*	73.2++ (17.5)+	91.1 (21.1)	25.0**	90.4++	7.5
Asymptomatic interval		433.5	42.5** (9.9)	211.8 (48.8)	90.2 (21.0)*	90.1 (20.7)	17.3	139.4*	5.3
Normal young adults (n = 10)	21.1	456.3	64.5 (14.5)	234.8 (51.2)	67.4 (14.8)	89.6 (19.5)	6.1	94.8	3.5

Only mean values and mean percentages in parentheses are shown here.

[1] Sleep onset REM periods was observed in 2 patients during the hypersomniac episode. Mean REM latency during hypersomniac episode was obtained from the findings in the remaining 9 patients.

* Different from the finding in the normals ($0.05 < p < 0.1$); ** significantly different from the finding in the normals ($p < 0.05$); + different from the finding in asymptomatic interval ($0.05 < p < 0.1$); ++ significantly different from the finding in asymptomatic interval ($p < 0.05$).

sleep time were significantly larger in the hypersomniac episode than in the asymptomatic interval (p<0.05). The amounts of stage 2 and stage 3+4 in the patients tended to be smaller in the hypersomniac episode than in the asymptomatic interval (0.05<p<0.1). The amount of stage REM and its percentage in the patients did not differ significantly from those in the normals.

Sleep latency to the first episode of stage 1 was significantly longer in the patients during the hypersomniac episode than in the normals (p<0.05).

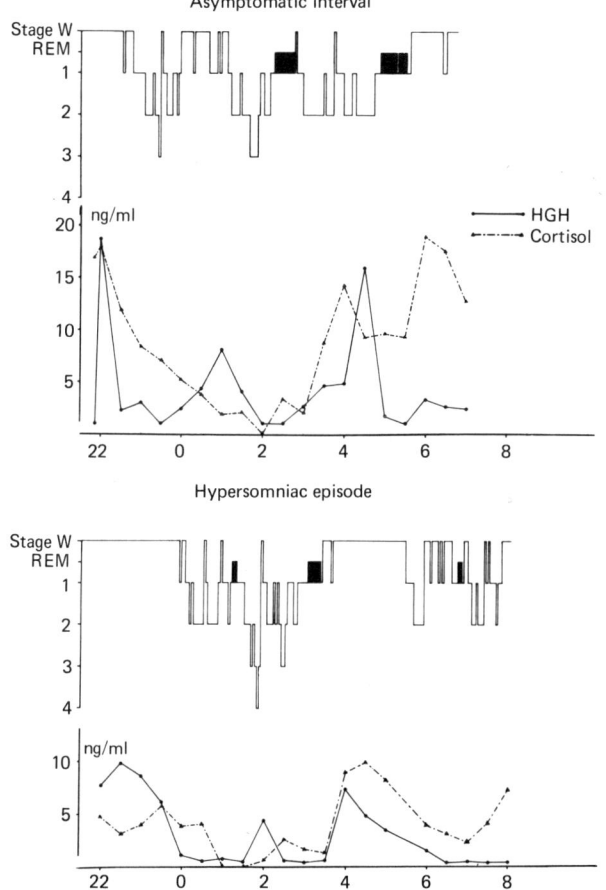

Fig. 3. Sleep diagrams and serum levels of GH (——) and cortisol (–··–) in a patient with periodic hypersomnia (female, 23 years old). The patterns of serum cortisol levels were normal, but the patterns of serum GH levels were abnormal. See text for further explanations.

Sleep onset REM period occurred in 2 patients only during the hypersomniac episode. When these findings of sleep onset REM period were excluded, REM latency from the onset of stage 1 in the patients was significantly shorter during the hypersomniac episode than in the asymptomatic interval ($p<0.05$), but it did not differ significantly from that in normals. The number of awakenings in the patients was larger during the hypersomniac episode than that in their asymptomatic interval and that in normals, though without statistical significance.

Endocrinological Findings

The sleep diagrams and serum levels of GH and cortisol in 1 patient both in the hypersomniac episode and in the asymptomatic interval are shown in figure 3. Serum cortisol level showed a normal pattern with increased secretion in the later part of the night and towards awakening in the morning. GH secretion pattern showed 3 apparent peaks in the asymptomatic interval and 2 peaks in the hypersomniac episode. The pattern of GH secretion seemed apparently abnormal, since the GH peak was absent or very small during the initial period of deep NREM sleep. In the normal subjects, the pattern of serum GH level showed an apparently large peak only during the initial period of deep NREM sleep. In the other 3 patients, the pattern of GH secretion was also abnormal during the hypersomniac episode and/or in the asymptomatic interval, because one or two apparent large peaks of serum GH level were observed during REM sleep or stage 2, which was not just preceded by an episode of deep NREM sleep.

Discussions and Conclusions

Based on the findings obtained by polysomnographic examination, we consider that the primary disturbance in the patients with periodic hypersomnia is not increased activity in the neural mechanism for inducing and maintaining sleep. Considering the clinical symptoms and slowing of alpha rhythm in the EEG during the hypersomniac episode, which may indicate slight disturbance of consciousness, and our polysomnographic findings, we rather suggest that the primary disorder in the patients with periodic hypersomnia is some functional disturbance of the neural structure for inducing and maintaining wakefulness in the brain. Probably because of disturbed wakefulness and decreased activity, the patients lie in bed most of the time during the hypersomniac episode and, as the results of lying in bed, they secondarily spend much longer time in sleep than they do sleep in their usual daily life in the asymptomatic interval.

Recently, the similarity of periodic hypersomnia to primary depression has been suggested by *Reynolds et al.* [37] on the basis of the polysomnographic findings of short REM latency, decreased REM percentage, and unstable nocturnal sleep in the patients with either of these diseases. But this suggestion cannot be supported because of our present findings that both REM latency and percentage of REM sleep in total sleep time in the patients with periodic hypersomnia did not differ significantly from those in the normals, although sleep onset REM period was observed in 2 patients in the hypersomniac episode.

The secretion pattern of serum cortisol was considered to be within normal limits. The secretion pattern of serum GH in the patients showed apparent deviation from the normal pattern either during the hypersomniac episode or in the asymptomatic interval or in both states. The peak of serum GH level during the initial period of deep NREM sleep was very small or absent, or there was one or two large peaks later in the night without being preceded by deep NREM sleep. A similar abnormality of GH secretion has also been found in narcoleptic patients [5]. The mechanism producing these abnormal GH patterns is unknown. But it is possible that the abnormal pattern of GH secretion is due to some disturbance in the neural mechanism regulating the secretion of GH in the hypothalamus.

Sleep 1980. 5th Eur. Congr. Sleep Res., Amsterdam 1980, pp. 133–137
(Karger, Basel 1981)

Hypersomnia and Manic Depressive Illness[4]

J. Montplaisir, J. L. de Liry, A. Dardenne, Montréal, Canada

Among 74 hypersomniacs seen in our laboratory over the past 2 years, 2 patients presented periodic changes of their symptomatology and symp-

[4] For reference see compound list on page 138.

toms of manic depressive illness (MDI). Their case history will be summarized and the results of polygraphic recordings will be reported.

Case No. 1

M.L. is a 62-year old woman presenting for 31 years recurrent episodes of hypersomnia. The first episode started shortly after the delivery of her third child, a premature baby requiring constant care, day and night for the first month. One night she went to bed as usual but she had 3 days of almost uninterrupted sleep. She was irritable and confused when aroused. Since then, approximately every 4 months she goes to bed 'normally' and wakes up 20 h later and the same pattern repeats itself for 4–30 consecutive days. Several hypersomniac episodes but not all were accompanied or followed by depressive symptoms such as lack of motivation, crying spells and suicidal thoughts which have led to 13 hospitalizations and to treatment with ECT and tricyclic antidepressant.

Over the past 20 years, she also presented approximately 15 psychotic episodes with severe insomnia lasting 2 or 3 days followed by auditory hallucinations, paranoid delusion and increased psychomotor activity for 1–6 weeks and treated with various neuroleptics. Lithium carbonate given for 2 years prevented both hypersomniac and psychotic manifestations but was discontinued because of suspected nephrotoxicity. For the last 2 years she was treated with haloperidol which suppressed psychotic episodes but had no effect on hypersomnia. Her sleep was recorded for 2 nights and 4 naps when she was asymptomatic. Her sleep was characterized by long sleep latency, several awakenings and stage shifts, and rapid eye movements were seen during stage 2 sleep.

Discussion (Case No. 1)

Previous studies [23] have shown the frequent occurrence of mental symptoms in classical cases of the Kleine-Levin syndrome (KLS) such as motor unrest, irritability, difficulty in thinking, incoherent speech and hallucinations. Some of Kleine's cases had affective symptoms at the end of the hypersomniac phase. Lithium carbonate effective to control sleep and mental symptoms in this patient was also successfully used to treat a few cases of KLS. In addition to these clinical observations there are several other similarities between periodic hypersomnia and MDI. Similar EEG sleep findings were reported for KLS and unipolar depression and there is much evidence that both conditions are related to disruption of biological rhythms.

Case No. 2

M.Q. is a 39-year old patient referred for narcolepsy. Since age 16 she had experienced hypersomnolence and sleep attacks in various inappropriate situations and severe nocturnal

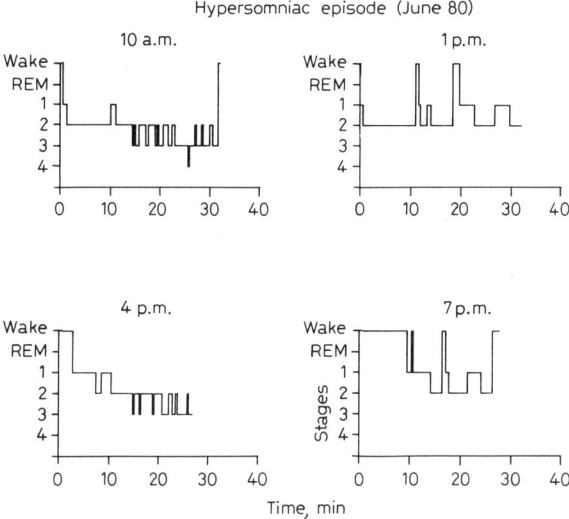

Fig. 4. Results of 4 polygraphic nap recordings of patient M.Q. during a period of hypersomnia.

sleep disruption with frequent awakenings and nightmares. She was treated with low dose of dextroamphetamine (5 mg/die). At age 20, she had her first 'nervous breakdown' described as a period of severe insomnia and restlessness followed within 2 or 3 days by auditory hallucinations and persecutory delusion. This first psychotic episode lasted approximately 2 months. After cessation of psychotic symptoms hypersomnia reappeared. At age 25, typical cataplectic attacks appeared, triggered by laughter or anger. Since age 20, she experienced hypersomnia interrupted about 5 times a year by acute psychotic episodes, as described above, lasting 1–3 weeks. She is currently treated with haloperidol and has developped symptoms of tardive dyskinesia.

M.Q. had 2 nights and 4 naps polygraphic recording during a hypersomniac and a psychotic episode. Figure 4 shows the result of nap recording during hypersomnia. Sleep latency was short and only non REM sleep was seen which is surprising considering the clinical picture of narcolepsy. On the contrary during the psychotic phase with insomnia she fell asleep during 2 naps and only REM sleep was recorded (fig. 5).

Nocturnal sleep recording confirmed the complaint of insomnia when she was psychotic. Numerous and long wakes were seen (fig. 6). Total sleep time was reduced and the cyclic organization of sleep was disrupted with delta sleep (stages 3 and 4) in large amounts towards the end of the night. REM sleep in this hypnogram is underscored since rapid eye movements were abundant in all stages, but muscle tone showed only phasic suppression (intermediate sleep).

Nocturnal sleep was also disrupted during hypersomnia (fig. 6) with numerous wakings and increased stage 1 sleep. REM sleep distribution was not clearly periodic and delta sleep was seen at the end of the recording. Total sleep time was normal.

Workshop A. Periodic Hypersomnia

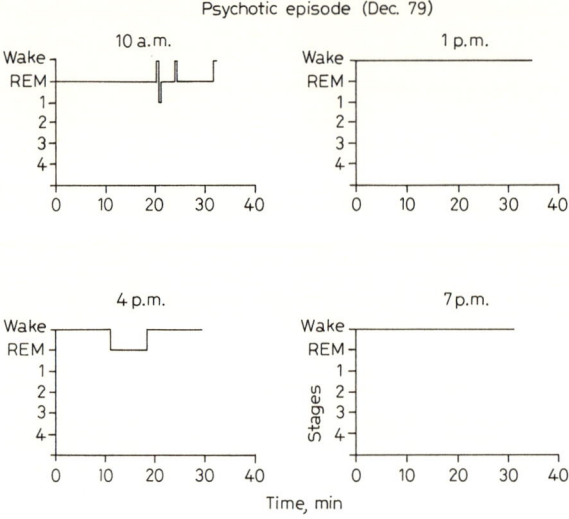

Fig. 5. Result of 4 polygraphic nap recordings of patient M.Q. during a period of acute psychosis with insomnia.

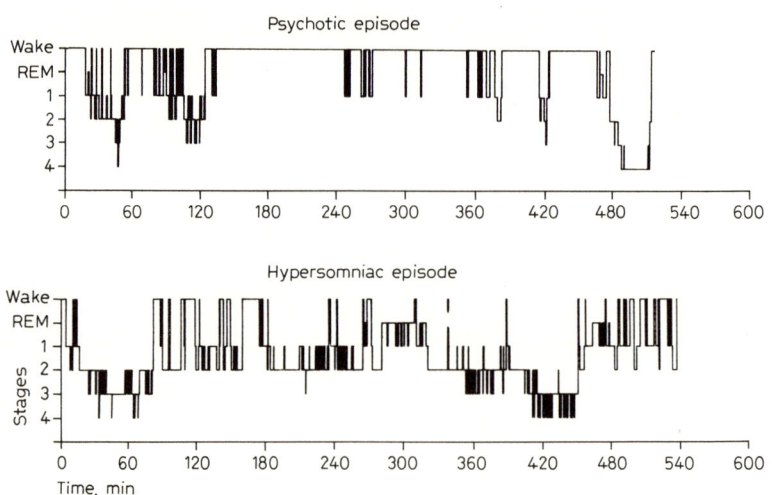

Fig. 6. Nocturnal polygraphic recordings of patient M.Q. during a period of hypersomnia and a period of acute psychosis with insomnia.

Cases 1 and 2 were remarkable for being free of any psychiatric symptoms on examination between psychotic episodes. They could recall the psychotic episodes in detail and manifest good insight and judgement when reporting it.

Discussion (Case No. 2)

M.Q. had symptoms of MDI in addition to typical symptoms of narcolepsy. However, nap recording during hypersomnia was not typical of narcolepsy, so she remains a diagnostic problem. Both cases had severe disruption of their nocturnal sleep with abnormal distribution of REM and delta sleep. Further studies in these patients should investigate the circadian variation of other biological parameters.

Conclusion

2 cases were reported where hypersomnia and recurrent episodes of insomnia with paranoid psychosis were seen. The periodicity is not clearly shown in these patients as in most cases of so-called 'periodic hypersomnia'.

The association of hypersomnia and affective symptoms is illustrated by these 2 cases. It is a common finding not only in periodic hypersomnia but even in narcolepsy. In a recent study, we found high values in the depression and mania scales of the MMPI (Minnesota multiphasic personality inventory) of 15 narcoleptics.

On many occasions depression occurs at the end of the hypersomnia and mania after 1 or 2 days of insomnia, which may suggest that mood changes would be secondary to disruption of the sleep-wake cycle. However, both conditions may have a common origin, involving the same brain region and neurotransmitter. Several observations raise the possibility of a defect in the control of dopamine (DA) activity in the CNS of these patients. (1) DA was found to be of low concentration in the CSF of hypersomniacs. (2) Medications blocking the reuptake of DA, such as amphetamine, are effective for the treatment of hypersomnia but may induce insomnia and acute psychotic episodes with higher dosage. (3) On the contrary haloperidol and other neuroleptics block DA activity and are effective for the treatment of insomnia and psychotic symptoms.

List of References of Workshop A

1 Abe, K.: Lithium prophylaxis of periodic hypersomnia. Br. J. Psychiat. *130:* 312–316 (1977).
2 Andreiev, B.V.: To the question of periodic hypersomnia (in Russian). Zh. Nevropatol. Psychiat. im. Korsakova. *19:* 49–53 (1950).
3 Anfimoff, J.A.: Quoted by Kaplinsky and Schulman (1935).
4 Barontini, F.; Zappoli, R.: A case of Kleine-Levin syndrome. Clinical and polygraphic study. Proc. 20th Eur. Meet. on Electroencephalography, pp. 239–245 (Gaggi, Bologna 1967).
5 Besset, A.; Bonardet, A.; Billiard, M.; Descomps, B.; Craste de Paulet, A.; Passouant, P.: Circadian patterns of growth hormone and cortisol secretions in narcoleptic patients. Chronobiologia *6:* 19–31 (1979).
6 Billard, C.; Ponsot, G.; Lyon, G.; Arfel, G.: Syndrome de Kleine-Levin, à propos d'une observation. Archs. fr. Pédiat *35:* 424–431 (1978).
7 Billiard, M.; Guilleminault, C.; Dement, W.C.: A menstruation linked periodic hypersomnia, Neurology Minneap. *25:* 436–443 (1975).
8 Bonkalo, A.: Hypersomnia: a discussion of psychiatric implications based on three cases. Br. J. Psychiat. *114:* 69–75 (1968).
9 Bucking, P.H.; Palmer, W.R.: New contribution to the clinical aspects and pathophysiology of the Kleine-Levin syndrom. Münch. med. Wschr. *120:* 1571–1572 (1978).
10 Chiles, J.A.; Wilkus, R.J.: Behavioral manifestations of the Kleine-Levin syndrome. Dis nerv. Syst. *37:* 646–648 (1976).
11 Critchley, M.: Periodic hypersomnia and megaphagia in adolescent males. Brain *85:* 627–656 (1962).
12 Critchley, M.; Hoffman, H.L.: The syndrome of periodic somnolence and morbid hunger (Kleine-Levin syndrome). Br. med. J. *1:* 137–139 (1942).
13 Davidova, F.B.: On the syndrome of periodic hypersomnia (in Russian). Zh. Nevropatol. Psychiat. im. Korsakova. *74:* 361–363 (1974).
14 Duffy, J.P.; Davison, K.: A female case of the Kleine-Levin syndrome. Br. J. Psychiat. *114:* 77–84 (1968).
15 Earle, B.V.: Periodic hypersomnia and megaphagia (the Kleine-Levin syndrome). Psychiat. Q. *39:* 79–83 (1965).
16 Fresco, H.; Giudicelli, S.; Poinso, Y.; Tatossian, A.; Mouren, P.: Le syndrome de Kleine-Levin. Annls Méd.-psychol. *1:* 625–668 (1971).
17 Furuya, E.; Hishikawa, Y.; Wakamatsu, H.; Kinoshita, R.; Doi, T.: A case of Kleine-Levin syndrome (in Japanese). Clin. Psychiat. *15:* 503–509 (1973).
18 Gallinek, A.: The Kleine-Levin syndrome: hypersomnia, bulimia and abnormal mental states. Wld Neurol. *3:* 235–241 (1962).
19 Gallinek, A.: The Kleine-Levin syndrome. Dis. nerv. Syst. *28:* 448–451 (1967).
20 Gilligan, B.S.: Periodic megaphagia and hypersomnia. An example of the Kleine-Levin syndrome in an adolescent girl. Proc. Aust. Ass. Neurol. *9:* 67–72 (1973).
21 Gran, D. von; Begemann, H.: Neue Beobachtungen bei einem Fall von Kleine-Levin-Syndrom. Münch. med. Wschr. *115:* 1098–1102 (1973).
22 Green, L.N.; Cracco, R.Q.: Kleine-Levin syndrome. Archs Neurol., Chicago *22:* 166–175 (1970).

23 Jeffries, J.J.; Lefebvre, A.: Depression and mania associated with Kleine-Levin-Critchley syndrome. Can. Psychiat. Ass. J. *18:* 439–444 (1973).
24 Kaneda, H.; Sugita, Y.; Masaoka, S.; Iijima, S.; Tanaka, K.; Wakamatsu, H.; Hishikawa, Y.: Red blood cell concentration and growth hormone release in periodic hypersomnia. Waking Sleeping *1:* 369–374 (1977).
25 Kleine, W.: Periodische Schlafsucht. Mschr. Psychiat. Neurol. *57:* 285–320 (1925).
26 Kupfer, D.J.; Foster, F.G.: EEG sleep and depression; in Williams, Karacan, Sleep disorders: diagnosis and treatment (Wiley & Sons, New York 1978).
27 Lavie, P.; Gadoth, N.; Gordon, C.R.; Goldhammer, G.; Bechar, M.: Sleep patterns in Kleine-Levin syndrome. Electroenceph. clin. Neurophysiol. *47:* 369–371 (1979).
28 Levin, M.: Narcolepsy (Gelineau's syndrome) and other varieties of morbid somnolence. Archs Neurol. Psychiat., Chicago *22:* 1172–1200 (1929).
29 Lewis, N.D.C.: The psychoanalytic approach to the problems of children under twelve years of age. Psychoanal. Rev. *13:* 424–443 (1926).
30 Lobzin, V.S.; Shamrei, R.K.; Churilov, I.K.: Pathophysiological mechanisms of periodic hypersomnia and the Kleine-Levin syndrome (in Russian). Zh. Nevropatol. Psychiatr. im Korsakova. *73:* 1719–1724 (1973).
31 Markman, R.A.: Kleine-Levin syndrome: report of a case. Am. J. Psychiat. *123:* 1025–1026 (1967).
32 Maxion, H.; Jacobi, P.: Klinische und polygraphische Untersuchungen bei periodischer Schlafsucht. Dt. Z. Nerv. Heilk. *197:* 192–202 (1970).
33 Messimy, R.; Weil, B.; Safar, J.: Sur un cas d'hypersomnie avec troubles des conduites alimentaires, excitation sexuelle et troubles du comportement. Sem. Hôp. Paris *49:* 3100–3105 (1967).
34 Ogura, C.; Nakaza, K.; Kishimoto, A.; Okuma, T.: A case of periodic somnolence improved by lithium carbonate (in Japanese). Clin. Psychiat. *17:* 59–63 (1975).
35 Popoviciu, L.; Corfariu, O.: Etude clinique et polygraphique au cours du nycthémère d'un cas de syndrome de Kleine-Levin-Critchley. Revue roum. Neurol. *9:* 221–228 (1972).
36 Prebhakaran, N.; Murthy, G.K.; Mallya, U.L. A case of Kleine-Levin syndrome in India. Br. J. Psychiat. *117:* 517–519 (1970).
37 Reynolds, C.F.; Black, R.S., Coble, P.; Holzer, B.; Kupfer, D.J.: Similarities in EEG sleep findings for Kleine-Levin syndrome and unipolar depression. Am. J. Psychiat. *137:* 116–118 (1980).
38 Ronald, J.: Hypersomnia associated with abnormal hunger. The Kleine-Levin syndrome. Br. med. J. *ii:* 326–327 (1946).
39 Sagripanti, P.: Sull'associazione di ipersonnia e bulimia (sindromi di Kleine-Levin). Cervello *28:* 194–205 (1952).
40 Sallares Dillon, C.: Sindrome de Kleine-Levin. Acta psiquiat. psicol. Am. lat. *19:* 148–151 (1973).
41 Suwa, K.; Toru, M.: A case of periodic somnolence whose sleep was induced by glucose (in Japanese). Folia psychiat. neurol. jap. *23:* 253–262 (1969).
42 Takahashi, Y.: Clinical studies of periodic somnolence. Analysis of 28 personal cases (in Japanese). Psychiatria Neurol. jap. *5:* 853–889 (1967).
43 Takahashi, K.; Takahashi, S.; Azumi, K.; Honda, Y.; Utena, H.: Changes of plasma growth hormone during nocturnal sleep in normals and in hypersomniac patients (in Japanese). Adv. Neurol. Sci. *14:* 743–754 (1971).

44 Vein, A.M.: Disturbances of sleep and wakefulness, (in Russian), p. 383 (Meditsina, Moscow 1974).
45 Vlach, V.: Periodická somnolence, bulimie a psychické poruchy (syndrom Kleineuv-Levinuv) (in Czech). Čslká Neurol. *25:* 401–405 (1962).
46 Wilkus, R.J.; Chiles, J.A.: Electrophysiological changes during episodes of the Kleine-Levin syndrome. J. Neurol. Neurosurg. Psychiat. *38:* 1225–1231 (1975).

Adresses of Authors

Dr. M. Billiard, Service de Physiopathologie des Maladies Nerveuses, Centre Gui de Chauliac, Cliniques St-Eloi, F-34059 Montpellier Cedex (France)

Dr. Y. Hishikawa, Department of Neuropsychiatry, Osaka University, Medical School, Osaka 553 (Japan)

Dr. J. Montplaisir, Département de Psychiatrie, Université de Montréal, Montréal, Quebec (Canada)

Dr. B. Roth and Dr. S. Nevšímalová, University Karlova, Klinika Neurologicka, Praha (Czechoslovakia)

B. Hypnotics and Insomnia Models in Animal and Man

Chairmen: *R. Scherschlicht*, Basel; *W. P. Koella*, Basel
Participants: *H. Bittiger*, Basel; *W. Haefeli*, Basel; *J. Jaekel*, Basel; *I. Oswald*, Edinburgh; *R. Spiegel*, Basel

Introduction

W. P. Koella, Basel, Switzerland

Hypnotics are among the most frequently prescribed, used, as well as misused and abused drugs. Hypnotics (should) induce sleep; in practice, they ought to be effective in the treatment of *insomnia* by shortening the long sleep latency, and suppressing early and/or frequent awakening. Or, to put it quite simply, they should turn a (subjectively) bad sleep, with little 'detiring' efficacy, into a good sleep.

The presently available hypnotic drugs, although some good progress has been made in recent years, are as yet not 'ideal'. They oftentimes lack full efficacy. They rarely, if at all, have a protracted effect. And they are plagued by some undesirable, often serious, side effects.

In part due to these various drawbacks there is, at present, *little unifying concept* among the medical practitioners as to when, how often, and how long, what particular hypnotic should be given in anyone of the various forms of sleep disturbance. With this background we thought it to be worthwhile to discuss, during the course of a 'Sleep Congress', some practical and theoretical aspects of the anti-insomnia drugs and to make an attempt to redefine, on the basis of today's knowledge about the physiology, pharmacology, biochemistry and clinic of sleep, the set of problems that make it desirable to develop still better hypnotics.

In the first paper a researcher who has been in the forefront of the *clinical testing* of new hypnotics tackles the problem of how to go about obtaining the information necessary to designate a new drug a good hypnotic. In the second paper an investigator with a good deal of experience in the preclinical development of hypnotic drug discusses *new animal models of insomnia,* a problem of paramount importance for the detection of still better anti-insomnia remedies. The third paper deals mainly with today's knowledge (and interpretation) about the *mechanism(s) of action of, in particular, the benzodiazepine-type hypnotics*. The last paper in this series dwells on *side effects* of hypnotic drugs and, using this information, offers some suggestions as to what can still be improved in future generations of hypnotics and as to some strategies that may be helpful in reaching this goal. To avoid too much 'subjectivity' the chairmen have asked some 'official discussants' to follow the presentations with some experiences and opinions of their own. These remarks are included in the write-up of this workshop.

Sleep 1980. 5th Eur. Congr. Sleep Res., Amsterdam 1980, pp. 142–147
(Karger, Basel 1981)

Testing an Hypnotic in Man[1]

I. Oswald, Edinburgh, England

The first and most important thing in testing a new hypnotic is subjective assessment of the effect of the drug on sleep [61]. The method is cheap, but not very precise for any one individual. There is therefore a high scatter in the data, and so a large number of subjects is needed if one is going to achieve a significant result. An inadequate number of subjects will simply fail to give an answer.

I will illustrate by describing a recent study that my colleagues and I have done in Edinburgh, in which 100 people, aged 40–68, participated with a view each to completing a 32-week study. They were selected because they themselves believed that they were poor sleepers. The fact that they were of an older age group, and therefore relatively reliable, and the fact that they were paid and were seen weekly for the first 2 months and thereafter every 2 weeks, so that their morale and interest was sustained, contributed to the very low drop-out rate. 97 of the subjects completed the entire study. In the first four weeks each individual subject took placebo pills at bedtime, and then for the next 24 weeks, 50 subjects took lormetazepam 2 mg nightly, 25 subjects took nitrazepam 5 mg nightly, and 25 subjects continued on placebo. A final 4-week period was one in which placebos were taken at bed-

---

[1] For references see compound list on page 164.

time by all of the subjects. The study was double-blind, except that the experimenters knew of the initial and final periods, each of 4 weeks' duration on placebos.

The subjects made many subjective ratings each day. In the mornings, for example, they made visual analogue scale ratings of the quality of their sleep the previous night and, on a modified log scale, they rated how many minutes it had taken them to fall asleep the previous night. Each evening they made other ratings and completed a 25-item checklist of somatic and nervous symptoms.

The analysis of data was a big task but, in brief: the self-ratings of sleep quality showed no change among the 25 subjects who continued on placebo throughout. In those who took nitrazepam there was a very clear and significant improvement of the quality of sleep when they started the drug and this was sustained throughout the drug intake period, to be followed by a sharp rebound worsening in the quality of sleep on withdrawal. Likewise those 50 subjects who took lormetazepam 2 mg nightly had a very clear and significant improvement in the quality of sleep that was sustained throughout the 24 weeks of intake, to be followed, once again, by a sharp rebound worsening of sleep quality on withdrawal.

The results answered the common question of whether the effectiveness of benzodiazepine hypnotics is lost during months of intake, and the answer is that effectiveness continues largely undiminished.

The results of the analysis of sleep latency were very similar. Lormetazepam caused a shortening of sleep latency that was sustained throughout the drug intake period, and there was a rebound increase of sleep latency after withdrawal. Nitrazepam also improved sleep latency at first, but this effect was not sustained. Again there was a rebound worsening in the time taken to fall asleep after withdrawal.

When the rebound period was looked at more closely, in the case of both sleep quality and sleep latency, it was found that the impairment of sleep, to a degree worse than baseline, reached its peak on the second night after the withdrawal of lormetazepam and on the fourth night after the withdrawal of nitrazepam, and that in the case of both drugs, return to normal took 2–3 weeks. The longer period to the peak of the withdrawal abnormality after nitrazepam may be attributed to a longer time taken to eliminate active products of the drug.

The 25-item checklist showed no effect of the two active drugs on such symptoms as lethargy, dizziness, fatigue or irritability, in so far as these were subjectively rated. In addition we made an examination of the gross total of all possible symptoms by adding together all the symptoms complained of and found no change as a consequence of either of the drugs. It was a conclusion that was possible only because of the inclusion in the design of a large group of persons who continued on placebo throughout.

We concluded that lormetazepam 2 mg and nitrazepam 5 mg at bedtime both improved the subjective quality of sleep and reduced the latency to falling asleep. These effects were slightly better sustained in the case of lormetazepam. Neither drug caused side-effects and the two drugs did not differ significantly in their effects.

At the outset of the study and again in the last 6 weeks on active drug (or continued placebo) each subject had a comprehensive general medical examination, including tests of liver and renal function and including measurement of body weight in underclothes. Whereas the placebo group of 25 subjects did not change in their body weight during the study (a loss of only 0.02 kg), the nitrazepam group lost a mean of just over 1 kg and the lormetazepam group lost a mean of just over 1.6 kg. The groups had not differed in age, weight or sex distribution at the start and the weight loss was not related to initial weight or to age or sex. The small loss of body weight on nitrazepam was significantly greater than the loss on placebo, and likewise the loss on lormetazepam was significantly greater than the loss on placebo. There was no significant difference between the two drugs in their effect of causing a small loss of body weight [62].

Equally important in the investigation of a new hypnotic is to examine its effects when taken at bedtime, but on daytime skills, especially during the chronic intake of the drug by persons of the age group who characteristically take such drugs. Too many of the published studies have been single-dose studies with healthy young adults.

In illustration I will again draw upon a recent study by my colleagues and me in Edinburgh, of 12 poor sleepers of mean age 53, each of whom took part in a study having a Latin square design in which there were four sequence conditions. At one time of year this meant that each subject took placebo continuously for 6 weeks, at another time of year each subject took placebo for 2 weeks and then flurazepam 30 mg for 3 weeks, followed by placebo for 1 week. At another time of year each volunteer took placebo for 2 weeks, then lormetazepam 1 mg nightly for 3 weeks, and then placebo again for 1 week. At another time of year each volunteer took placebo nightly for 2 weeks, lormetazepam 2.5 mg nightly for 3 weeks, and then placebo for a further week. The study was double-blind, except that the experimenters knew of the initial and final placebo periods in each sequence.

There were 4-week breaks between each sequence and therefore each subject was involved in the study for a total period of 37 weeks (there having been one extra placebo week during the very first week of attendance, purely for extra practice).

Each subject attended a laboratory for the testing of skilled performance each week during the various sequences, attending for 2 h in the morning, 2 h in the middle of the day, and 2 h in the early evening. Among the tasks was a 1 h modification of the Wilkinson auditory vigilance task and in addition there were tests of manual dexterity, cardsorting and digit-symbol substitution, all tasks being carried out three times during the testing day.

The results have already been described [63]. The subjects made subjective ratings of the quality of their sleep throughout and they considered that all three drug regimes were effective hypnotics, each giving a significant improvement over placebo. By day the objective testing failed to show any adverse effects of lormetazepam, either in the 1 mg or in the 2.5 mg doses. On the other hand, flurazepam 30 mg nightly caused a significant impairment by day in all of the tests, the impairment tending to increase between the first and the third weeks of testing, and the impairment being sustained throughout the morning, afternoon and evening sessions of the day. Subjects made ratings on visual analogue scales each morning of whether they felt their vigilance was impaired, and in the first week of receiving flurazepam they recognized the presence of impairment, but by the last couple of weeks they considered themselves to have returned to normal, in sharp contradistinction to what the objective tests showed. It is evident that people's own ratings of whether they are impaired by an hypnotic drug are not to be relied upon.

It is not only psychomotor skills that are important, but social skills also. It is a commonplace that the most traditional of hypnotics, alcohol, can lead to social disasters. In the course of the whole study, which included a sleep laboratory study with another 9 subjects, to be mentioned below, there were 450 subject weeks and in the course of that time there were seven crises, involving such things as major quarrels, threats of murder, or, in one case, a car crash. When the code was broken it was found that all seven of these crises had been associated with the intake of flurazepam, which had itself occupied only 63 out of the 450 subject weeks, as a consequence of which the high rate of crises cannot be attributed to chance but must be attributed to the intake of flurazepam. Flurazepam is a drug that is metabolized to long-life compounds that are cumulative to such an extent that the intake of the drug on a regular basis at bedtime gives rise to a sustained plateau, with tissue concentrations of active product that can vary little during the 24 h.

I have thus far scarcely mentioned the EEG laboratory, but this is by now established as providing the means for a precise, though financially costly, investigation of the effects of hypnotics on human sleep. Owing to the fact that it is so precise, the variance in the observation is reduced, and one can expect to obtain significant effects by the use of relatively small numbers of subjects, though one should never use less than about 9 subjects in a satisfactory design. Again, my colleagues and I have carried out a Latin square design comparison of flurazepam 30 mg, lormetazepam 1 mg and lormetazepam 2.5 mg, using a group of poor sleepers of mean age 61 years. The drugs were each effective as hypnotics by EEG criteria, there was a sharp rebound worsening of sleep on withdrawal of lormetazepam, whereas on withdrawal of flurazepam there were signs of persistence of the drug [63].

In the long-term we shall also want to know how drugs might affect the restorative value of sleep [1]. First explorations in this field have involved examination of the anabolic and catabolic hormones during sleep, by means of indwelling catheters in forearm veins and the removal of small quantities of blood at regular intervals through the night without disturbing the sleeper. We found, for example, that the intake of benzoctamine, a modern minor tranquillizer, significantly lowered plasma corticosteroids sampled during sleep. Following withdrawal, and subsequent to 5 weeks of intake, there was a rebound increase in the level of plasma corticosteroids during sleep to a degree significantly above baseline levels, with return to normal after a couple of weeks [60]. These are catabolic hormones. Growth hormone is an anabolic hormone released concurrently with human slow wave sleep. Although nitrazepam 5 mg nightly significantly diminished slow wave sleep, it had no effect on human growth hormone in a group of 7 people of mean age 59 [2].

Summary

In the investigation of a new hypnotic drug in man we therefore ask: What are the subjective effects by night and by day? What are the long-term effects, for example, on body weight? What are the objective effects on sleep using the EEG? What are the effects on psychomotor skills by day and on social skills? And increasingly we shall want to know if the drug in some way alters the restorative value of sleep and the balance of metabolism between the contrasting states of wakefulness and sleep.

Discussion

In his discussion remarks on Dr. *Oswald's* paper, *Spiegel* (Basel) emphasized the following two points: (1) polygraphic parameters of sleep, if used in isolation as indicators of a compound's clinical activity, can be misleading; (2) benzodiazepines, in particular some of the more recent compounds, are extremely well suited drugs for a majority of patients complaining of insomnia, and it is difficult to find better hypnotics outside of this chemical class.

The first point was illustrated by the example of mesoridazine, which, according to polygraphic criteria appeared to be an ideal hypnotic: it reduced the number of sleep interruptions, slightly increased the percentage of REM sleep and had no influence on SWS [3, 72]. It was also active in hospitalized patients with insomnia, as found in a controlled study with subjective assessment of sleep [71]. However, when subsequently tested in a controlled field trial in outpatients, it was clearly inferior to nitrazepam, as the subjective quality of sleep and aftereffects clearly favoured the latter. Thus, the polygraphic data, as well as the subjective data from an in-patient population, were not predictive of the drug's performance in the real target population of outpatient insomniacs.

The second point, the superiority of benzodiazepine compounds in most insomniacs, is illustrated by today's sales figures of sleeping pills. They are not only a consequence of superior marketing techniques, but due to the real advantages of these compounds: low toxicity, efficacy, few interactions with other drugs, few and reversible side-effects, no cardiotoxicity, relatively low abuse potential. Having worked as a project leader for 8 years and entrusted with the task of finding a non-benzodiazepine hypnotic, *Spiegel* feels he knows how difficult this task is. During this time 10 different compounds from diverse chemical classes were clinically and polygraphically assessed and none of them was superior to, or even equal to, the more recent benzodiazepines. Thus, any future hypnotic, its chemistry and biology as fanciful as may be, will have to be measured against benzodiazepines. Good luck!

Sleep 1980. 5th Eur. Congr. Sleep Res., Amsterdam 1980, pp. 147–155
(Karger, Basel 1981)

Model Insomnia in Animals[2]

R. Scherschlicht, J. Marias, J. Schneeberger, M. Steiner, Basel, Switzerland

Most hypnotics presently known were either found by chance or in screening procedures that included the study of various animal behavioural items, but not true sleep. The effects mostly looked for in the latter case

---

[2] For references see compound list on page 164.

were a decrease of spontaneous locomotor activity and a prolongation of 'sleepingtime' (e.g. time of loss of righting reflex) after combined administration of the unknown substance and a standard hypnotic, usually a barbiturate. Both parameters were only indirectly linked to true sleep and the substances found in this way were consequently called 'sedatives' or 'sedativehypnotics'. With the possible exception of thalidomide [53] they all induced severe motor impairment and anaesthesia in high doses, indicating an unspecific central depressant action rather than specific induction of sleep in these doses. The early benzodiazepines were discovered by the same screening approach, and it was only by designing and consequent development of a new methodology that the hypnogenic, i.e. sleep-promoting, activity in this chemical class could be optimized in its modern representatives and that now concepts of their mode of action exist [31].

Among these new methods the cable-bound or wireless telemetric [68] recording of central and peripheral signals from unrestrained animals of different species opened the way to direct measurements of hypnotic-induced changes in the sleep pattern. The results of sleep studies based on polygraphic recordings from normal animals, examples of which are shown in table I, remained, however, unsatisfactory. Table I shows that several well-known hypnotics as well as the sleep-promoting oligopeptide DSIP [58] un-

Table I. Influences of standard hypnotics and DSIP on the sleep-wakefulness cycle of normal cats

	Dose	W	NREMS	REMS	REMS/TS
Pre-drug control (saline)		100	100	100	100
Nitrazepam	0.01 i.p. (7)	94	105	111	101
Nitrazepam	0.03 i.p. (3)	132	86	56	61
Flunitrazepam	0.001 i.p. (8)	66	119	169	134
Flunitrazepam	0.1 i.p. (5)	175	54	33	67
Phenobarbitone	1.0 i.p. (4)	102	107	48	44
Phenobarbitone	10.0 i.p. (3)	71	134	17	12
Glutethimide	1.0 i.p. (7)	100	105	65	60
Glutethimide	10.0 i.p. (3)	121	102	45	38
Thalidomide	1.0 i.p. (4)	77	108	140	125
DSIP	0.025 i.v. (5)	70	116	177	147
DSIP	0.25 i.v. (3)	98	104	79	79

Number of animals in parentheses. Doses in mg/kg. The values given are percent of pre-drug controls.

doubtedly affected the sleep-wakefulness cycle of normal cats, but that they either did not augment sleep or had opposite effects in different doses or produced unexpected effects. Of the two benzodiazepines shown in table I, for instance, only flunitrazepam in the low dose of 0.001 mg/kg positively influenced sleep by decreasing wakefulness (W) and increasing both non-REM sleep (NREMS) and REM sleep (REMS). Higher doses of both flunitrazepam and nitrazepam had an opposite effect, namely increased W. Phenobarbitone and glutethimide in low doses did not affect the amount of W, but decreased REMS. The dose of phenobarbitone which increased NREMS, decreased REMS and induced ataxia, thus showing unspecific central depression. The high dose of glutethimide had a paradoxical effect similarly to the high doses of the benzodiazepines. The active dose of DSIP in cats mainly augmented the amount of REMS, in contrast to its effect in rabbits, which led to the name 'delta sleep inducing peptide'.

It was, of course, not surprising that differing effects were found in animal experiments with substances of different chemical classes. It was, however, disappointing that, within the broad spectrum of visible effects, the most important one, i.e. induction of sleep, was found rather infrequently. To some extent, differences in the physiology of man as a primate and the cat as a carnivore might have accounted for the lack of a clear-cut hypnogenic activity of the clinically approved substances in the latter species. However, species differences cannot be the only explanation.

The goal of today's search for better hypnotics is, of course, no longer just 'sedation', i.e. a general decrease of the excitability of the CNS to various stimuli, but a selective hypnogenic effect. Since another communication of this workshop [50] deals with the desired properties of future hypnotics in detail, it is not necessary to define such a substance here. One should, however, keep in mind, that a hypnotic is a drug capable of alleviating human insomnia. Different types of insomnia may have manyfold origins and different manifestations, but they have in common that falling asleep is prevented at a time of day at which a given human being should be asleep according to his personal rhythm of life and to social rules. Such states of sleeplessness may occur with and without a preceding period of sleep. The hypnogenic property of a substance correspondingly is its ability to terminate, in appropriate dosage, a punctual state of sleeplessness by inducing and thereafter maintaining sleep in its meanwhile well-known natural composition.

Such a property of a drug can hardly be found in normal animals under so-called controlled conditions, because a state of sleeplessness that could be terminated cannot occur. The reasons for this are that normal animals, by definition, must not suffer from a disease possibly leading to a sleep disturbance, and that controlling the conditions leads to an environment that is poor in external stimuli and, hence, to a monotonous life. Furthermore, the animals have free access to food and water, i.e. they are not compelled to go in search of prey, and are free to sleep whenever they are inclined to do so. It is obvious that animals under such conditions within the usual observation period produce the maximum amount of sleep as determined by species, age and individuality. A further increase from this maximum, by definition, cannot be physiological sleep. On the other hand, controlled experimental conditions and good health are a prerequisite for reproducible results. In order to achieve sleeplessness, the condition required to detect a hypnogenic effect, a state of insomnia, therefore, has to be induced in a reproducible manner too.

Numerous behavioural (stress), physical (shocks, noise, etc.) and pharmacological means have been used to realize this condition. It is necessary to induce a disturbance of sleep of moderate degree, and not a complete suppression, the counterbalance of which would require the high and unspecifically depressant doses, which have to be avoided. A specific problem in behavioural and physical models is the habituation of the animals to the stressful conditions, inherent problems in pharmacologic model insomnia are the choice of a suitable substance in its appropriate dose and the avoidance of false positive results with compounds that are not hypnotics, but antagonists of the drug used to induce insomnia. It is mainly with respect to this last-mentioned point that morphine, proposed as a model substance in cats by *Echols and Jewett* [19], has advantages as compared with stimulants, e.g. methylphenidate or LSD, which also reduce sleep in this animal species [67]. Specific opiate antagonists can easily be recognized with relatively simple screening procedures. Thus, when a compound is found to antagonize morphine insomnia in cats, it can subsequently be tested for specific antagonism and thus can be recognized as a false positive.

The experiments reported here with intraperitoneally administered morphine were done in male cats with chronically implanted electrodes on the surface of the parietal cortex, within the hippocampus and the lateral geniculate body. During the recovery period of at least 2 weeks and the whole period of the experiments the animals were housed in sound-attenuated, artificially illuminated (12 h L/D-schedule) and ventilated boxes with food and water *ad libitum*. Bipolar EEGs were transmitted telemetrically by means of a four-channel FM emitter [68], fixed on the animals's head, and recorded continously from 0900 to 1500 hours. Sleep scoring was done on 1 min epochs, distinguishing between W, NREMS and REMS. Drugs were administered at 0855 hours. Saline and standard hypnotics were given i.p. as separate contralateral injections immediately after morphine. DSIP was given i.v. through an indwelling catheter in one jugular vein. Student's t-test for paired values was used for statistical comparisons, since each animal was its own control. Differences were regarded as significant when $p < 0.05$.

The sleep depressant effect of morphine was found to be dose-dependent: 0.1 mg/kg i.p. increased the amount of W to 147% of the pre-drug control value on the preceding day, reduced NREMS to 73%, REMS to 32% and the ratio of REMS to total sleep (REMS/TS) to 50%. The higher dose of 0.3 mg/kg i.p. increased W to 198%, reduced NREMS to 20% and abolished REMS. The increase of W was mainly due to prolonged sleep latency (SL): increase to 295 and 404%, respectively.

On the basis of these results 0.1 mg/kg morphine i.p. was considered to be sufficient to induce the desired moderate sleep reduction and was selected as standard dose to test the hypnotics listed in table I. The results (table II) show that the amounts of W and NREMS were fully normalized by all substances tested. The values after combined administration of morphine and a hypnotic were significantly different from the morphine values, but not from the saline control values. The relative potency of the standard hypnotics, furthermore, correlated very well with their relative potency in clinical practice. All substances, except phenobarbitone, also normalized the

Table II. Reversal of morphine-induced insomnia in cats by standard hypnotics and DSIP

	Dose			W	NREMS	REMS	REMS/TS
Pre-drug control (saline)				100	100	100	100
Morphine-HCl	0.1	i.p.					
+ saline			(6)	147	73	32	50
+ nitrazepam	0.03	i.p.	(4)	93	106	82	79
+ flunitrazepam	0.01	i.p.	(5)	98	105	73	71
+ phenobarbitone	3.0	i.p.	(6)	112	99	46	46
+ glutethimide	3.0	i.p.	(3)	124	86	95	110
+ thalidomide	1.0	i.p.	(3)	117	87	90	105
+ DSIP	0.025	i.v.	(4)	89	114	81	75

Number of animals in parentheses. Doses in mg/kg. The values given are percent of pre-drug controls.

amount of REMS. No explanation can be given at present for the failure of phenobarbitone to normalize the amount of REMS decreased by morphine; this finding might, however, be relevant for theories about different modes of action of barbiturates and benzodiazepines. Another important result was that nitrazepam normalized both NREMS and REMS with 0.03 mg/kg i.p., a dose which in normal animals had a paradoxical effect (table I). Finally, it should be stressed that DSIP also proved its activity by normalizing disturbed sleep and that the activity of this peptide could be better shown under conditions of disturbed sleep than of normal sleep.

After the usefulness of the model had been established by these experiments, initial steps were done in the direction of possible mechanisms of action by which morphine suppresses sleep in cats. Preliminary results revealing an antagonistic activity of naltrexone, an agonistic activity of dihydromorphine, but inactivity of ethylketazocine and (D-Ala; D-Leu)-enkephalin point to a possible link with μ-receptors [51].

Compared with primates, rats and cats which are the most frequently used animal species in sleep research, the rabbit played a less important role in this research area. Interest in this species is, however, increasing because two of the putative sleep-inducing peptides, DSIP and 'factor S', either have been isolated from this animal [58] or, at least, tested in it [52]. An insomnia model in rabbits, therefore, was felt to be of special interest for evaluating the effects of these peptides, whose activity and physiological role is not yet generally accepted. The morphine model was not applicable, because morphine in rabbits induces a slowing of the corticogram accompanied by a behavioural state resembling superficial NREMS. Being a herbivorous animal and prey of predatory mammals and birds, the rabbit seemed to be predestinated for a behavioural model insomnia based on anxiety and stress. One possibility to evoke fear appeared to us to be the confrontation with a natural enemy, e.g. a dog.

The experiments in rabbits reported here were done in male animals with chronically implanted brain (sensorimotor cortex, dorsal hippocampus) and muscle electrodes. During the 6-h (0900–1500) recording sessions the animals were housed in sound attenuated and electrically shielded boxes. EEGs and EMGs were recorded by means of flexible cables suspended, with their rotating contacts, from a balanced beam. Sleep scoring was done on epochs of 5 s duration, distinguishing between W, NREMS and REMS. In certain experiments NREMS was subdivided in light NREMS (l-NREMS, spindle sleep) and deep NREMS (d-NREMS, delta sleep). The percentages of the stages of vigilance were calculated for each whole session as well as parts of it, down to 30-min periods (HHPs). In order to disturb the sleep-wakefulness cycle of the animals, they were, 5 min before the beginning of the recording, confronted in a separate room with an agitated, barking beagle dog for 1 min at a distance of about 50 cm. Drugs or saline were administered intravenously immediately before the confrontation.

As may be seen from the time-effect functions of figure 1, the confrontation with a barking dog for only 1 min resulted in a distinct, moderate decrease of both NREMS and REMS. It took about 3 h before this disturbance was normalized again, as is shown on the right side of the figure: compared to normal control values, the confrontation increased the amount of W and decreased the amounts of both NREMS and REMS particularly in the first 3-h period, and hence, of course, also in the total session. Expressed as percent of controls the values for hours 0–3 after the confrontation became as follows: W was increased to 123% while NREMS was decreased to 88% and REMS to 24%. Hence it became obvious that REMS was relatively far more affected than NREMS. In another series of experiments (fig. 2a), total NREMS in fact was not diminished significantly in the first 2-h period after the confrontation, but further analysis revealed that this

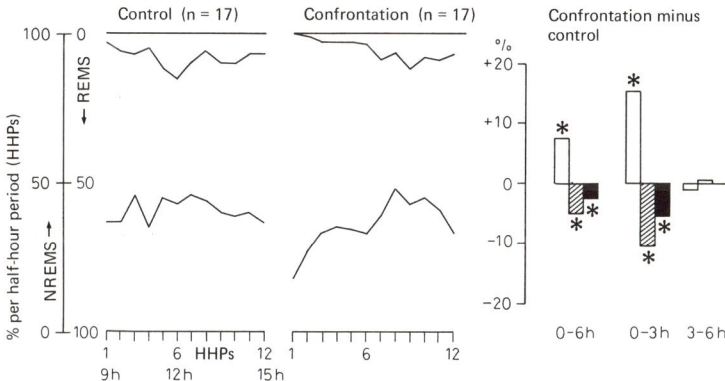

Fig. 1. Confrontation of rabbits (n = 17) with an agitated dog. The two diagrams on the left side show the percentages of NREMS (lower line, read from bottom to top) and REMS (upper line, read from top to bottom) as well as W (area between the lines) between 0900 and 1500 hours in half-hour periods (HHPs) before (left-most) and after (middle) confrontation. On the right side are shown the differences between the percentages of W (white bars), NREMS (hatched) and REMS (black) before and after confrontation for the whole recording sessions (∅–6 h) as well as their first (∅–3 h) and second (3–6 h) halves.

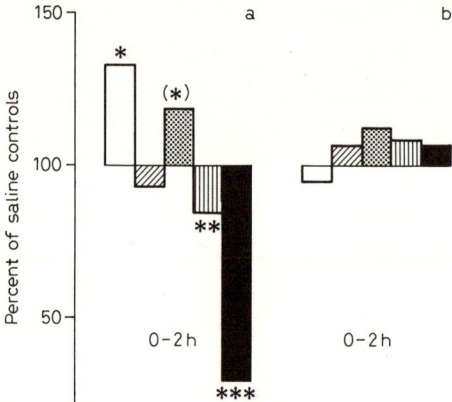

Fig. 2. Effect of DSIP on the confrontation insomnia in rabbits (n = 5). *a* Confrontation without DSIP. *b* Confrontation after 0.05 mg · kg^{-1} i.v. DSIP. The values after both confrontations are expressed as percentages of the values of the same animals without confrontation on the day preceding *a*. W white, total NREMS obliquely hatched l-NREMS dotted, d-NREMS vertically hatched, REMS black. Asterisks indicate significant differences to nonconfronted controls.

was due to a shift from d-NREMS to l-NREMS, the latter having been augmented on account of the former.

At present our data are not sufficient to discuss the value of the confrontation model in detail. It may, however, be stated that flunitrazepam and phenobarbitone in doses not effective in normal rabbits (0.01 and 5 mg/kg i.v., respectively) prevented the sleep disturbance induced by stress. Even more important, perhaps, is the fact that DSIP was effective too (fig. 2). If this peptide was given intravenously in the dose of 0.05 mg/kg immediately before the confrontation, the values of W, d-NREMS and REMS remained normal in the following 2 h.

It remains to be shown whether this behavioural insomnia model is specific for 'pure' hypnotics or whether hypothetical anxiolytics without sleep-inducing properties will also be detected in this procedure.

Discussion

J. Jaekel Basel, discussing and supplementing *R. Scherschlicht's* paper, presents another 'insomnia model'. When kept in groups of from 5 to 8, rhesus monkeys develop a rank order in which the 'boss' quite clearly dominates the ongoing individual and social activities (and inactivities) within the experimental territory. The lower ranked monkeys, and in particular the 'omega' animal, reveal signs of 'anxiety' and/or 'depression' if not suppression. Also quite typically the lower ranked members of the group succumb to an occasional nap significantly less frequently – if at all – than do the higher ranked ones. The thought is close at hand that the lower ranked animals reveal a 'natural' insomnia akin to the sleep disturbance due, in man, to anxiety. Characteristically, benzodiazepines and some of their modern derivatives as well as anxiolytics of the adrenergic β-blocker type are capable to markedly enhance sleeping time in these lower ranked 'insomniacs'. It seems that this 'insomnia' model may be quite useful to test putative hypnotic drugs.

Mechanism of Action of Benzodiazepines[3]

W. Haefely, Basel, Switzerland

The availability of such potent and safe drugs as the benzodiazepines for the treatment of various sleep disorders is not only a great help for the insomniac, but also a challenge for scientists interested in the mechanism of action of neuropsychotropic agents. The present state of knowledge on the mechanism of action of benzodiazepines is briefly reviewed.

A first breakthrough in the understanding of these drugs has been the finding that they affect in a highly selective manner one type of chemical synapse in the central nervous system. In fact, as first proposed by *Haefely et al.* [32] and since then confirmed in many laboratories, the primary effect of benzodiazepines is clearly an enhancement of GABAergic neurotransmission. This has been shown by electrophysiological and, in part, by biochemical methods, to occur both on presynaptic inhibition of primary afferent neurones in spinal cord and dorsal column nuclei and on postsynaptic inhibition in all brain structures studied so far, i.e. cerebral and cerebellar cortex, hippocampus, hypothalamus, Deiter's nucleus, substantia nigra, raphé nuclei and locus coeruleus (for references see *Haefely et al.* [33]). In their effect on GABAergic transmission benzodiazepines differ quite clearly from propandiol carbamates which do not potentiate GABAergic trans-

[3] For references see compound list on page 164.

mission; they resemble in some way barbiturates and ethanol which also enhance GABAergic transmission, albeit by mechanisms distinct from those of benzodiazepines and which, in addition to their effect on the GABAergic synapse, also affect a great number of other synaptic and nonsynaptic events in the central nervous system and in the periphery.

The characteristic profile of activity of benzodiazepines and their unmatched influence on GABAergic synapses are accounted for on a molecular level by the existence of binding sites with a very high affinity and specificity for active benzodiazepines [11, 56] and which undoubtedly include the receptors to which benzodiazepines have to bind as a first step to initiate their effects. These receptors are distributed unevenly in the body; they are highly concentrated in those central nervous structures in which benzodiazepines have been shown to affect neuronal activity and they are absent in peripheral tissues, where benzodiazepines are well-known to be inactive at pharmacologically relevant concentrations. The presence of specific benzodiazepine-binding sites in GABAergic synapses has recently been confirmed by combined electron microscopic immunohistochemistry and photoaffinity labelling [57]. The central benzodiazepine-binding sites are an inhomogenous class of proteins differing both in their molecular weight [70] and in their affinity for various ligands [46]. The meaning of this heterogeneity of benzodiazepine-binding sites is at present unknown. The radioreceptor-binding technique has opened several new lines of research. One is the search for possible endogenous ligands that might act either as agonists or antagonists at benzodiazepine receptors. So far no convincing evidence has been found for the existence of a 'natural benzodiazepine' in the brain. The radioreceptor technique makes it possible to study the relation between receptor occupation and brain or blood concentration of benzodiazepines, which should be essential in future relevant interpretations of plasma levels of drugs with respect to their concentration at the sites of action and, hence, their effect.

At present basic research in the benzodiazepine field is mainly concerned with two problems. One is to elucidate the chain of events initiated by the binding of these drugs to their receptors and which result in an enhanced activity of GABAergic synapses. One hypothetical explanation of the link between receptor occupation and increased GABAergic transmission has been proposed [15] and assumes the interaction of benzodiazepines with an endogenous modulatory protein in the GABA receptor-benzodiazepine receptor-chloride channel complex, resulting in a shift of the GABA receptor from a low-affinity state to a high-affinity state for

GABA. Alternative mechanisms, e.g. an enhanced release of GABA from GABAergic nerve terminals in response to an action potential, require more experimentation.

An equally important link which remains to be resolved is that between enhanced GABAergic synaptic effectiveness and the various somatic and psychic effects of benzodiazepines. Benzodiazepines are, of course, not specific hypnotics, as they produce also potent anticonvulsant, anxiolytic, sedative and anti-stress effects. The unrivaled effectiveness and acceptance of benzodiazepines in sleep disorders might well be related to their broad-spectrum activity. As a clue to the understanding of the link between the effects of these drugs on the GABAergic synapse and complex somatic and psychic functions, it is suggested to concentrate on the role of the feedback inhibitory GABAergic loops in various structures and the consequences of an enhanced operation of these pathways. A stabilization of principal neurone activity – by prevention of excessive responses to stimuli without blockade of excitatory impulse transmission – might well underline the various pharmacological effects of these drugs. Although some central nervous structures could be particularly important for some of their actions, the great therapeutic value of benzodiazepines might well result from an essentially similar effect on neuronal activity in most areas of the central nervous system.

Discussion

H. Bittiger, Basel, in discussing *W. Haefely's* paper, made some comments on a new concept in neurobiology; namely *receptor modulation*. The term *receptor* here refers not to sensory transducers but rather to those molecular entities which recognize specific transmitters or other active compounds of either endogenous or exogenous origin. *Modulation* in this context means a change in the *number* of receptors, or in the *affinity*, between receptors and a specific agent interacting with them. Both mechanisms can enhance or attenuate *amplification* of neurotransmission. Changes in numbers and in affinity of receptors can be measured, namely by radioreceptor assays. A typical example of a consequence of receptor modulation for the physiological response, is the well-known effect of *hyperthyreosis* where the number of β-receptors is increased and, consequently, heart rate and metabolism are increased and sleep is disturbed.

Receptor modulation, in fact, may play an important role in the organization and regulation of sleep, and in the mechanism of action of hypnotic drugs. *Guidotti* and co-workers [1978] suggested that benzodiazepines enhance the affinity of GABA receptors for GABA. This would mean that, at a given GABA concentration, during the 'high-affinity state' more receptors would be occupied than during the 'low-affinity state'. This would offer a reasonable explanation – at a molecular level – for the benzodiazepine-induced increase in GABA transmission, as demonstrated by electrophysiological procedures. This concept certainly could be of great value. However, it has not yet been proven in vivo, i.e. under physiological conditions; all experimental data were obtained under very special, in vitro, conditions and some important questions are still open.

The mechanisms involving *modulation of the benzodiazepine receptors* are considerably better established. One can state that GABA agonists are prominent such modulators; yet, agents that do not interact with GABA receptors directly are also active in modulating benzodiazepine receptors. Still other compounds are liable to modulate both GABA *and* benzodiazepine receptors without interacting directly with the recognition sites of these compounds.

There is an ever-increasing number of findings, indicating modulation of still other receptors and it seems a highly promising task to consider these proposed mechanisms in the context of the organization of sleep. At this meeting of ESRS, *A. Wirz-Justice* has reported on circadian changes in receptor sensitivity; a 'modulation' concept which may prove to be of paramount importance for the understanding of that most important circadian business: the alternation of waking and sleep.

Side Effects of Today's Hypnotics and the Hypnotic of the Future[4]

W. P. Koella, Basel, Switzerland

The presently available hypnotics not always reveal good efficacy in all forms of insomnia. They often lose their hypnotic effect in the course of longer treatments and they are plagued by often severe and serious side effects. Thus, we should continue our efforts and try to produce remedies of still better efficacy, tailor-made for the various sleep disturbances and not afflicted by undesired side effects.

In this paper we present and discuss first some of the more common side effects, typical, but not unique, for hypnotics. In the second part we shall have a closer look at what we consider the desirable qualities of the *'ideal hypnotic'* and try to offer some ideas as to strategies and approaches that may be helpful in producing this 'wonder drug'.

[4] For references see compound list on page 164.

The Side Effects of Hypnotic Drugs

It is convenient to list the side effects according to the following grouping: (1) 'commonplace' side effects; (2) interaction with other drugs; (3) dependence liability including tolerance and withdrawal symptoms; (4) dyssomnia.

Among the more banal *'commonplace' side effects* one has to list the characteristic 'morning-after' symptoms such as *difficulties to wake up, thick head, slight headache, vertigo* – altogether the 'hangover'. Dry mouth, gastric disturbances and mild paresthesias may be added to the symptomatology. Depending on the half-life, type, and dosage of the drug, and on sensitivity of the patient – to a large degree a function of age, state of health, *and* tolerance – these hangover symptoms may be more severe and turn into real disturbances of equilibrium and into a disruption of sensory-motor coordination. Performance in 'psychomotor' tests can be severely impaired in the early parts of the 'dayafter'; so is performance of 'practical' activities such as automobile driving, piloting and delicate manipulations.

Disturbed *equilibrium function* and *sensory-motor coordination* as 'morning-after' effects were observed after *secobarbital* [25], *oxazepam* [26], *nitrazepam* [54], *diazepam* [66], *and methaqualone* [23].

In the card-sorting test impairment of performance was noticed 13 and 17 h after ingestion of *nitrazepam* and *amylobarbitone*. Nitrazepam (5 or 10 mg), but not amylobarbitone (100 or 200 mg) prolonged 'motor-time' component; both drugs in higher dosage affected the 'decision-time' component [54]. *Butobarbitone* in doses of 100 or 150 mg did not affect 'simple reaction time' 8 h after ingestion [9]. *Nitrazepam* (10 mg) and *flurazepam* (30 mg) impaired performance in this task 12 or 8 h post administration [8]. The 'symbol-copying test' was not affected after *butobarbital* (150 mg) but after 15 or 30 mg of *flurazepam* [10].

Dry mouth has been reported after ingestion of *methaqualone, oxazepam, diazepam* and *nitrazepam* [28]. *Methaqualone* produced paresthesias in the face and on the extremities [23, 36, 45]. Diminished libido was observed after *oxazepam* [23, 28]. Also allergic reactions in response to *glutethimide, methaqualone* and a variety of *benzodiazepines* have been reported [28].

Many hypnotics, in particular those of the barbiturate type affect respiration; they are well known to reduce – already in clinical doses – respiratory rate and alveolar ventilation; they increase alveolar pCO_2. In higher doses they diminish the sensitivity of the medullary respiratory centers to

CO_2. These effects on respiration may have serious consequences in elderly people and in cases of sleep disturbances attended by apnea.

Many drugs, if given together, may influence each other in their (effective) blood and/or tissue concentrations or in their biological activity. The 'mutual potentiation', for instance, of analgesics and hypnotics, and of ethyl alcohol and hypnotics is well established. As to the 'mutually inhibiting' phenomena we concentrate here on only one, namely the so-called *enzyme induction*. A drug may accelerate its own metabolization by enhancing the activity of the metabolizing enzymes and thus reduce its own 'effective' blood and tissue concentration. This enhanced enzyme activity may accelerate the metabolization of other drugs, reducing their effective concentration as well. Barbiturates are capable of enzyme induction affecting various types of coagulants [17, 38, 74]. Glutethimide and *chloralhydrate* have similar effects [16]; but *not* nitrazepam and diazepam [3, 12, 55]. *Phenobarbital* accelerates the demethylation of *methadon* [5]; barbiturates 'induce' enzymatic metabolization of *nortryptilin* [4] and of a variety of anticonvulsants [69]. Barbiturates through enzyme induction accelerate the metabolization of steroid hormones including hormonal contraceptive drugs [6, 14].

Dependence liability is well documented for hypnotics of the barbiturate type and for glutethimide [18, 20, 24, 40, 73]. A steeply increasing trend toward abuse with methaqualone has been noted about 15 years ago [13, 39, 65]. Although less pronounced and relatively less frequent, dependence develops also with benzodiazepines [22, 30, 35]. With barbiturates, and to a lesser extent with other hypnotics, tolerance – a drop in potency and efficacy with continued use (or abuse) – typically develops together with dependence.

Withdrawal symptomatology after fully developed dependence with barbiturates and methaqualone is quite dramatic: anxiety, irritability, insomnia, hyperreflexia, blepharospasm, tachycardia, hyperthermia, orthostatic hypotonia, nausea, vomiting and severe abdominal cramps constitute the typical symptomatology. During withdrawal from barbiturate dependence 75% of the patients develop convulsions [21]. The withdrawal symptomatology after chronic (ab-)use with benzodiazepines is said to be not unlike the one observed after barbiturates [37].

It is the crux, and one of the serious 'side effects', of practically all of the presently available hypnotics that they are unable to produce and maintain a sleep which in structure and time course is comparable to natural sleep. Most, if not all, of today's hypnotics are liable to shorten sleep latency, reduce intermittent awakenings and increase its duration. But they all, to a major or lesser degree, reduce slow wave (stages NREM 3 + 4) and REM sleep and favor stages 2 (and 1). Chloralhydrate may be the only hypnotic as yet that does not follow this 'rule'. With long treatment regimes this side effect cannot be tolerated.

A few examples should illustrate these facts (for a detailed compilation of relevant data, see ref. [48]). *Secobarbital* (100 mg) and *pentobarbital* (100 mg) were shown to reduce REM sleep in normals from the about 21 to 15 and 18%, respectively [44]. In a newer study *secobarbital* in insomniacs was found to lower REM sleep to 17.9% (by the 30th day of treatment) and to eliminate stage 4 completely [41]. *Glutethimide* suppressed stage 3 and REM sleep in insomniacs to about 5/6 of their original values; *methaqualone* (300 mg) reduced stage 4, increased REM sleep latency and greatly diminished the number of eye movements [29].

With *benzodiazepines* such dyssomnic side effects are – in general – less pronounced but by far not absent. In normal sleepers a reduction of REM sleep by 15% (from the original 100%) after 10 mg of *diazepam* was observed [59]. *Oxazepam* (0.77 or 2.31 mg/kg) was found to reduce REM sleep in normals from 24.7 to 18.8 and 15.1%, respectively [26], with a reduction of average eye movements from 3.49 to 1.44 and 0.59, respectively. Stage 4 in this study was suppressed from 48 to 32 and 27 min respectively.

Flurazepam in insomniacs showed only little effect on REM time (26–21%); but it increased REM latency by roughly 100%, reduced REMs by 50% and all but completely suppressed stage 4 [43]. *Nitrazepam* in doses of 10 or 20 mg reduced REM sleep by 50 and 65% respectively, during the first 6 h of the night and doubled REM latency without much of an effect on stage 3 and 4 [34]. In another study, little effect of 10 mg of *nitrazepam* on REM sleep, but an about 65% reduction of stage 4 was observed [27].

Flunitrazepam (2 mg) in normal sleepers was found to reduce REM sleep from 23 to 16%; the 6-mg dose REM sleep to 6% of TST [27]. In insomniacs 1 mg of this hypnotic was not effective as far as sleep prolongation was concerned; but it caused a reduction of stage 4 by 90%. An 'effective' (anti-insomnia) dose (2 mg) cut down REM sleep by about 70% and all but suppressed stage 4 [7].

In the course of longer treatments one can notice with practically all hypnotics that the sleep-inducing (anti-insomnia) as well as the 'dyssomnic' effects are attenuated; total sleep after an initial increase tends to become shorter again, and REM and SW sleep, to whatever extent they have been suppressed by the hypnotic, tend to increase and to move toward the status quo. With *acute withdrawal of medication* some of these tendencies are greatly accelerated, total sleep becomes greatly reduced – the *'rebound insomnia'*. REM sleep increases to above control – the *REM rebound*. SWS moves toward control in a somewhat accelerated fashion.

REM rebounds of often extremely long duration are seen after extended treatments with *barbiturates* [64]; after cessation of treatments with the latter, one observes a pronounced insomnia rebound [64, 42]. REM rebound is pronounced also after *glutethimide* and *methaqualone* [29, 44]. In turn, no pronounced REM and insomnia rebound is seen after *benzodiazepines* [7, 27, 44]. There are not as yet enough hard data available about the 2nd (or 3rd) generation of benzodiazepines to make statements similar to the ones just discussed.

The Hypnotic of the Future

Ideally a hypnotic should reinstitute normal sleep, with short treatments. It should be a 'healing' device, with which the patient 'relearns' to sleep and not a 'crutch' that substitutes for a lost function. It should be free of adverse effects – poor performance in the 'morning after', enzyme induction, tolerance, dependence liability, overt effects on autonomic functions.

How does one develop something that approaches these ideal properties? 'Morning-after' disturbances can be eliminated by proper pharmacokinetic characteristics, i.e. short half-life. High efficacy for the whole period of the projected duration of induced sleep may be maintained by using compounds with ultrashort half-lives and by maintaining constant and adequately high blood levels via galenic manipulations: slow or controlled release for predetermined times. Such techniques are available.

As to *enzyme induction*, it seems that most or all of the benzodiazepines are devoid of this side effect. Future developments may make use of this fortuitous beneficial characteristic; intensive research should try to clarify as to what particular basic properties make this class of drug so different from the older compounds in this respect.

Dependence liability, be it of psychical or of the physical nature, may prove to be a more difficult problem to solve. If one envisions short treatment regimes, time may be too limited to form a habit and/or to develop the chemical changes in the organism responsible for the physical dependence. Still, the opportunity to produce drugs with *no* liability in this direction should not be missed. Again, the benzodiazepines which definitely have a lower liability, by degree and by number, than most of the older hypnotics may lead us in the right direction – if we can find out what exactly makes them drugs of reduced efficacy in producing habit and dependence. In addition, proper legal controls and novel attitudes in prescription and distribution may help to complete what the pharmacologist cannot accomplish. Development of tolerance should not prove to be a problem if the new drugs are tailor-made for short-term applications.

As concerns *side effects* in the autonomic sphere – respiration, circulation, sexual functions, etc. – one may predict that drugs that really 'zero in' on the 'sleep and waking systems' may prove to have little chance of affecting other functions, except for indirect and quantitatively proper influences; during natural sleep, respiratory and circulatory activity are reduced but (usually) not to an extent that endangers the well-being of the organism.

As to the *primary therapeutic effect,* – the reinstitution of normal sleep – there can be little doubt that the proper remedy should attack at the 'pathology'. *It must correct the abnormal function;* if possible for good; if not, at least for extended periods. To achieve this, one must know the pathogenesis of the insomnia. To understand the pathogenesis, the underlying cause of the faulty behavior, one needs a full understanding of the normal physiology of the 'function'; i.e. deep insight into the 'dry' and 'wet' neurophysiological organization of sleep.

We have to be aware that sleep, including its various phases and stages is the resultant of organizing and regulating activity of a *'sleep system'* supplemented – in a permissive fashion – by reduced activity of a 'waking system'. Sleep is not only an actively induced increase in output of *sleep-specific functions,* it is also *dewaking*.

With the steeply increasing knowledge about neurohumoral transmitters we have come to learn that serotonergic pathways are probably an all-important part of the 'sleep system' and catecholaminergic pathways are part of the 'waking system'. Additional organizational links are probably represented by such transmitter systems as γ-aminobutyric acid, acetylcholine, polypeptides (such as DSIP, vasotocin and substance S) in the former and (again) acetylcholine, γ-aminobutyric acid, as well as glutamate, enkephalins and the thyrotropic hormone-releasing factor (TRH) in the latter [cf. 49].

Sleep disturbances, such as insomnia, are quite probably the faulty behavioral manifestation, of a faulty activity pattern in anyone or several, of these transmitter systems. These are the 'weak spots' in the various central information channels. Thus, it seems that hypnotic drugs should in fact be tailor-made to be able to correct (only) the faulty transmission link(s).

Again, the introduction of the benzodiazepines is one right step in this direction; they appear to enhance (deactivating) GABA transmission and thus have a 'dewaking' effect. Future research should establish ways and means to produce drugs that in addition – and this in a permanent fashion – enhance, at the right time, the activity in the 'pro-sleep' systems most of all in serotonergic but also in polypeptidergic pathways. The introduction of L-tryptophan into the armamentarium of 'hypnogenesis' constitutes – as we think – another step in the right direction. So does the novel use of β-adrenergic blockers as another 'dewaking' principle [47].

List of References of Workshop B

1 Adam, K.: Do drugs alter the restorative value of sleep? in Passouant, Oswald, Pharmacology of the states of alertness, pp. 105–111 (Pergamon, Oxford 1979).
2 Adam, K.; Adamson, L.; Brezinova, V.; Hunter, W. M.; Oswald, I.: Nitrazepam: lastingly effective but trouble on withdrawal. Br. med. J. *i:* 1558–1560 (1976).
3 Adam, K.; Allen, S.; Carruthers-Jones, I.; Oswald, I.; Spence, M.: Mesoridazine and human sleep. Br. J. clin. Pharmacol. *3:* 157–163 (1976).
4 Alexanderson, R.; Svans, D. A. P.; Sjöquist, F.: Steady state plasma levels of nortriptyline in twins; influence of genetic factors and drug therapy. Br. med. J *iv:* 764–768 (1969).
5 Alvares, A. P.; Kappas, A.: Influence of phenobarbital on the metabolism and analgesic effect of methadone in rats. J. Lab. clin. Med. *79:* 439–451 (1972).
6 Azarnoff, D. L.; Hurwitz, A.: Interacciones medicamentosas Farmacol. Med. *4:* 1–7 (1970).
7 Bixler, E. O.; Kales, A.; Soldatos, C. R.; Kales, J. D.: Flunitrazepam, an investigational hypnotic drug: sleep laboratory evaluations. J. clin. Pharmacol. *17:* 569–578 (1977).
8 Bixler, E. O.; Scharf, M. B.; Leo, L. A.; Kales, A.: Hypnotic drugs and performance; in Kagan, et al., Hypnotics (Spectrum, New York 1975).
9 Bond, A. J.; Lader, M. H.: Residual effects of hypnotics. Psychopharmacologia *25:* 117–132 (1972).
10 Bond, A. J.; Lader, M. H.: The residual effects of flurazepam. Psychopharmacologia *32:* 223–235 (1973).
11 Braestrup, C.; Squires, R. F.: Specific benzodiazepine receptors in rat brain characterized by high-affinity ^{3}H-diazepam binding. Proc. natn. Acad. Sci. USA *74:* 3805–3809 (1977).
12 Breckenridge, A.; Orme, M.: Clinical implications of enzyme induction. Ann. N. Y. Acad. Sci. *179:* 421–431 (1971).
13 Bridge, T. P.; Ellinwood, E. H.: Quaalude alley: a one-way street. Am. J. Psychiat. *130:* 217–219 (1973).

14 Conney, A.H.: Pharmacological implications of microsomal enzyme induction. Pharmacol. Rev. *19:* 317–366 (1967).
15 Costa, E.; Guidotti, A.: Molecular mechanisms in the receptor action of benzodiazepines. Annu. Rev. Pharmacol. Toxicol. *19:* 531–545 (1979).
16 Cucinell, S.A.; Odessky, L.; Weiss, M.; Dayton, P.G.: Effect of chloralhydrate on bishydroxycoumarin metabolism. A fatal outcome. J. Am. med. Ass. *197:* 366–368 (1966).
17 Dayton, P.G.; Tarcan, Y.; Chenkin, T.; Weiner, M.: Influence of barbiturates on coumarin plasma levels and prothrombin response. J. clin. Invest. *40:* 1797–1802 (1961).
18 Deniker, P.; Loo, H.; Colonna, L.; Cottereau, M.J.: Les agents psychotropes utilisés dans les toxicomanies actuelles en France. Leurs effets et leurs risques. Toxicomanie *6:* 135–148 (1973).
19 Echols, S.D.; Jewett, R.E.: Effects of morphine on sleep in the cat. Psychopharmacology *24:* 435–448 (1972).
20 Eddy, N.B.; Halbach, H.; Isbell, H.; Seevers, M.H.: Drug dependence: its significance and characteristics. Psychopharmacol. Bull. *3:* 1–23 (1966).
21 Eskenazy, J.; Stamate, A.: La prébarbituromanie et la barbituromanie. Revue roum. Neurol. *8:* 81–88 (1971).
22 Fabre, L.E.; Harris, R.T.: Clinical considerations in the abuse of antianxiety and antidepressant drugs; in Kagan et al., Hypnotics (Spectrum, New York 1975).
23 Fournier, E.; Piva, C.; Diamant-Berger, O.: Chronic intoxications in France. Int. Congr. Toxicology, Montreal 1974.
24 Fraser, H.F.; Shaver, M.R.; Maxwell, E.S.; Isbell, H.: Death due to withdrawal of barbiturates. Ann. intern. Med. *38:* 1319–1325 (1953).
25 Fregly, A.R.; Smith, M.J.; Wood, C.D.; Cramer, D.B.: Effects of some antiemetic sickness drugs and secobarbital on postural equilibrium functions at sea level and at 12,000 feet. Aerospace Med. *44*: 145–150 (1973)
26 Gaillard, J.-M.; Aubert, C.: Specificity of benzodiazepine action on human sleep confirmed. Another contribution of automatic analysis of polygraph recordings. Biol. Psychiat. *10:* 185–197 (1975).
27 Gaillard, J.-M.; Schulz, P.; Tissot, R.: Effects of three benzodiazepines (nitrazepam, flunitrazepam and bromazepam) on sleep of normal subjects, studied with an automatic sleep scoring system. Pharmakopsychiatr. Neuro-Psychopharmakol. *6:* 207–217 (1973).
28 Ginestet, D.; Poirier-Littre, M.F.; Cuche, H.: Les effets indésirables des hypnotiques. Thérapie *31:* 77–103 (1976).
29 Goldstein, L.; Graedon, J.; Willard, D.; Goldstein, F.; Smith, R.R.: A comparative study of the effects of methaqualone and glutethimide on sleep in male chronic insomniacs. J. clin. Pharmacol. *10:* 258–268 (1970).
30 Gordon, E.B.: Addiction to diazepam (valium). Br. med. J. *i:* 112 (1967).
31 Haefely, W.: Mechanism of action of benzodiazepines; in Koella, Sleep 1980 (this volume).
32 Haefely, W.; Kulcsár, A.; Möhler, H.; Pieri, L.; Polc, P.; Schaffner, R.: Possible involvement of GABA in the central actions of benzodiazepines; in Costa, Greengard, Mechanisms of action of benzodiazepines, pp. 131–151 (Raven Press, New York 1975).
33 Haefely, W.; Pieri, L.; Polc, P.; Schaffner, R.: General pharmacology and neuropharmacology of benzodiazepine derivatives; in Hoffmeister, Stille, Handbook of experimental pharmacology, vol. 55, part II (Springer, Berlin, in press, 1981).

34   Haider, I.; Oswald, I.: Effects of amylobarbitone and nitrazepam on the electrodermogram and other features of sleep. Br. J. Psychiat. *118:* 519–522 (1971).
35   Hanna, S.M.: A case of oxazepam (serenid D) dependence. Br. J. Psychiat. *120:* 443–445 (1972).
36   Hauberg, B.: The chlordiazepoxide HCl (librium) and analogue nitrazepam (mogadon) in the treatment of epilepsy in children. Devl. Med. Child Neur. *10:* 302–309 (1968).
37   Hollister, L.E.; Motzenbecker, F.P.; Degan, R.O.: Withdrawal reactions from chlordiazepoxide (librium). Psychopharmacologia *2:* 63–68 (1961).
38   Hunninghake, D.B.; Azarnoff, D.L.: Drug interactions with warfarin. Archs intern. Med. *121:* 349–352 (1968).
39   Inaba, D.S.; Gay, G.R.; Newmeyer, J.A.; Whitehead, C.: Methaqualone abuse, 'luding out'. J. Am. med. Ass. *224:* 1505–1509 (1973).
40   Isbell, H.: Addiction to barbiturates and the barbiturate abstinence syndrome. Ann. intern. Med. *33:* 108–121 (1950).
41   Kales, A.; Hauri, P.; Bixler, E.O.; Silverfarb, P.: Effectiveness of intermediate-term use of secobarbital. Clin. Pharmacol. Ther. *20:* 541–545 (1976).
42   Kales, A.; Kales, J.D.; Bixler, E.O.: Insomnia: an approach to management and treatment. Psychiat. Ann. *4:* 28–44 (1974).
43   Kales, A.; Kales, J.D.; Bixler, E.O.; Scharf, M.B.: Effectiveness of hypnotic drugs with prolonged use: flurazepam and pentobarbital. Clin. Pharmacol. Ther. *18:* 356–363 (1975).
44   Kales, A.; Malmstrom, E.J.; Scharf, M.B.; Rubin, R.T.: Psychophysiological and biochemical changes following use and withdrawal of hypnotic; in Kales, Sleep: physiology and pathology (Lippincott, Philadelphia 1969).
45   Kessel, A.; Williams, A.G.; Youngs, T.D.S.: Side effects with a new hypnotic drug potentiation. Med. J. Aust. *ii:* 1194–1195 (1967).
46   Klepner, C.A.; Lippa, A.S.; Benson, D.I.; Sano, M.C.; Beer, B.: Resolution of two biochemically and pharmacologically distinct benzodiazepine receptors. Pharmacol. Biochem. Behav. *11:* 457–462 (1979).
47   Koella, W.P.: Beta-Blockers and sleep; in Kielholz, Beta-blockers and the central nervous system. Int. Symp., St. Moritz 1976, pp. 174–183 (Huber, Bern 1977).
48   Koella, W.P.: Nebenwirkungen der Hypnotika. In: XVII Internationaler Fortbildungskurs für praktische und wissenschaftliche Pharmazie. Schriftenreihe der Bundesapothekerkammer zur Wissenschaftlichen Fortbildung, Band VII/Gelbe Reihe, Meran (1979, pp. 77–101).
49   Koella, W.P.: The neuropharmacology of sleep (the role of neurotransmitters in the making of sleep); in Wheatley, Nicholson, Psychopharmacology of sleep. (Raven Press, New York, in press 1981).
50   Koella, W.P.: Side effects of today's hypnotics and the hypnotic of the future; in Koella, Sleep 1980 (this volume).
51   Kosterlitz, H.W.; Lord, J.A.H.; Paterson, S.J.; Waterfield, A.A.: Effects of changes in the structure of enkephalins and of narcotic analgesic drugs on their interactions with mu- and delta-receptors. Br. J. Pharmacol. *68:* 333–342 (1980).
52   Krüger, J.M.; Pappenheimer, J.R.; Karnovsky, M.L.: Sleep-promoting factor S: purification and properties. Proc. natn. Acad. Sci. USA *75:* 5235–5238 (1978).
53   Kunz, W.; Keller, H.; Mückter, H.: *N*-Phthalyl-glutaminsäureimid. Drug Res. *6:* 426–430 (1956).

54 Malpas, A.; Rowan, A.J.; Joyce, C.R.B.; Scott, D.F.: Persistent behavioral and electroencephalographic changes after single doses of nitrazepam and amylobarbitone sodium. Br. med. J. *ii:* 762–764 (1970).
55 Matis, P.: Toleranzänderungen (Effekte von Nebenmedikamenten) bei langzeitiger Antikoagulanzienbehandlung. Thromb. Diath. haemorrh. *12:* suppl. pp. 33–38 (1964).
56 Möhler, H.; Okada, T.: Benzodiazepine receptors: demonstration in the central nervous system. Science *198:* 849–851 (1977).
57 Möhler, H.; Richards, J.G.; Wu, J.-Y.: Autoradiographic localization of benzodiazepine receptors in immunocytochemically identified GABAergic synapses. Proc. natn. Acad. Sci. USA (in press, 1980).
58 Monnier, M.; Schönenberger, G.A.: Characterization, sequence, synthesis and specifity of a delta-EEG sleep-inducing peptide; in Koella, Levin, Sleep 1976, pp. 257–263 (Karger, Basel 1977).
59 Nicholson, A.N.; Stone, B.M.; Clarke, C.H.: Effect of diazepam and fosazepam (a soluble derivative of diazepam) on sleep in man. Br. J. clin. Pharmacol. *3:* 533–541 (1976).
60 Ogunremi, O.O.; Adamson, L.; Brezinova, V.; Hunter, W.M.; MacLean, A.W.; Oswald, I.; Percy-Robb, I.W.: Two anti-anxiety drugs: a psychoneuroendocrine study. Br. med. J. *ii:* 202–205 (1973).
61 Oswald, I.: Sleep studies in clinical pharmacology. Br. J. Pharmacol. *10:* 317–326 (1980).
62 Oswald, I.; Adam, K.: Benzodiazepines cause a small loss in body weight. Br. med. J. *281:* 1039–1040 (1980).
63 Oswald, I.; Adam, K.; Borrow, S.; Idzikowski, C.: The effects of two hypnotics on sleep, subjective feelings and skilled performance; in Passouant, Oswald, Pharmacology of the states of alertness, pp. 51–63 (Pergamon, Oxford 1979).
64 Oswald, I.; Priest, R.G.: Five weeks to escape the sleeping pill habit. Br. med. J. *ii:* 1093–1099 (1965).
65 Pascarelli, E.F.: Methaqualone abuse, the quiet epidemic. J. Am. med. Ass. *224:* 1512–1514 (1973).
66 Pignataro, F.P.: Experience with chemotherapy in refractory psychiatric disorders. Curr. ther. Res. *4:* 389–398 (1962).
67 Polc, P.; Schneeberger, J.; Haefely, W.: Effects of several centrally active drugs on the sleep-wakefulness cycle of cats. Neuropharmacology *18:* 259–267 (1979).
68 Polc, P.; Wolfgang, H.: Telemetry of the EEG and EMG in the cat under the influence of psychotropic drugs. Biotelemetry *1:* 264–272 (1974).
69 Rubin, E.; Lieber, C.S.: Alcohol, other drugs and the liver. Ann. intern. Med. *69:* 1063–1078 (1968).
70 Sieghart, W.; Karobath, M.: Molecular heterogeneity of benzodiazepine receptors. Nature, Lond. *286:* 285–287 (1980).
71 Spiegel, R.: Controlled clinical investigation with a phenothiazine derivative used as a sleep-inducer in nonpsychotic insomniacs; in Koella, Levin, 1st Eur. Congr. on Sleep, Basel 1972, pp. 91–101 (Karger, Basel 1973).
72 Spiegel, R.: Effect of mesoridazine on sleep in normal subjects. Sleep Res. *4:* 119 (1975).
73 Wang, R.I.H.: Dependence liability of sedatives and hypnotics; in Kagan, et al., Hypnotics, pp. 297–307 (Spectrum, New York 1975).
74 Welch, R.M.; Harrison, Y.; Conney, A.H.; Burns, J.J.: An experimental model in dogs for studying interaction of drugs with bishydroxycoumarin. Clin. Pharmacol. Ther. *10:* 817–825 (1969).

Addresses of Authors and Discussants

Dr. H. Bittiger, Pharma Research Department, Ciba-Geigy Ltd, CH-4002 Basel (Switzerland)
Dr. W. Haefely, F. Hoffmann-La-Roche & Co. Ltd, Pharma Research Department, Grenzacherstrasse 124, CH-4002 Basel (Switzerland)
Dr. J. Jaekel, Pharma Research Department, Ciba-Geigy Ltd, CH-4002 Basel (Switzerland)
Prof. W. P. Koella, Friedrich Miescher Institute, CH-4002 Basel (Switzerland)
Prof. I. Oswald, Royal Edinburgh Hospital, Edinburgh EH10 5HF (Scotland)
Dr. R. Scherschlicht, F. Hoffmann-La-Roche & Co. Ltd, Pharma Research Department, Grenzacherstrasse 124, CH-4002 Basel (Switzerland)
Dr. R. Spiegel, Department of Experimental Therapeutics, Sandoz Ltd, Postfach, CH-4002 Basel (Switzerland)

C. Special Symposium on Dreams

Chairmen: *D. Lehmann,* Zürich; *I. Strauch,* Zürich
Participants: *D. Foulkes,* Atlanta, Ga.; *M. Kramer,* Cincinnati, Ohio; *P. Salzarulo,* Paris

Introduction

I. Strauch, Zürich, Switzerland

Experimental dream research has resulted in relatively few publications over the last decade: only 4% of all papers listed in *Sleep Research* was concerned with this field, for example 48 of 1700 papers in 1978. Similarly, this present symposium on dreams is only the second one at a European Sleep Research Meeting. This may reflect a number of developments: For one, the naive hopes of finding simple straight-forward physiological correlates in the sense of a catalogue or taxonometry of dream contents have not been fulfilled. On the other hand, the traditional experts in the field, the psychoanalysts, have not become excited about possible potentials of experimental dream research. Finally, hard-nosed scientists still are apalled by the fact that dreams – because of their very nature – necessarily constitute a re-worked version of what really happened.

As to the last point, internal private experiences are once again becoming an admissable tool in psychological research. In addition, more sophisticated ideas and methods are being applied to physiological and psychological data collection and evaluation, which might remedy the disappointment coming out of unfulfilled early expectations.

In the present symposium we attempted to discuss some of the multifaceted concurrent approaches to dreams and their physiology, by inviting four workers in the field who recently discussed different aspects of mentation during sleep.

Dr. *Lehmann* and Dr. *Koukkou* attempt to combine what is known from electrophysiology, experimental psychology and work with dreams into a model of dream formation. Within the framework of different functional states of the brain they draw analogies between EEG and cognitive phenomena during wakefulness in childhood and during functional sleep states in the adult.

Dr. *Foulkes* approaches dream development in longitudinal studies of children's dreams. He discusses the analyses of an extensive data base, relating to cognitive processes in dreams and cognitive maturation during wakefulness, where different developmental stages are reflected in both types of activity.

Dr. *Salzarulo* investigates how memory processes are involved in the recall of mental experiences during REM and NREM sleep. Following the criteria of the generative-transformational grammar, a psycholinguistic analysis provides information of how memory processes operate with regard to consolidation and long-term memory storage.

Dr. *Kramer* considers the various still hypothetical functions of dreaming and by doing this he proposes a conceptional and methodological clarification of the alleged types of function. He develops a functional approach to dreaming which takes into account the relationship between dreaming and mood during preceding and subsequent wakefulness.

Sleep 1980. 5th Eur. Congr. Sleep Res., Amsterdam 1980, pp. 170–174
(Karger, Basel 1981)

Dream Formation in a Psychophysiological Model: the State-Change Theory[1]

D. Lehmann, M. Koukkou, Zürich, Switzerland

This psychophysiological model of dream formation accounts for the characteristics of dream mentation, as incoherence [15], discontinuity, lack of reality testing (acceptance of physically impossible phenomena, appearance of deceased or far-away persons), non-logical conclusions, etc. In psychoanalytic theories, these characteristics are the result of 'dream work' (e.g. condensation, displacement, censorship), i.e. mechanisms which operate on the result of a postulated primary process predominantly or specifically in dreams.

Dreams are thinking, mentation during sleep, the result of recall and cognitive treatment of old or recently acquired experiences. Dream ex-

[1] For references see compound list on page 186.

periences as different from wakeful thinking need to be considered by the model. Awake-type mentations which might have been collected in awakenings during laboratory nights require no specific attention in a model of dreaming, as one may readily assume that they are generated during the awakening. Arguments which are specific for REM sleep [24] are not satisfying for a model of dream formation, because dream reports are also obtained at sleep onset or during the first non-REM period prior to the first REM period [18, 39]. However, a model will have to account for the high yield of typical dream reports out of REM phases as opposed to the lower yield out of non-REM phases.

The present model [29] uses brain mechanisms established in experimental psychology, behavioral research, psychopharmacology, electroneurophysiology and developmental studies; it does not postulate dream-specific processes, but views mentations during sleep as a set of mentation classes which occur during different functional states of the sleeping brain.

The pivoting point of the model is the conception of functional states of the brain with state-dependent information processing, as for instance sleep and wakefulness, or infancy and adult age [30]. Functional states are observable in studies on electrophysiology, learning and recall, cognitive strategies, behavior, etc. The phenomena measured with the different methods vary depending on the functional state, or vice versa, may serve as a classifier of state. The functional state at each moment in time depends on heredity, structural maturation and development, past experiences, circadian changes, type and relevance of new information input, metabolic conditions, and possibly, pathological factors as intoxication or disease. Minor, short-lasting, changes (fluctuations) of the functional state are observed within gross states, and are associated with changes in sensory threshold, cognitive strategies, and EEG patterns [40]. Changes of functional state as a reaction to relevance of new incoming information will be determined and constrained by the state which was present at the moment of arrival of the information, in addition to the factors above.

Our functional model of state-dependent information processing in the brain is shown in figure 1. The characteristics of the operations of all subcomponents of the model change depending on the functional state which is continuously reset by the controller as a result of the adaptational decisions of the logic to input. The frequent changes of short-lasting EEG patterns (see segmentation analysis of EEG [47]) during sleep indicate numerous brief changes of the functional state within conventional sleep stages.

Basically, predominance of slower EEG waves indicates states during

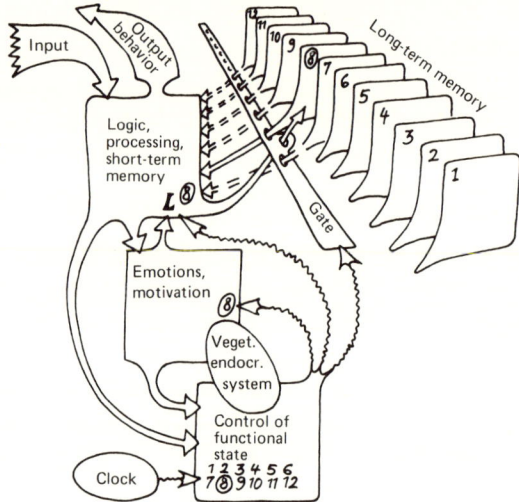

Fig. 1. Functional model of the brain. Double lines: open information channels; dashed lines: partially open information channels; solid thin lines with black arrowheads: control channels: The functional state of the system (illustrated is state '8') is set by the control according to the clock input, and information received from the logic/processing (analysis) subsystem with its short-term memory, and from the motivation/emotion subsystem. The control in return adjusts the operations of logic and motivation to the current functional state, and opens the gate to the associated long-term memory space for deposit and recall (in the picture, space '8'). Memory of higher states ('9' to '12') might be partially recalled; of the lower states, only the next neighbor space '7' is partially open for recall. Changes of the functional state put different memory spaces and processing strategies into use.

which simpler organizational levels become operative (e.g. sleep, childhood, narcosis, intoxication), whereas higher EEG frequencies are associated with more complex functional organization. Examples are wakefulness from childhood to adult age, and vigilance in the adult from sleep to wakefulness (see also higher EEG frequencies as result of increased connectivity in models of neural nets, [44]). Thus, during the night repeated functional regressions to lower levels occur in the functional state governed by the circadian and ultradian clock, but under finer control according to the requirements which are set by the results of cognitive operations on information treated by the logic. As to the proposed functional regression we note that children's cognitive strategies during symbolic play have been described as comparable to those during adult dreaming [50]; in fact, emotional experiences in children were reported to be associated with slow EEG (theta) waves [46] which are also observed during sleep.

Different states are associated with access to different memory storages, according to the studies on learning and recall; learning and recall are state-dependent [49, 61]. Material stored in storages of states of higher complexity (wakefulness, or no-drug state) can be recalled in lower states (sleep, or drug state). Accordingly, selective awakening by relevant (new or important) stimuli is possible during sleep after comparison of the new with old information, or automatized responses might be executed correctly during sleep without awakening [28] or during sleepwalking [25].

Recall of information from lower state storages is limited to the closely neighboring storage; thus, recall of early childhood material is impossible for awake adults; drug state information cannot be recalled in no-drug states. This asymmetry of state-dependent recall accounts for the difficulty of recall of mentation out of non-REM sleep phases, which by definition show more EEG slow wave activity than wakefulness epochs, and are thus organizationally remote from wakeful states. Indeed quality of awake recall of new information which is presented during slow wave sleep is a function of the amount of EEG activation (slow wave reduction) induced by the presentation [41]. The relatively superior recall of dreams out of REM sleep follows from the closer functional proximity of wakefulness states to REM sleep (which contains more wakefulness-near, faster EEG patterns) than to non-REM sleep (with slow waves). On the other hand, the laws of state-dependent learning and recall make it possible that during periods with EEG slow waves, memory storages containing childhood experiences might become re-accessible for partial recall, accounting for childhood material in dreams.

Different functional states also are associated with different cognitive strategies, state-innate (reflecting the complexity level), or acquired. Thus, a change of state will bring new strategies into operation (e.g. in daydreaming, night dreams, or in different maturational levels [50]).

Information which is evaluated by the logic during sleep might be recognized as important (e.g. threatening or exciting) to the subject. This result of evaluation will produce an orienting reaction, whose EEG component is an increase in wave frequency. The orienting reaction is a larger or smaller change of state to higher levels, thus interrupting automatized responses and enabling the system to work out new solutions.

The information which induced a change of state via an orienting reaction will remain in short-term memory (processor memory), but will, in the newly attained state, be further processed with strategies which are different from those with which it was originally acquired; search for context material will also be made in the newly available state-dependent memory. In this way, material from one storage area becomes distorted by unrelated treatment strategies, and gets connected with originally unrelated material, these connections being made with rules having possibly still different origins. This interplay of spontaneous and induced state changes, with the resulting changes of accessible memory and available cognitive strategies, and the following recombinations and reinterpretations of memories from different sources, while at the same time considering recently acquired information, are the mechanisms which lead to the end result of the typical dream experience, as far as it is recallable, i.e. if it has been stored eventually during a state which was close enough to wakefulness for later recall.

This model of dream formation implies the biological significance of dreaming: The utilization of earlier experiences for the review and reappraisal of new experiences, and the storage of such treatments in a form which might be accessible for wakeful usage.

Sleep 1980. 5th Eur. Congr. Sleep Res., Amsterdam 1980, pp. 174–178
(Karger, Basel 1981)

Dreams and Cognitive Development[2,3]

D. Foulkes, Atlanta, Ga., USA

In two concurrent 5-year longitudinal studies, male and female children served 8 or 9 non-consecutive laboratory nights per year. The children were studied from, respectively, ages 3–4 to 8–9 and ages 9–10 to 14–15. Of the original study groups, all 14 younger subjects, and 12 of 16 older subjects, remained in the study for its duration. Both samples were supplemented cross-sectionally. Here I report some data on *correlates* of children's rates of REM dream reporting and of the kinds of content which they reported.

[2] Supported by NIMH grant 32063.
[3] For references see compound list on page 186.

Extra-dream variables ranged in number, at different age-levels, between 176 and 400. They included behavior observations and ratings and numerous family life, sleep, physical-maturational, personality, and cognitive measures.

Normatively, REM recall was low in early childhood, but increased to near-adult values by middle childhood. Significant intra-group increases were observed from ages 5–7 to 7–9 and significant inter-group increases were observed from ages 7–9 to 9–11, periods in which Piagetian analysis might lead us to expect the increasing systematization and autonomy of cognitive processing.

At ages 3–5, there were no positive cognitive correlates of REM reporting. High rates of dream reporting were associated with an expansive-expressive behavioral style and with approaches to adults for instrumental and emotional support. These traits clearly suggest the possibility of both false positives and false negatives. However, there were no negative cognitive correlates of REM reporting, and all 14 children had higher REM than non-REM report rates, suggesting that at least some of the children's reports were veridical at ages 3–5. The problem, however, is to know which ones. A possible solution is to move forward to ages where children's reports did seem generally credible, and to use correlational data at those ages for backward extrapolation.

At ages 5–7, a relatively different set of correlates emerged, and there was a somewhat different set of high- and low-rate REM reporters. No longer were the high-rate reporters particularly active or obsequious children. High-rate reporters came from democratically organized, higher social-class backgrounds. From the perspective of actual dream construction, the most interesting correlates came from among the few cognitive measures available for correlation. Absolute level of proficiency on WPPSI block designs, and proficiency in a visuospatial recognition task – Kagan's matching familiar figures test – were significant predictors of REM recall. *Increases* in REM recall from ages 3–5 to 5–7 were predicted by similar parent and behavioral variables, and by still another visuospatial cognitive measure, the children's embedded figures test.

The 5- to 7-year-old children's REM correlates deserve particular stress because, from these ages forward in our two longitudinal series, three facts always were observed: (1) there was intra-subject consistency in REM report rate; (2) cognitive variables were correlated with REM report rates, and (3) the direction of these correlations always was positive. Thus, the children's REM reports seemed, from ages 5–7 on, generally credible.

The nature of the cognitive correlates which appeared at ages 5–7 also set a pattern that held at later ages: namely, so-called visuospatial or performance measures correlated more strongly with REM report rates than did either measures of verbal proficiency or measures of memory span. Given the nature of the cognitive tasks correlated with REM reporting, it is difficult to believe that we are dealing simply with report phenomena rather than with experience phenomena as well. Whatever block designs measures, it cannot be simply verbal facility or memory or reporting skill. It is tempting to believe, and I believe, that block designs was measuring some skill(s) essential to the production of dream experiences themselves. Elsewhere [19], I have reported still more impressive evidence in favor of this hypothesis: 2 boys who were studied from ages 11–12 on, and who had atypically low REM report rates (e.g. 0% and 7% at ages 13–14) also had specific cognitive defects in block designs performance.

What seems to be needed to experience dreams, then, is not merely some pattern of brainstem-initiated EEG or eye-movement activity, but specific cognitive skills that enable one to employ an occasion of general cortical activation for the symbolic reconstruction of reality in the form of dreamlike narratives. Our data speak to the question of what those skills might be and to when they first generally reach a level of functioning permitting the synthesis of dreams in REM sleep.

How does dreaming begin as an organized psychological process? There was a definite 'preoperational' slant to the 3- to 5-year-olds' REM reports. (a) They were short, 11 of 14 children describing a typical dream in 18 words or less, as if the children were unable to sustain an internally scripted line of thought. (b) They contained many animal characters, most often cast in humanoid roles. (c) They included little active self-representation, consistent with children's waking deficiencies in self-knowledge and self-awareness. (d) They contained remarkably little movement imagery, consistent with *Piaget and Inhelder's* [51] demonstration of the absence of kinematic imagery in preoperational children. (e) They contained practically no affect, consistent with children's waking deficiencies in knowledge and self-awareness of emotions. (f) Their most frequent theme – the self sleeping somewhere (for instance, on top of a car, in a barber-shop window, at a refreshment stand) – would seem to reflect egocentrism, an inability of symbolization to decenter from the child's own current situation. Note that, from a cognitive perspective, the prediction of dream features does not follow from what children can *do* in wakefulness, but rather depends on how well children symbolically can understand and reconstruct the texture of their waking lives. In this latter respect, children's waking thought is notably deficient, in now well-documented ways, and so too, our data suggest, is their dream thinking.

But, are these so-called preoperational features of children's reports *credible* ones? For features that are almost totally absent in the 3- to 5-year-olds' dream reports, the correlational data obviously can shed no light on this issue. Let us focus, therefore, on the two features of the children's REM reports which are peculiar in the degree to which they are present, rather than absent, as compared to adult dreams. These features are sleep themes and animal characters. At ages 3–5, the presence of sleep themes was correlated significantly with block designs raw scores, the best general predictor in later years of dream-making, or at least of dream-reporting. Children high in such themes also were high in the *accuracy* of their waking memory and were relatively less animistic than were their peers. Dreamers who dreamed in an *exclusive* way of self-fatigue themes emerged as even more competent cognitively, and were older than peers within their study group. By inference, not only is the sleeping self a relatively credible dream feature, but also the moving other is not a credible dream feature. Thus, the construction of highly static dream episodes in which no social interaction occurs is a cognitive achievement that only the oldest and most talented children can accomplish during the preschool years. Such episodes, closely focused on the child's own current body state, may be prominent among the ways that dreaming begins as an organized psychological process.

Our correlational data suggested that animals also were genuine dream features. The *exclusive* reporting of animal characters was associated positively with cognitive skill but negatively with two measures of dependency. By inference, children's animal characters are credible dream features, while their human characters are not.

The general postulate of egocentrism suggests that children's animal characters may constitute a form of self-expression. We also found that exclusive animal dreamers were vigorous, assertive, and aggressive children. If animals *are* self-representations, they probably are selected not as crafty disguises of the self, but because of representational inability to construct any more veridical form of self-representation. At ages 7–9, where dreams first seemed, by correlational criteria, reliably to contain a genuinely participating self character, animal characters began their longitudinal decline, and the residual use of animals at those ages was associated with cognitive immaturity, including a measure of animism.

From their preoperational base of static imagery, body-state content, animal characters, and no affect, dreams develop toward adult form in predictable ways. At ages 5–7, kinematic imagery, as reflected in movement by other characters, is present and is, by cognitive correlational criteria, credible. Self-representation and the experience of affect first appear in substantial degree and with positive cognitive correlates, at ages 7–9.

The influence of cognitive maturation on dreaming does not, from our data, cease in middle childhood. Even in early adolescence, when new representational modalities appeared in REM dream content, e.g. the analytic detachment of objects both from a larger setting or background and from any strong narrative context, their appearance was correlated with visuospatial measures.

In summary, our data suggest that, throughout childhood, the possibility of dream experiencing, and the particular kinds of dreams that can be dreamed, are greatly constrained by children's cognitive maturation. Conversely, dreams offer us a wide window through which to view the development of children's cognitive representations and mental operations.

Sleep 1980. 5th Eur. Congr. Sleep Res., Amsterdam 1980, pp. 178–182
(Karger, Basel 1981)

Memory Processing of the Mental Sleep Experience[4]

Piero Salzarulo, Carlo Cipolli, Paris, France

What are the effects of sleep on the memory and what are the functioning characteristics of the latter in terms of the cognitive processes? These problems have been discussed for many years with often contradictory results [57]. Several theories have been advanced without one of them allowing us to better explain all the phenomena [13]. Here we will not discuss the processes of memory involved in recall of stimuli presented before, during or after sleep; we will instead talk about those memory processes concerned in the mental sleep experience (MSE) related either after provoked awakening at a particular sleep stage or after spontaneous awakening in the morning.

[4] For references see compound list on p. 186.

In the field of dream psychophysiology, *Foulkes* [17] was among the first to underline the importance of the memory processes in explaining the absence of dream report. In much research the reference framework was the consolidation theory [23]. The absence of recall has been attributed to an absence or to a difficulty in consolidating the contents of what has been elaborated during sleep: this is how the low frequency of recall after awakening in NREM sleep has been explained. We do not think that this theory best explains the memory processing of the contents of mental activity during sleep. In fact, this theory has mainly been used for explaining the effects of sleep on the memory in learning experiments in the animal [2]. This situation is in every way radically different to that where man memorizes his mental sleep experience. Firstly, the period considered necessary for the consolidation of the trace in the experiments of learning in the animal is longer than in man, where unlearnt stimuli are concerned. In addition, in the learning situation there is a repeated presentation of the stimuli, contrary to what occurs in the memory processing of MSE contents. Moreover, in these studies reference is made to a neurobiological conception of the consolidation according to which the trace is either completely consolidated or not at all. In studies on humans, we should use more sophisticated criteria of recall, allowing an evaluation of several levels of retention. Thus, we will use the notion of consolidation in accordance with present psychological theories [10], in order to indicate a process which transfers the material from short-term to long-term store. This material will be organized in memory according to the different levels of analysis developed above all when it was in short-term store.

It is obvious that the information created during sleep should be recorded and then stored at short-term during sleep. Is there already during sleep a transfer of information into long-term store? The model proposed by *Koulack and Goodenough* [31] presupposes that in almost all cases information will be stored during sleep in short-term: the passage to long-term would take place after awakening. These conclusions do not seem to be entirely justifiable to us. In fact, if the model of *Atkinson and Shiffrin* [1] is referred to (it was created to explain the memory processes during waking) information does not stay in short-term store for more than 30 s: consequently the contents obtained more than 30 s after awakening (the usual situation in research where the subject relates the report after provoked awakening) come above all from long-term store. In addition, it has been shown that there is a correlation between the length of the report (which roughly gives the amount of contents in the memory) and the length of

REM sleep [12]; this correlation would be inexplicable only in terms of short-term storage during sleep, since short-term store has only a limited capacity. Other arguments resulting from personal research will follow.

How would the passage from short-term to long-term store take place? If the model of *Atkinson and Shiffrin* [1] is referred to again, it would be possible to imagine a buffer functioning in accordance with the modality 'first-in first-out': it would follow that long-term storage would have a serial organization; the contents 'first-in', would be the first to be forgotten if there were a limit in the capacity of long-term store or in the persistence of the trace: the contents 'first-in' would no longer be accessible when the subject is retrieving. The period following the awakening is also very important, even crucial. What occurs during this particular moment can lead to important differences in the material recalled.

It should be remembered, in particular, the instructions that are given to the subject after awakening. *Foulkes* [16] modified the instructions used before by *Dement and Kleitman* [11] and obtained a larger amount of reports after NREM sleep; moreover, one particular instruction allowed *Foulkes and Pope* [20] to obtain more contents, and their equal distribution between tonic and phasic periods of REM sleep, than in a previous research [48]; we obtained more hallucinatory verbal recall [56] after an instruction focused on this type of content [7].

The importance of this factor resides essentially in the modification of the processes of retrieval of the mental sleep experience, that is to say the cognitive processes that the subject must use to give a report of his MSE. These processes are essentially those of retrieval of the contents in the memory and of verbal encoding. We shall see that the processes of retrieval and of verbal encoding are important since they give us a mean of having information about the consolidation and the organization in memory of the contents of the MSE. In order to study these problems we undertook the psycholinguistic analysis of the reports of the MSE. Methods and general principles that have guided us in our research can be found in *Salzarulo and Cipolli* [56, 58].

It is assumed that strategies (sequential reproduction, evaluation, abstraction) and the difficulties of retrieval are expressed by the length of the sentences and by the frequency of the waking-related kernel-sentences. Moreover, the development of semantic planning is shown by the frequency and the location of the pauses within the sentence. The effectiveness of the processes of consolidation in memory can be evaluated by the amount of content present in the reports (= number of sleep-related kernel-sentences); the degree of consolidation of single contents can be appreciated by their presence or non-presence in successive reports; the degree of consolidation of all the contents can be deduced from the data concerning retrieval: the fewer the cues to the consolidated contents, the harder their retrieval. The organization of the contents in memory can be obtained by comparing the serial position of the sentences including the best consolidated contents and the stability of this position in successive reports.

To obtain data for studying these processes, we did two experiments. In the first [58] following a classical paradigm, we awakened the subjects either after REM sleep or after NREM sleep; after the awakening, the subject was asked: 'What was passing through your mind before awakening?'. In the second experiment [8] one provoked awakening was performed as in the previous experiment; after their verbal report was completed, the subjects were asked to go back to sleep; in the morning, upon spontaneous awakening, the subjects were asked to relate again the mental experience that had preceded the previous night's experimental awakening.

The data obtained in the first experiment indicated that the contents of MSE are stored in long-term memory after the two types of sleep. This is shown, in addition to the high frequency of contentful reports, by the remarkable length of the reports made up of numerous kernel-sentences that encode the MSE. In fact, only short-term consolidation in one or both types of sleep would have led to the recovery of only very few contents and the verbal encoding of few sleep-related kernel-sentences.

Information regarding the degree of consolidation in long-term memory can be deduced from the data on retrieval. Since the frequency of between-kernel-sentence pauses in complete waking-related kernel-sentences, and of within-kernel-sentence pauses in incomplete waking-related kernel-sentences are greater in NREM than in REM reports, the degree of consolidation is supposed to be lower in NREM reports than in REM reports. That there should be better consolidated contents than others in long-term memory, comes from the comparison between night and morning reports. In fact, only a ¼ of sleep-related kernel-sentences previously encoded in night reports are also reported in the morning. It can be added that the best consolidation concerns single contents rather than sequences of contents. Sentences similar in length are organized around the better consolidated contents on both types of reports, but there are also encoded contents that are less consolidated and therefore less easily accessible in memory. From this it may be argued that the relationships of coreference and of thematic progression which link the better consolidated to the less consolidated, and therefore less accessible, contents are more numerous than can be deduced from the analysis of the night report alone.

What can we say about the organization of the contents in the memory? Some answers can be had, first of all, by analyzing the reports obtained after awakening in REM and NREM, and then by a comparison between these reports and those of the following morning. We found that the strategy of retrieval adopted is of the sequential type (see great length of sentences and high frequency of intrusions), which in turn allows it to be supposed that the storage in memory of the contents is also of a serial type. This is confirmed by the fact that following the comparison between night and morning reports, the sentences in which the same contents appear in both reports occupy very similar serial positions: sequential organization is thus stable in time. This is also additional proof that the contents preceding the night awakening were consolidated at long-term. This data can be considered as partial proof of the fact that the transfer from short-term to long-term storage takes place according to the 'first-in, first-out' procedure, relatively independent from the degree of consolidation of the contents.

Though the data we presented does not allow general conclusions to be drawn about the memory processes during sleep (for example, the limit of the capacity of long-term store is still unknown), we feel that interferences between the numerous contents produced during sleep and inappropriate retrieval procedures could explain, in part, difficulties in recall.

There is evidence from clinical and experimental studies on dreams which shows that the dream failure is not necessarily due to the lack of long-term storage of contents during sleep.

Sleep 1980. 5th Eur. Congr. Sleep Res., Amsterdam 1980, pp. 182–185
(Karger, Basel 1981)

The Function of Psychological Dreaming: a Preliminary Analysis[5]

Milton Kramer, Cincinnati, Ohio, USA

The present essay is directed at exploring some of the conceptual issues related to demonstrating a function for psychological dreaming. It will attempt (a) to clarify the concept of dream function, (b) to define the nature

[5] For references see compound list on page 186.

of adaptive dream function, and (c) to specify the criteria necessary to demonstrate a function for psychological dreaming.

The Functions of Dreaming: a Distinction

It is necessary to distinguish two senses in which the concept function can be applied to psychological dreaming. The concept function can be used either in the sense of what does the dream do or how is the dream constructed [3]. If one explores the consequences of dreaming, one is interested what the dream does. What is the effect or result of dreaming? In this sense of function, one is concerned about the dream as an independent variable and waking as the subsequent dependent variable. *Leveton* [43] has suggested that dreams can be viewed as the 'night residue' that helps precipitate the consequent waking behavior.

Function can be used in another sense, to describe how things are made. In the case of psychological dreaming this would relate to the construction and organization of the dream. What contents are selected by what rules and how are they melded together? These are the questions the how function of dreaming addresses. Studies which examine the mode by which pre-sleep elements are selected [52] and enter, directly or by transformation [62] into the dream, which describe the interconnection among the parts of the dream, [27] and those which delineate the sequential development of dreams across the night [32, 38] and across the REM period [36] are all examples of the how function of dreaming.

What the dream achieves, its effects or consequences, is often related to how the dream is constructed and organized. *Freud* [15] clearly related the two. The dream, in *Freud's* conception, was organized as an attempt at a disguised gratification of a disturbing infantile wish in order to protect sleep.

The distinction between the dream as independent and dependent variable has been explored by others [4, 22]. Unfortunately, the distinction has not been consistently applied and evidence for the what of dream function has often been confused with evidence for the how of dream formation.

Theories of Adaptive Dream Function

Those theories of psychological dreaming which specify a consequent function may be considered adaptive in nature. Such theories espouse either an *assimilative* or *accommodative* function for dreaming [50].

Theories of dreaming which view it as *assimilative* are more likely to be able to account for the totality of the dream experience. In these assimilative theories the dreaming process functions automatically outside of conscious awareness, i.e. without recall and secondary reworking. These theories generally have the dream achieve some corrective [34] or reductive goal [14]. In the *accommodative* view of dream function the dreamer is altered as a result of the dream experience [26]. The dream is a special experience which has the potential for a significant and often decisive impact on the dreamer's life. The dream must become conscious, enter awareness, and be 'understood' in order to have its transforming effect.

It is possible, if not probable, that dreams serve both an assimilative and accommodative function. It may be that those dreams that are recalled are related to more significant events of the moment [9, 60], and carry the potential for the transforming experience either in themselves or in conjunction with a waking exploration. Meanwhile, the bulk of dreaming would be continuing, outside of awareness, to achieve the more modest assimilative goals.

Criteria for Demonstrating a Function of Dreams

In order to study the adaptive function of dreaming one must examine the relationship between dreaming and the consequent state of the dreamer. One must be able to manipulate the dream experience and quantitatively observe the consequent change in the dreamer. The function of psychological dreaming can only be demonstrated by treating the dream as the independent variable.

It has been shown, to a greater or lesser degree, that dream content can be directly manipulated [59]. For example, *pre-sleep suggestion* to dream about certain topics or to have specific dreams has been effective. *Hypnosis* with post-hypnotic suggestion has also met with some success. Paradigms which have *conditioned* subjects to respond when certain type dreams are occurring have been accomplished [5]. Stimuli *introduced during REM* sleep have been shown to effect dreaming systematically [6]. And lastly, dream content can probably be altered by *prior REM deprivation* with intensity increases in the rebound condition [21]. The systematic exploration of some if not all of these manipulations might well permit us to gain sufficient control over the dreaming process to manipulate it effectively.

If one can manipulate various aspects of the dream experience, the possibility for demonstrating an effect on subsequent wakefulness then exists. However, one does need to have some idea in what aspect of wakefulness one might expect to see some change as a result of a change in dream content.

The dependent variables, to be examined in wakefulness, which are likely to be effected by changes in dream content, are those of an affective rather than cognitive nature [13]. The waking experiences which are most likely to appear in dreams are those with affective charge. The link of waking life to dreams then is more likely to be in the affective than cognitive realm [37]. The research done to date would suggest that it is the waking affective life of the dreamer, as captured in adjective check lists [34], verbal samples [53] and in aspects of the TAT [37] and Rorschach [42] which is the most promising area to explore.

My laboratory has examined the links between sleep and waking affects in a number of studies. We have shown that how you feel before and after going to sleep is systematically different [55] in and out of the laboratory and in both men and women [45]. Depriving subjects of sleep altered this relationship [54]. We were also able to show that this subjective state difference between night and morning was related both to the intervening physiology of sleep (sleep stages) [33] and to the psychology of sleep (dream content) [34]. This lead us to postulate a differential mood regulatory function for sleep [35].

Conclusion

The function or functions of psychological dreaming remain to be demonstrated. The current essay is an effort to clarify some of the conceptual and methodological problems involved in searching for a function for psychological dreaming. The evidence seems hopeful that a relationship between dreaming and subsequent wakefulness will be elucidated. The dream needs to be systematically explored as an independent variable if its function is to be established.

List of References of Workshop C

1. Atkinson, R.C., Shiffrin, R.M.: Human memory: a proposed system and its control processes; in Spence, Spence, The psychology of learning and motivation. Advances in research and theory, vol. 2, pp. 189–195 (Academic Press, New York 1968).
2. Bloch V., Fishbein W.: Sleep and psychological functions: memory, in Lairy, Salzarulo, The experimental study of human sleep: methodological problems, pp. 157–173 (Elsevier, Amsterdam 1975).
3. Breger, L.: The functions of dreams. J. abnorm. Psychol. Monogr. *72:* 1–18 (1967).
4. Cartwright, R.D.: A primer on sleep and dreaming; MA, Reading (Addison-Wesley, London 1978).
5. Cartwright, R.D.: Personal communication (1979).
6. Castaldo, V.; Holzman, P.: The effects of hearing one's voice on dreaming content: a replication. J. nerv. ment. Dis. *148:* 74–82 (1969).
7. Cipolli, C.; Salzarulo, P.: Effect of a memory retrieval task on recall of verbal material obtained after awakening from sleep. Biol. Psychol. *3:* 321–326 (1975).
8. Cipolli, C.; Salzarulo, P.; Calabrese, A.: Memory processes involved in morning recall of mental REM-sleep experience: a psycholinguistic study. Percept. Mot. Skills (in press).
9. Cohen, D.B.: Toward a theory of dream recall. Psychol. Bull. *81:* 138–154 (1974).
10. Craik, F.I.M.; Lockhart, R.S.: Levels of processing: a framework for memory research. J. verb. Learn. *11:* 671–684 (1972).
11. Dement, W.; Kleitman, N.: The relation of eye movements during sleep to dream activity: an objective method for the study of dreaming. J. exp. Psychol. *53:* 339–346 (1957).
12. Dement, W.C.; Wolpert, E.: The relation of eye movements, body motility and external stimuli to dream content. J. exp. Psychol. *55:* 543–553 (1958).
13. Ekstrand, B.; Barrett, T.; West, J.; Maier, W.: The effect of sleep on human long-term memory; in Drucker-Colin, McGaugh, Neurobiology of sleep and memory, pp. 419–438 (Academic Press, New York 1977).
14. French, T.M.: The integration of behavior, vol. I–III (University of Chicago Press, Chicago 1952, 1953, 1958).
15. Freud, S.: The interpretation of dreams (Basic Books, New York 1955).
16. Foulkes, D.: Dream reports from different stages of sleep. J. abnorm. soc. Psychol. *65:* 14–25 (1962).
17. Foulkes, D.: The psychology of sleep (Scribners, New York 1966).
18. Foulkes, D.: A grammar of dreams (Basic Books, New York 1978).
19. Foulkes, D.: Home and laboratory dreams: four empirical studies and a conceptual reevaluation. Sleep *2:* 233–251 (1979).
20. Foulkes, D.; Pope, R.: Primary visual experience and secondary cognitive elaboration in stage REM: A modest confirmation and an extension. Percept. Mot. Skills *37:* 107–118 (1973).
21. Firth, H.: Sleeping pills and dream content. Br. J. Psychiat. *124:* 547–553 (1974).
22. Fiss, H.: Current dream research: a psychobiological perspective; in Wolman, Handbook of dreams, pp. 20–75 (Van Nostrand Reinhold, New York 1979).
23. Goodenough, D.R.: Dream recall: history and current status of the field; in Arkin et al., The mind in sleep, pp. 113–142 (Erlbaum, Hillsdale 1978).

24 Hobson, J.A.; McCarley, R.W.: The brain as a dream state generator: an activation-synthesis hypothesis of the dream process. Am. J. Psychiat. *134:* 1335–1348 (1977).
25 Jacobson, A.; Kales, A.; Lehmann, D.; Zweizig, J.: Somnambulism: all night electroencephalographic studies. Science *148:* 975–977 (1965).
26 Jung, C.G.: in Hull, Dreams (Princeton University Press, Princeton 1974).
27 Klinger, E.: Structure and functions of fantasy (Wiley-Interscience, New York 1971).
28 Koukkou, M.; Lehmann, D.: EEG and memory storage in sleep experiments with humans. Electroenceph. clin. Neurophysiol. *25:* 455–462 (1968).
29 Koukkou, M.; Lehmann, D.: Psychophysiologie des Träumens und der Neurosentherapie: Das Zustands-Wechsel-Modell, eine Synopsis. Fortschr. Psychiat. Neurol. *48:* 324–350 (1980).
30 Koukkou, M.; Lehmann, D.: Brain functional states: determinants, constraints and implications; in Koukkou, Lehmann, Angst, Functional states of the brain: their determinants, pp. 13–20 (Elsevier, Amsterdam 1980).
31 Koulack, D.; Goodenough, D.R.: Dream recall and dream recall failure: an arousal-retrieval model. Psychol. Bull. *83:* 975–984 (1976).
32 Kramer, M.; McQuarrie, E.; Bonnet, M.: Dream differences as a function of REM period. Proc. Meet. of the Association for the Psychophysiological Study of Sleep, Mexico City 1980.
33 Kramer, M.; Roehrs, T.; Roth, T.: Mood change and the physiology of sleep. Compreh. Psychiat. *17:* 161–165 (1976).
34 Kramer, M.; Roth, T.: The mood-regulating function of sleep. Proc. 1st Eur. Congr. Sleep Res., Basel 1972, pp. 563–571 (Karger, Basel 1972).
35 Kramer, M.; Roth, T.: The dream as selective affective modulator. Acad. For. *22:* 13 (1978).
36 Kramer, M.; Roth, T.; Czaya, J.: Dream development within a REM period. Proc. 2nd Eur. Congr. Sleep Res., Rome 1974, pp. 406–408 (Karger, Basel 1975).
37 Kramer, M.; Roth, T., Palmer, T.: The psychological nature of the 'REM' dream. I. A comparison of the REM dream report and T.A.T. stories. Psychiat. J. Univ. Ottawa *1:* 128–135 (1976).
38 Kramer, M.; Whitman, R.M.; Baldridge, B.J.; Lansky, L.M.: Patterns of dreaming: The interrelationship of the dreams of a night. J. nerv. ment. Dis. *139:* 426–439 (1964).
39 Kuhlo, W.; Lehmann, D.: Das Einschlaferleben und seine neurophysiologischen Korrelate. Arch. Psychiat. NervKrankh. *205:* 687–716 (1964).
40 Lehmann, D.: Fluctuations of functional state: EEG patterns and perceptual and cognitive strategies; in Koukkou, Lehmann, Angst, Functional states of the brain: their determinants, pp. 189–202 (Elsevier, Amsterdam 1980).
41 Lehmann, D.; Koukkou, M.: Computer analysis of EEG wakefulness-sleep patterns during learning of novel and familiar sentences. Electroenceph. clin. Neurophysiol. *37:* 73–84 (1974).
42 Lerner, B.: Rorschach movement and dreams: a validation study using drug induced dream deprivation. J. abnorm. Psychol. *71:* 75–86 (1966).
43 Leveton, A.T.: The night residue. Int. J. Psycho-Analysis *42:* 506–616 (1961).
44 Lopes da Silva, F.H.; Hoeks, A.; Smits, H.; Zetterberg, L.H.: Model of brain rhythmic activity. Kybernetik *15:* 27–37 (1974).

45 Lysaght, R.; Kramer, M.; Roth, T.: Mood differences before and after sleep: a test of its generalizability. Proc. at the Meet. of the Association for the Psychophysiological Study of Sleep, Tokyo 1979.
46 Maulsby, R. L.: An illustration of emotionally evoked theta rhythm in infancy: hedonic hypersynchrony. Electroenceph. clin. Neurophysiol. *31:* 157–165 (1971).
47 Michael, D.; Houchin, J.: Automatic EEG analysis: a segmentation procedure based on the autocorrelation function. Electroenceph. clin. Neurophysiol. *46:* 232–235 (1979).
48 Molinari, S.; Foulkes, D.: Tonic and phasic events during sleep: psychological correlates and implications. Percept. Mot. Skills *29:* 343–368 (1969).
49 Overton, D. A.: Major theories of state-dependent learning; in Ho, Richards, Chute, Drug discrimination and state dependent learning, pp. 283–318 (Academic Press, New York 1978).
50 Piaget, J.: Play dreams and imitation in childhood (Norton, New York 1962).
51 Piaget, J.; Inhelder, B.: Mental imagery in the child (Basic Books, New York 1971).
52 Piccione, P.; Jacobs, G.; Kramer, M.; Roth, T.: The relationship between daily activities, emotions and dream content. Sleep Res. *6:* 133 (1977).
53 Roth, T.; Kramer, M.; Arand, D.: Dreams as a reflection of immediate psychological concern. Sleep Res. *5:* 122 (1976).
54 Roth, T.; Kramer, M.; Lutz, T.: The effects of sleep deprivation on mood. Psychiat. J. Univ. Ottawa *1:* 136–139 (1976).
55 Roth, T.; Kramer, M.; Roehrs, T.: Mood before and after sleep. Psychiat. J. Univ. Ottawa *1:* 128–135 (1976).
56 Salzarulo, P.; Cipolli, C.: Spontaneously recalled verbal material and its linguistic organization in relation to different stages of sleep. Biol. Psychol. *2:* 47–57 (1974).
57 Salzarulo, P.; Cipolli, C.: Verbal memory in relation to sleep; in Levin, Koella, Sleep 1974, pp. 145–149 (Karger, Basel 1975).
58 Salzarulo, P.; Cipolli, C.: Linguistic organization and cognitive implications in REM and NREM sleep-related reports. Percept. Mot. Skills *49:* 767–777 (1979).
59 Tart, C. T.: From spontaneous event to lucidity: a review of attempts to consciously control nocturnal dreaming; in Wolman, Handbook of dreams, pp. 226–268 (Van Nostrand Reinhold, New York 1979).
60 Trinder, J.; Kramer, M.: Dream recall. Am. J. Psychiat. *128:* 76–81 (1971).
61 Weingartner, H.: State-dependent learning; in Ho, Richards, Chute, Drug discrimination and state-dependent learning, pp. 361–380 (Academic Press, New York 1978).
62 Witkin, H. A.; Lewis, H. B.: Presleep experiences and dreams; in Witkin, Lewis, Experimental studies of dreaming, pp. 148–201 (Random House, New York 1967).

Addresses of Authors

Dr. D. Foulkes, Georgia Mental Health Institute, 1256 Briarcliff Road,
   Atlanta, GA 30306 (USA)
Dr. M. Kramer, V.A. Hospital, University of Cincinnati, 3200 Vine Street,
   Cincinnati, OH 45229 (USA)
Dr. D. Lehmann, Department of Neurology, University Hospitals, Rämistrasse 100,
   CH-8091 Zürich (Switzerland)
Dr. P. Salzarulo, INSERM U.3, 47 Boulevard de l'Hôpital, F-75651 Paris (France)
Dr. Inge Strauch, Department of Clinical Psychology, Institute of Psychology,
   University of Zürich, Schmelzbergstrasse 40, CH-8044 Zürich (Switzerland)

D. Chronobiology, Shiftwork and Sleep

Chairman: *O. Benoit*, Paris
Participants: *T. Åkerstedt*, Stockholm; *J. Foret*, Paris; *P. Knauth*, Dortmund; *J. Zulley*, Andechs

---

Sleep 1980. 5th Eur. Congr. Sleep Res., Amsterdam 1980, pp. 190–195
(Karger, Basel 1981)

Age, Sleep and Adjustment to Shiftwork[1]

T. Åkerstedt, L. Torsvall, Stockholm, Sweden

Introduction

From studies of normal night sleep it is apparent that sleep is shortened and more broken, and contains less slow wave sleep (SWS) with increasing age [15, 34, 35]. One might then expect that great irregularity of sleep hours would add to this deterioration. However, though the shiftwork literature often mentions increased difficulties of adjustment with increasing age, very few studies have explicitly focussed on this issue [1, 5, 9, 32]. Moreover, it has seldom been made clear whether the disturbances merely reflect the normal deterioration of night sleep as age increases or whether displaced sleep hours exacerbate the effects. We have touched upon these issues in four studies, particularly emphasizing day sleep difficulties in contrast to the 'normal' aging problems of night sleep.

Study 1

Study 1 consisted of a questionnaire to 300 3-shift working steel workers [4]. Shift change hours were 0445, 1300, 2115 with weekly or 2 to 3-day rotation. The dependent variables were sleep length and a sleep quality index. The latter contained questions on difficulties going to sleep, difficulties remaining asleep, disturbances during sleep, and (not) feeling well rested after awakening. Answers ranging from 'never' to 'often', were scored from 4 to 1, and averaged to form the index score. Homogeneity using the coefficient alpha [12] was equal to 0.79.

---

[1] For references see compound list on page 208.

Analysis of variance showed that age was, by far, the most important predictor of day sleep difficulties. Also, length of experience of shiftwork and diurnal type [33] were rather good predictors. Notably, type of housing did not account for any variance. Figure 1 shows the relation between age and sleep length and sleep quality, respectively. For day sleep (after night shift), both were sharply reduced. For normal night sleep (after afternoon shift), the trend was similar, but less pronounced for sleep length and not significant for sleep quality. For early night sleep before the morning shift the age trend was *reversed*, i.e. the effects of age on sleep depended on the time of day to which sleep was displaced.

Since age is closely correlated with length of exposure to shiftwork, we tried to establish the importance of experience by controlling for age. This was done by computing the correlations between length of exposure and the dependent variables within 6 age groups. Below the age of 45 most correlations were slightly positive, i.e. there was a weak trend towards fewer problems with increasing experience. At 45, however, correlations turned significantly negative, i.e. problems increased with increasing exposure. The results were similar for sleep length, except for the significant negative correlation in the 23- to 29-year interval. Similar results have recently been presented by *Foret et al.* [18], and it appears that, in any case, adjustment to shiftwork does not improve with practice but rather the reverse in higher age groups.

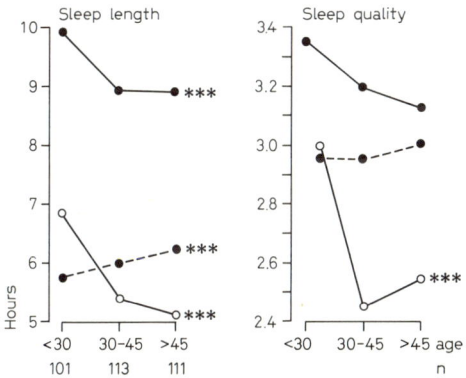

Fig. 1. Sleep length and sleep quality for three age groups of workers in rotating shiftwork – in connection with day sleep (after night work ○—), normal night sleep (after afternoon work ●—), and early night sleep (before morning work ●---). *=p<0.05; ***=p<0.001. Higher values = better quality. n for <30=101; n for 30–45 = 113; n for >45=111.

Study 2

In the second study we wanted to identify the particular types of disturbance connected with different work hours. To this end we presented a questionnaire to 1000 train drivers who were asked to describe sleep length and sleep quality in connection with work in the morning (starting 0300–0500 hours), during the day (0800–1000 hours), and during the night (2200–2400 hours). The questions were the same as in study 1. Furthermore, we also asked more global questions about general difficulties falling asleep and remaining asleep, and also about medical treatment for insomnia and the use of hypnotics.

The answers to the global questions (fig. 2) show that the reports of premature awakenings, medical treatment, and use of hypnotics, but not difficulties falling asleep, increased with age.

With respect to reactions to different work hours (fig. 2) all four variables showed a significant interaction with increasing age: (1) for night sleep there was a moderate decrease of sleep length and ability to remain asleep, but no change of recuperation or ease of falling asleep; (2) for day sleep (after night work) all four variables indicated considerably worse sleep as age increased, out of proportion with the moderate deterioration of night sleep; (3) for early night sleep (before morning work) only premature awakenings showed an increase.

Thus, it appears that the negative effects of day sleep grow progressively worse with increasing age, while those of night sleep are far less affected – early night sleep may even show a reversed trend. Thus, as in study 1, it seems that age and the temporal position of sleep interact.

Study 3

In order to further elucidate the relation between age and day sleep difficulties, we carried out EEG recordings of the sleep of 16 train drivers from study 2, 8 between 50 and 60 years of age, and 8 between 20 and 40 (reported in detail by Torsvall elsewhere in these proceedings). Recordings were made in the subjects' homes with unobtrusive Medilog equipment.

With respect to the composition of sleep stages, normal night sleep did not differ between the age groups except for the expected low amounts of stage 4 sleep in the older group. This difference was retained in day sleep, but in addition, the older group showed more time awake, and more time in stage 1. The older groups also showed more frequent awakenings and more

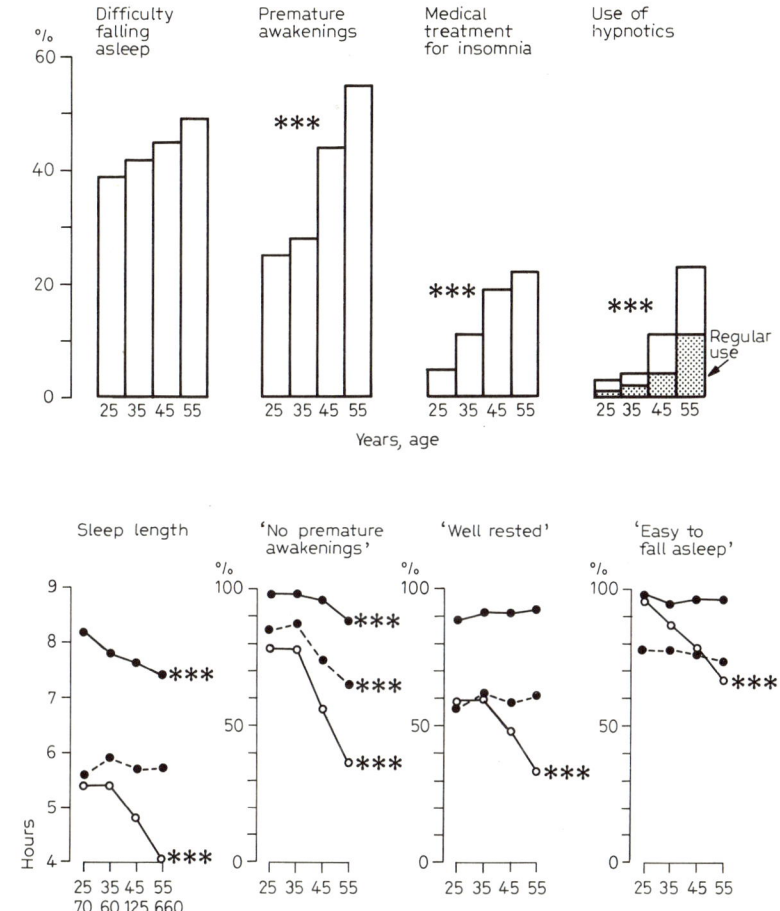

Fig. 2. Results from a questionnaire to 1000 train drivers, divided into four age groups: 21–30 years ('25', n=70); 31–40 years ('35', n=60); 41–50 years ('45', n=125), and 51–60 years ('55', n=660). *a* Percent of subjects: who 'often' or 'rather often' experience difficulties falling asleep, awaken prematurely, who have been under medical treatment for sleep difficulties, and who use hypnotics 'sometimes' or regularly (at least once a week = shaded). *b* Sleep length; percent of subjects who 'never' or 'seldom' experience premature awakenings, percent of subjects who always or almost always feel well rested after sleep, percent of subjects who never or almost never experience difficulties falling asleep. Separate values for night sleep (●—) after normal work hours, early night sleep (●---) before morning work, and day sleep (○—) after night work (see fig. 1 for levels of significance).

frequent stage changes during day sleep but did not differ in total sleep length or sleep latency. Apparently then, the day sleep of older subjects is more restless than that of younger ones. The present data did not show a longer day sleep for the younger group, which has been found in the questionnaire studies. Possibly, the reason may be that the upper age limit for the younger group was too high.

Study 4

The previous data were all collected from experienced shiftworkers who had been through the selection process. To obtain information about the reactions of the normal day-working subject, we reanalyzed the diary data from a 'natural experiment' [2] in which day workers were temporarily exposed to night work (2230–0730 hours). We obtained two groups of 7 subjects each, the point of separation being 29 years.

The sleep length data showed that for the younger group the day sleep after night work did not differ from sleep during days off and was actually *longer* than normal night sleep in connection with day work (fig. 3). The only exception was on Friday when many skipped the day sleep and turned to a day-oriented life. The sleep length for the older group was very close to that of the younger group on weekends and day work. In connection with night work, however, sleep length dropped 2–3 h below that of the younger groups. Thus, as well as in inexperienced night workers, age strongly influenced the effects of displacing sleep to day time.

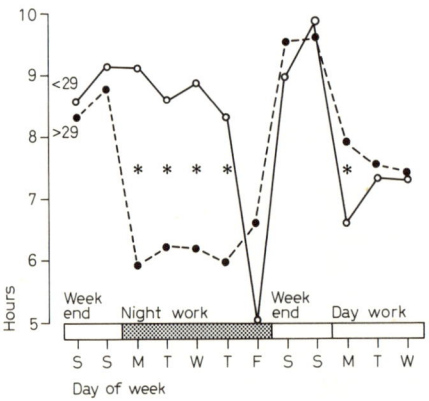

Fig. 3. Diary records of sleep length for 7 younger (<29 years) and 7 older (>29 years) day workers who temporarily were assigned to night work (see fig. 1 for levels of significance).

Conclusion

In conclusion then, it appears that with increasing age day sleep becomes progressively more difficult – in a way that is not merely an extrapolation of the moderate deterioration of night sleep. As a matter of fact, on schedules which require a phase advance of sleep, higher age may be of an advantage! The reason for the interaction between age and displacement of sleep is not clear, but one might speculate that circadian rhythmicity is involved, e.g. the tendency to become a behavioral 'morning type' increases with age [33]. This could be due to a biological phase advance or to a weakening of the sleep mechanism allowing weaker arousal levels to terminate sleep earlier in the arousal cycle. Another possibility is the increased tendency to desynchronize which *Wever* [43] has found to exist in older subjects.

Incidentally, one might wonder whether age may be involved in the discrepancies between day sleep in shiftwork studies on the one hand [14, 16, 28] and laboratory phase-shift experiments on the other [7, 39, 42]. The latter generally show very minor effects on total sleep time. The laboratory studies have been carried out with students or similar age groups, while shiftworkers have been considerably older. In any case, both for methodological and applied reasons the issue of age and phase-shifted sleep needs further research.

Sleep 1980. 5th Eur. Congr. Sleep Res., Amsterdam 1980, pp. 195–197
(Karger, Basel 1981)

Individual Factors, Sleep Characteristics and Circadian Evolution of Body Temperature[2]

J. Foret, O. Benoit, Paris, France

The ability of an individual to work on shift is practically impossible to predict. Who will be able to adapt to shiftwork and how long he will be able to sustain it remain unsolved questions.

[2] For references see compound list on page 208.

Two critical levels can be differentiated: the global capability to adapt to shiftwork and the ability to sleep well. These two levels are closely related, and the beginning of an inability to continue to work on shift is often signaled by sleep difficulties.

It is now well established that whatever the circumstance shiftwork results in statistical differences of sleep durations as a function of the type of shift. It also causes a new composition of diurnal sleep that is different from that of nocturnal sleep. The duration as well as the composition are affected by both inter- and intraindividual factors.

Emphasis has been placed for the past few years on the large interindividual variance found with regards to wake/sleep patterns, and more generally, with regards to the temporal organization of the set of circadian variables such as length of spontaneous period, phase relationship, stability of temporal system, etc. To date, studies have documented spontaneous sleep duration [19, 36], spontaneous sleep schedules, regularity and irregularity of schedules [30], 'morningness' [20, 22, 29, 40], and personality traits [8, 11, 27]. All these characteristics are likely to play a role in the degree of tolerance to shiftwork, and they are probably interdependent.

We have hypothesized that the relative dependance of these characteristics is related to the extent to which each of them is coupled with different 'internal oscillators'. *Aschoff and Wever* [6] showed that to explain the whole system of circadian variations in physiology (and also in psychophysiology), at least two classes of 'internal oscillators' must be assumed: one that controls the sleep/activity alternation and another that regulates the circadian rhythm of body temperature and other physiological rhythms.

We have studied some individual parameters such as spontaneous sleep duration and arising and bed times which are clearly related to the sleep/ wake oscillators and how they vary with different circadian organization (temperature and vigilance peak times, morningness-eveningness scores). Spontaneous sleep duration is correlated only with bedtime not only during vacations but even more so during working periods: the greater the need for sleep, the earlier one goes to bed. There seems to be little or no correlation with the circadian temperature curves.

Significant correlations have been found between arising and bed times and temperature peak and vigilance times (table I). Thus, temperature peak time is closely related to arising time and less so to vigilance peak time ($p < 0.10$).

A number of studies have utilized the degree of 'morningness' as an index of both activity schedules and subjective feelings of activation [20]. This

Table I. Correlations between arising and bed times and temperature peak and vigilance times

	Bedtime on vacation	Arising time on vacation	Vigilance peak time	Morningness-eveningness score
Temperature peak time	no	***	~	no
Vigilance peak time	*	***		no
Morningness-eveningness score	***	***	***	
Spontaneous sleep duration	**	no	no	no

Spearman rank correlation test results: $*p<0.05$; $**p<0.02$; $***p<0.01$; $\sim$ = trend. 50 medical students studied, age range 19–23 years.

index is strongly associated with bed and arising times during vacation. In the *Horne and Östberg* [20] study it was also correlated with temperature peak time, while in our results it was not. 'Larks' (morning types) have a less variable sleep and awakening time than 'owls' [40]. We found this to be also true for bedtime, arising time, and sleep duration. Thus, 'morningness' index seems to reflect the wake/sleep oscillator rather than the temperature oscillator of an individual.

In addition, the circadian evolution of the vigilance peak time seems more closely linked to the wake/sleep clock than to the temperature clock. This conclusion is supported by *Wever's* [44] results: in his free-running experiments, subjects, because they are at different levels of desynchronization, may awaken and fall asleep at times of the day significantly different from their usual arising and bed times.

These data strongly support the hypothesis that there are at least two classes of internal oscillators. The phase relationships between them vary by individual. It is also possible that the strength of the coupling forces between them vary as well as their degree of synchronization to the 24-h system. For example, the sleep/wake rhythm of morning type subjects seems to be more adequately attuned to the standard schedules of living than that of the evening people.

Day Sleep during the Morning and during the Afternoon between Experimental Night Shifts[3, 4]

P. Knauth, E. Kiesswetter, S. Bruder, H. P. Romberg, J. Rutenfranz,
Dortmund, FRG

Introduction

Many shiftworkers complain of sleeping difficulties in connection with day sleep after night shifts. These difficulties refer both to the duration of sleep and to the quality of sleep. The results of time budget studies have indicated that the average duration of day sleep between 2 night shifts is reduced to about 6 h [25]. The quality of day sleep seems to be dependent on many factors, e.g. noise [26], age [3], prior wakefulness and time of day. Not all shiftworkers have the same 'strategy of sleep': though most shiftworkers prefer a morning sleep after the end of the night shift, there are others who sleep during the afternoon or even divide their day sleep into a morning and an afternoon part with a lunch with the family in between.

According to several publications it could be expected that the quality of sleep would differ between morning and afternoon sleep. It was assumed that the morning sleep would have more REM sleep than the afternoon sleep [21, 35, 37]. Furthermore, the afternoon sleep after the first night shift with a longer prior wakefulness should have more stage 4 sleep than the morning sleep because some authors showed the strong determining effect of length of prior wakefulness on sleep [21, 31, 37].

The effect of the reversal of the sleep-waking cycle in man has been studied, for example, by *Weitzman et al.* [42] and by *Webb et al.* [39]. Comparing night and day sleep in both studies, an increase in time awake after sleep onset and a shift of REM sleep towards the early part of the sleep period was found.

---

[3] This work was supported by a grant from the 'Deutsche Forschungsgemeinschaft' Bonn-Bad Godesberg.
[4] For references see compound list on page 208.

Methods

Three experimental groups (A, B, C) had to work 3 consecutive day shifts and 3 consecutive night shifts. The 6 subjects of group A slept from 0800 to 1400 hours between the night shifts. Furthermore, 6 subjects of group B slept during the afternoon from 1500 to 2100 hours whereas the sleep of the 6 subjects of group C was divided into two parts: from 0800 to 1100 hours and from 1500 to 1800 hours. The subjects were male students, aged 20–29 years.

During the whole experiment the subjects had to work, sleep and spend their leisure time in the Institute. The work consisted of fitting and soldering electronic parts to printed circuit boards. In regular intervals psychological tests were performed. Heart frequency and rectal temperature were recorded continuously. The subjects slept in soundproof chambers and their sleep was disturbed by random traffic noise [24]. During sleep 3 EEGs, 2 EOGs, 1 EMG and bed movements were recorded. 30-s epochs were analyzed by two scorers independently of each other according to standard procedures.

To test differences of the scored data between experimental groups, between days and trends within the sleep periods analyses of variance including repeated measurements were used.

Results

Differences between Experimental Groups. Group B had on an average less stage REM sleep during the first and second day sleep and more stage 4 sleep during the first day sleep than group A. However, these and all other differences between the experimental groups A, B, and C referrring to the average amount of each sleep stage per 6-h sleep period were not significant. The only significant difference between groups concerned the scoring of 'movement time' (MT): group C had less MT (3 min) than the other two groups (A: 6 min; B: 8 min). In the mean stage onset latencies, again very few differences between groups were found; e.g. stage 3 sleep appeared later in group C than in the groups A and B.

Differences between Days. With help of the analysis of variance the last night sleep and the first and second day sleep were compared with each other. As every day sleep was limited to 6 h, only the first 6 h of night sleep were included in the analysis of variance. In figure 4 the average amount of REM sleep per hour and per sleep period is shown for the groups A, B, and C. The trend over the days is significant ($p<0.05$). Regarding the slow wave sleep (fig. 5), also a significant trend was found ($p<0.05$). Furthermore, there was a clear descending trend of stage 2 sleep over the days ($p<0.01$). All sleep onset latencies, i.e. awake–stage 1, stage 1–stage 2, stage 1–stage 3, stage 1–stage 4 and stage 1–REM, were shorter in connection with day sleep than with night sleep ($p<0.05$).

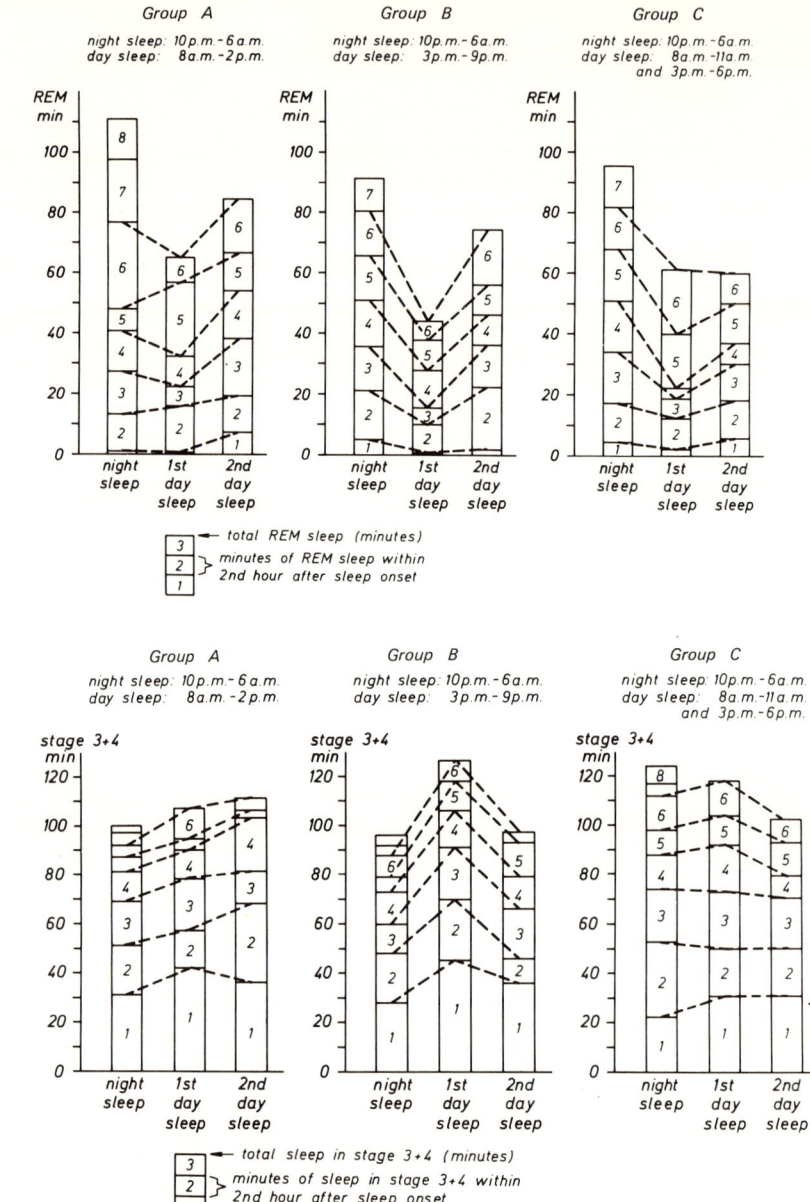

Fig. 4. Amount of REM sleep per hour and per sleep period in three experimental groups during 1 night and 2 days.

Fig. 5. Amount of slow wave sleep (stage 3 and 4) per hour and per sleep period in three experimental groups during 1 night and 2 days.

Distribution in the Amount of Slow Wave and REM Sleep within Sleep Periods. Slow wave sleep (stages 3 and 4) predominated in the first half of the sleep period both during night sleep and during day sleep. The distribution of REM sleep within the sleep period changed from night sleep to day sleep ($p<0.01$). During night sleep the minutes of REM on an average increased from the onset to the end of sleep. In contrast to this finding, during day sleep more REM sleep was found earlier, i.e. especially during the second and the third hour of sleep.

In all 36 recordings of day sleep only 2 sleep onset REM periods (SOREMPs) appeared.

Discussion

In contrast to the findings of some authors (see introduction) that stage 4 sleep increases as a function of prior wakefulness we did not find differences between the first morning and the first afternoon sleep with prior wakefulness of respectively 26 and 33 h. However, according to *Webb and Agnew* [38] there is evidence that any increase in stage 4 during the first 3 h of sleep is quite limited when time awake is increased beyond 21 h. As most of stage 4 sleep was observed within the first 3 h our results seem to confirm these findings.

The result that the total amount of REM sleep was not significantly different between morning and afternoon is not in agreement with the data of other authors cited in the introduction. *Hume and Mills* [21] defined a circadian rhythm of REM by cosinor procedure for experiments with a sleep period of 7 h. This best fitting cosinor curve had a maximum at 1100 hours and a minimum at 2300 hours. The time of mid-sleep of our group A was 1100 hours and of our group B 1800 hours. The best fitting cosine of *Hume and Mills* [21] showed much scatter due to the different scores of the subjects. We observed similar large interindividual variations. Therefore, our differences of REM sleep between morning and afternoon conditions did not achieve a statistical significance level.

Comparing night and day sleep *Weitzman et al.* [42] as well as *Webb et al.* [39] noted that there was an increase in the amount of awakenings during day sleep. However, these findings are not in agreement with our results. Furthermore, *Weitzman et al.* [42] made mention of a significant decrease of time of sleep onset to the first REM period during day sleep compared with night sleep. Beyond that we found this decrease in all other sleep onset latencies. Despite the long prior wakefulness before the first day sleep, *Foret and Benoit* [17] noted a decreased amount of REM sleep which is confirmed by our data. The slow wave sleep rebound at the first day sleep could be expected because of the long prior wakefulness. However, the data of *Foret and Benoit* [17] are not in agreement with this expectation and our results.

The striking change from night sleep to day sleep in the distribution of the amount of REM within the sleep period has already been observed several times [17, 23, 39, 42].

Summing up, our data show differences between day and night sleep but no essential differences of quality between morning and afternoon sleep during experimental shiftwork studies. Therefore, it is not possible to recommend a special strategy concerning the position of day sleep between night shifts.

Sleep 1980. 5th Eur. Congr. Sleep Res., Amsterdam 1980, pp. 202–206
(Karger, Basel 1981)

Comparison of Long and Short Sleep Durations in Free-Running Sleep-Wake Cycles[5]

J. Zulley, Andechs, FRG

Introduction

A consistent result in shift-work studies is the reduction in sleep duration during day sleep. It has been shown that this reduction is mainly due to the shift in the placement of sleep relative to the time of day [10]. Recent studies have shown that in the absence of external time cues, the duration of sleep, as well as the sleep stage structure, depends on the phase position of sleep in the circadian temperature cycle [13, 47, 48].

Another approach to this problem involves the study of sleep duration as a habitual factor in subjects who are short and long sleepers. By comparing the sleep stage structure of these two groups, long sleepers have, in ab-

[5] For references see compound list on page 208.

solute values, more stage REM, stage 1, stage 2 and the same amount of stage 3+4 as short sleepers. In relative values stage REM, stage 1, and stage 2 are the same or slightly higher and stage 3+4 is lower in long sleepers compared to short sleepers [36].

The question is whether these differences between long and short sleepers can also be seen in spontaneously long or short sleep episodes of a subject, who has normally a medium sleep duration.

Method

In this experiment 4 subjects (21–28 years old) spent 21–30 days individually in an underground room. Throughout the experimental period, the subject had no information about local time and a self-selected regime of lights on and off, bedrest-activity and meal-times. Body temperature was measured continuously by a rectal probe. Thus, the experimental conditions were the same as in the standard experiments for circadian research by *Aschoff and Wever* [6]. Additionally all bedrest episodes were polygraphically recorded. For the comparison of short and long sleep episodes of a subject, the 6 shortest and 6 longest sleep episodes of each subject were compared separately (fig. 6). These criteria were taken because of the different distributions of the sleep duration between the subjects. From the course of body temperature, the maxima and minima were scored by inspection, and for every sleep episode the distances from these values computed.

Results and Discussion

Sleep Duration. In isolation, all the subjects developed free-running circadian rhythms of body temperature and sleep-wake cycle with periods longer than 24 h [for further details see 46]. The distribution of the shortest and longest sleep episodes of each subject are shown in figure 6, where the sleep episodes are divided into three groups (indicated by different patterns of shading), according to the different phase position of the temperature minimum during sleep. On the right side of the figure, the positions of sleep in relation to the minima of body temperature are shown for two groups. Short sleep episodes always have an early temperature minimum (1.0 ± 0.93 h after sleep onset), while long sleep episodes have a late minimum (8.6 ± 3.1 h after sleep onset).

This difference in the course of body temperature is in accordance with the finding that the phase position of sleep onset in the circadian temperature cycle is related to the duration of sleep [48]. In those studies it was shown that the beginning of short sleep episodes is at the time of the cir-

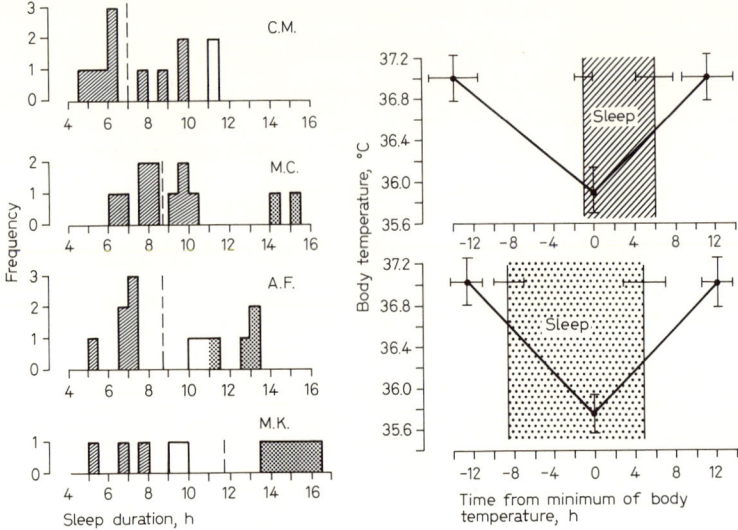

Fig. 6. Frequency distributions of the duration of the 6 shortest and 6 longest sleep episodes of the 4 subjects (left side). Additionally the phase position of the circadian temperature minimum is shown by different patterns of shading. Lined histograms represent sleep episodes where the temperature minimum occurred in the first half of sleep and hatched where it occurred in the second half. The non-shaded histograms represent sleep episodes where no definable minimum was found. On the right side for the groups with an early minimum and with a late minimum, the means and standard-deviation are given.

cadian minimum, while longer sleep episodes begin relatively earlier, with the minimum in the later part of sleep.

Sleep Stage Structure. The comparison of the absolute amounts of the different sleep stage structure showed more stage REM, stage 1, and stage 2 in long sleep episodes than in short sleep episodes while stage 3+4 was the same in most of the subjects (fig. 7). In relative amounts, stage REM and stage 1 were the same in short and long sleep episodes, while for stage 2 in long sleep episodes the amount was slightly higher. The relative amount of stage 3+4 was lower in long sleep episodes, compared with short sleep episodes in all the subjects. These results are in agreement with findings in habitually short and long sleepers [36]. Thus, the differences in the sleep stage structure, which were reported for the inter-individual comparison between long and short sleepers, are similar to those which are found in the

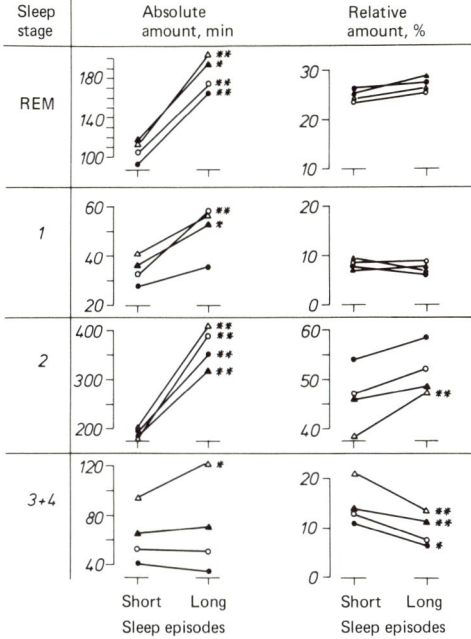

Fig. 7. Differences in the mean amount of various sleep stages between short and long sleep episodes for each subject. Asterisks denote statistical significance (t-test): *p<0.05; **p<0.01. Subjects: ● = CM; ▲ = MC; ○ = AF; △ = MK.

present study intra-individually between long and short sleep episodes of the same subject. A further analysis was done with regard to the relationship between the amount of REM sleep and the circadian temperature cycle cited above. Because of different phase positions in long and short sleep episodes in the temperature cycle, different distributions of REM sleep were expected. For this analysis, all the sleep episodes of the subjects were divided into those with an early temperature minimum and those with a late minimum as in figure 6. By comparing these two groups, sleep episodes with an early minimum have significantly more REM sleep in the first 3 h of sleep (56.67 ± 14.90 min) than the sleep episodes with a late minimum (42.75 ± 7.82 min).

This result agrees with the findings in studies of the relationship between REM sleep and the circadian temperature cycle [13, 41, 47]. By comparing sleep episodes with different durations, it has to be considered that these sleep episodes can be located at different phase positions in the temperature cycle, which results in a different distribution of REM sleep in the sleep episode.

In summary, the differences in the sleep stage structure of long and short sleep episodes in the same subject are comparable with those found in habitually long and short sleepers. In accordance with the finding that sleep duration is controlled by the temperature cycle, the long and short sleep episodes of an individual subject begin at different phases in the temperature cycle. From this it can be hypothesized that habitually long and short sleepers have different phase positions of sleep onset in the temperature cycle.

Sleep 1980. 5th Eur. Congr. Sleep Res., Amsterdam 1980, pp. 206–211
(Karger, Basel 1981)

Concluding Remarks[6]

O. Benoit, Paris, France

These four presentations underline the complexity of the biological aspects of shiftwork in relation to the adjustment of the sleep-wake system. They illustrate the role played by factors such as age, individual characteristics of the sleep-wake system, phase relationship between sleep period and body temperature, or time of sleep.

In most available studies, individual characteristics and body temperature have not been analyzed and in some cases not even mentioned. One may suppose that the subjects were heterogeneous at least in so far as these characteristics were concerned which might explain the great variability of the results as well as the lack of significant data concerning sleep after a schedule inversion.

Age appears to be a major factor, with the 40s as a critical period for

---

[6] For references see compound list on page 208.

sleep disturbances in shiftworkers. Day sleep of poor quality is one of the earliest critical symptoms of disadjustment, and middle age people seem to encounter more difficulties in adjusting their sleep after a phase delay than younger workers. These results, however, confirmed by our own [18], show that there is no long-term adaptation to shiftwork. However, one wonders if factors such as sleep length, regularity/irregularity of sleep schedules, or degree of morningness play a role in the ability or inability of an individual to continue on shiftwork as he grows older. Moreover, very little if anything is known about the affects of aging on the circadian rhythms. For example, does their relative phase change with age? Does either the coupling between the main internal oscillators or their sensitivity to zeitgebers change?

Does the time of the day when sleep occurs influence the recovery after a night shift? The results, when compared to those for non-shiftworkers, are surprising. Day sleep has distinctly different characteristics from night sleep mainly in a reduction of total sleep time and a displacement of the distribution of REM sleep. However, when sleep takes place in the afternoon rather than the morning, there is no clear difference. Two hypotheses for this discrepancy can be put forward: one that the lengthening of prior wakefulness overcomes the influence of the circadian rhythm normally acting on REM sleep. The other is that the interindividual variability obscures, on the average, the individual circadian changes induced by the phase shift and, for example, the change of body temperature.

As a matter of fact, the results obtained in people experiencing free-running conditions clearly illustrate the role of the phase position of sleep in relation to the temperature cycle for sleep length and REM sleep distribution. The change induced in the temperature rhythm by a phase delay of 8 h varies considerably from one individual to another. Thus, in shiftworkers the variable level of morning temperature before day sleep may interfere with its quality and thus recovery.

In order to understand better the influence of shiftwork on the sleep-wake rhythm it seems necessary to have further studies to consider, among other things, populations of workers who are relatively homogeneous with regards to parameters such as sleep, temperature, etc., as well as to obtain longitudinal studies of shiftworkers over a relatively long period of time.

One last point which has not been discussed here, but remains important none the less, is whether in the long-term it is better for the human organism to remain synchronized to the normal 24-h way of life (i.e. not to adjust to night shift) or to adjust quickly when exposed to a rapid rotation system.

List of References of Workshop D

1. Aaronsen, A.: Shift work and health (Universitetsforlaget, Oslo 1964).
2. Åkerstedt, T.: Inversion of the sleep wakefulness pattern: effects on circadian variations in psychophysiological activation. Ergonomics 20: 459–474 (1977).
3. Åkerstedt, T.; Torsvall, L.: Age, sleep and adjustment to shift work (this volume).
4. Åkerstedt, T.; Torsvall, L.: Shift-dependent well being and individual differences. Ergonomics (in press, 1981).
5. Andersen, J.E.: Three shift work. A socio-medical survey, vol. I+II (Teknisk Forlag, Copenhagen 1970).
6. Aschoff, J.; Wever, R.: Human circadian rhythms: a multioscillatory system. Fed. Proc. 35: 2326–2332 (1976).
7. Berger, R.J.; Walker, J.W.; Scott, T.D.; Magnusson, L.J.; Pollack, S.L.: Diurnal and nocturnal sleep stage patterns following sleep deprivation. Psychon. Sci. 23: 273–275 (1971).
8. Blake, M.J.F.; Corcoran, D.W.J.: Introversion-extraversion and circadian rhythms; in Colquhoun, Aspects of human efficiency, pp. 261–272 (English Universities Press, London 1972).
9. Bruusgaard, A.: Shift work as an occupation health problem. Stud. Laboris salutis 4: 6–8 (1969).
10. Bryden, G.; Holdstock, T.L.: Effects of night duty on sleep patterns of nurses. Psychophysiology 10: 36–42 (1973).
11. Colquhoun, W.P.; Folkard, S.: Personality differences in body temperature rhythm and their relation to its adjustment to night work. Ergonomics 21: 811–817 (1978).
12. Cronbach, L.J.: Coefficient alpha and the internal structure of tests. Psychometrika 16: 297–334 (1951).
13. Czeisler, L.A.; Zimmermann, J.C.; Ronda, J.M.; Moore-Ede, M.C.; Weitzman, E.D.: Timing of REM sleep is coupled to the circadian rhythm of body temperature in man. Sleep 2: 329–346 (1980).
14. Ehrenstein, W.; Schaffler, K.; Muller-Limmroth, W.: Die Wirkung von Oxazepan auf den gestörten Tagschlaf nach Nachtschichtarbeit. Arzneimittel-Forsch. 22: 421–427 (1972).
15. Feinberg, I.; Koresko, R.L.; Heller, N.: EEG sleep patterns as a function of normal and pathological aging in man. J. psychiat. Res. 5: 107–144 (1967).
16. Foret, J.; Benoit, O.: Structure du sommeil chez des travailleurs à horaires alternants. Electroenceph. clin. Neurophysiol. 37: 337–344 (1974).
17. Foret, J.; Benoit, O.: Shiftwork: the level of adjustment to schedule reversal assessed by a sleep study. Waking Sleeping 2: 107–112 (1978).
18. Foret, J.; Bensimon, G.; Benoit, O.; Vieux, N.: Quality of sleep as a function of age and shift work; in Reinberg, Studies on night and shift work: a multidisciplinary approach (Pergamon Press, New York 1981).
19. Hartmann, E.; Baekeland, F.; Zwilling, G.; Hoy, P.: Sleep need: how much sleep and what kind? Am. J. Psychiat. 127: 1001–1008 (1971).
20. Horne, J.A.; Östberg, O.: A self assessment questionnaire to determine morningness-eveningness in human circadian rhythm. Int. J. Chronobiol. 4: 97–110 (1976).

21 Hume, K.I.; Mills, J.N.: Rhythms of REM and slow-wave sleep in subjects living on abnormal time schedules. Waking Sleeping *1:* 291–296 (1977).
22 Kleitman, N.: Sleep and wakefulness (University of Chicago Press, Chicago 1963).
23 Knauth, P.; Rutenfranz, J.: Untersuchungen zum Problem des Schlafverhaltens bei experimenteller Schichtarbeit. Int. Arch. Arbeitsmed. *30:* 1–22 (1972).
24 Knauth, P.; Rutenfranz, J.: The effect of noise on the sleep of nightworkers; in Colquhoun, Folkard, Knauth, Rutenfranz, Experimental studies of shift work, pp. 57–65 (Westdeutscher Verlag, Opladen 1975).
25 Knauth, P.; Landau, K.; Dröge, C.; Schwittek, M.; Widynski, M.; Rutenfranz, J.: Duration of sleep depending on the type of shift work. Int. Archs occup. envir. Hlth *46:* 167–177 (1980).
26 Knauth, P.; Rutenfranz, J.; Schulz, H.; Bruder, S.; Romberg, H.P.; Decoster, F.; Kiesswetter, E.: Experimental shift work studies of permanent night, and rapidly rotating, shift systems. II. Behaviour of various characteristics of sleep. Int. Archs occup. envir. Hlth *46:* 111–125 (1980).
27 Lund, R.: Personality factors and desynchronisation of circadian rhythm. Psychosom. Med. *36:* 224–228 (1974).
28 Matsumoto, K.: Sleep patterns in hospital nurses due to shift work: an EEG study. Waking Sleeping *3:* 169–173 (1978).
29 Östberg, O.: Interindividual differences in circadian fatigue patterns of shiftworkers. Br. J. industr. Med. *30:* 341–351 (1973).
30 Taub, J.M.: Behavioral and psychophysiological correlates of irregularity in chronic sleep routines. Biol. Psychol. *7:* 37–53 (1978).
31 Taub, J.M.; Berger, R.J.: Sleep stage patterns associated with acute shifts in the sleep-wakefulness cycle. Electroenceph. clin. Neurophysiol. *35:* 613–619 (1973).
32 Thiis-Evensen, E.: Shift work and health. Stud. Laboris salutis *4:* 81–83 (1969).
33 Torsvall, L.; Åkerstedt, T.: A diurnal type scale; construction, consistency, and validation in shift work. Scand. J. Work, Environ. and Health (in press, 1981).
34 Tune, G.S.: The influence of age and temperament on the adult human sleep-wakefulness pattern. Br. J. Psychol. *60:* 431–449 (1969).
35 Webb, W.B.; Agnew, H.W.: Sleep cycling within twenty-four hour periods. J. exp. Psychol. *74:* 158–160 (1967).
36 Webb, W.B.; Agnew, H.W.: Sleep stage characteristics of long and short sleepers. Science *168:* 146–147 (1970).
37 Webb, W.B.; Agnew, H.W.: Variables associated with split period sleep regimes. Aerospace Med. *42:* 847–850 (1971).
38 Webb, W.B.; Agnew, H.W.: Stage 4 sleep: influence of time course variables. Science *174:* 1354–1356 (1971).
39 Webb, W.B.; Agnew, H.W.; Williams, R.L.: Effect on sleep of a sleep period time displacement. Aerospace Med. *42:* 152–155 (1971).
40 Webb, W.B.; Bonnet, M.: The sleep of 'morning' and 'evening' types. Biol. Psychol. *7:* 29–35 (1978).
41 Weitzman, E.D.; Czeisler, C.A.; Zimmerman, J.C.; Ronda, J.M.: Timing of REM and stages 3+4 sleep during temporal isolation in man. Sleep *2:* 391–407 (1980).
42 Weitzman, E.D.; Kripke, D.F.; Goldmacher, D.; McGregor, P.; Nogeire, C.: Acute reversal of the sleep-waking cycle in man. Archs Neurol., Chicago *22:* 483–489 (1970).

43 Wever, R.: Bedeutung der circadianen Periodik für das Alter. Naturw. Rdsch., Stuttg. *27:* 475–479 (1974).
44 Wever, R.: The circadian system of man (Springer, New York 1979).
45 Williams, R.L.; Karacan, I.; Hursch, C.J.: EEG of human sleep (Wiley, New York 1974).
46 Zulley, J.: Der Einfluss von Zeitgebern auf den Schlaf des Menschen (Fischer, Frankfurt 1979).
47 Zulley, J.: Distribution of REM sleep in entrained 24 hour and free running sleep-wake cycles. Sleep *2:* 377–389 (1980).
48 Zulley, J.: Timing of sleep within the circadian temperature cycle. Sleep Res. (in print).

Addresses of Authors

Dr. T. Åkerstedt, Laboratory for Clinical Stress Research, Karolinska Institute, Stockholm (Sweden)

Dr. O. Benoit, U.3 INSERM, 47 Boulevard de l'Hôpital, F-75634 Paris Cédex 13 (France)

Dr. J. Foret, Laboratoire de Physiologie du Travail du CNRS, 15 Rue de l'Ecole de Médecine, F-75270 Paris Cédex 06 (France)

Dr. P. Knauth, Institut für Arbeitsphysiologie an der Universität Dortmund, Ardeystrasse 67, D-4600 Dortmund (FRG)

Dr. J. Zulley, Max-Planck-Institut für Verhaltensphysiologie, D-8131 Andechs (FRG)

E. Traffic Noise, Sleep and Performance

Chairmen: *A. Kumar,* Amsterdam; *A. Muzet,* Strasbourg
Participants: *B. Griefahn,* Düsseldorf; *W. F. Hofman,* Amsterdam; *A. A. Jurriëns,* Delft; *M. Vallet,* Bron; *R. T. Wilkinson,* Cambridge; *H. L. Williams,* Oklahoma City, Okla.

Introduction

A. Kumar, Amsterdam, The Netherlands

This workshop reviewed data about the effects of noise during sleep on sleep and performance. Data from experiments performed in the laboratory and in the homes are presented. A special feature of the workshop is the presentation of an interim report of an international joint project.[1] *Muzet et al.* summarized their results on cardiovascular responses to simulated noises in the laboratory. *Griefahn* presented an introduction to the joint project. *Jurriëns* presented results on sleep stages, *Vallet et al.* presented visual analysis of arousals and transients in the EEG. *Wilkinson* studied the effects of noise on the EEG in detail by examining the changes in the EEG frequencies by automatic procedures. In addition, he presented results on performance and subjective sleep quality. *Hofman et al.* presented the results on cardiac changes due to changes in noise intensity. *Williams* presented critical remarks on these presentations and provided suggestions for future work.

---

[1] Supported by a grant from the Commission of European Communities.

Habituation and Age Differences of Cardiovascular Responses to Noise during Sleep[2,3]

A. Muzet, J. Ehrhart, R. Eschenlauer, J. P. Lienhard, Strasbourg, France

Introduction

During the last 20 years there has been an increasing number of studies concerned with the effects of noise on sleep. Whether these studies were done in the laboratory or at home, they were focused primarily on subjective assessments and/or sleep EEG modifications related to the noise. Habituation to noise has been studied within the same night [3, 6] and over several nights [5, 7, 10, 12, 18, 19]. From these studies it appears that habituation depends upon the nature of the stimulus and also of the type of criteria considered. Thus, if habituation seems to occur over several nights with respect to the number and the duration of EEG modifications, contradictory results were reported for autonomic responses to noise. Few studies considered the age factor since the study described by *Lukas and Kryter* [13]. Therefore, the purpose of the present study was to compare the habituation to traffic noises at three different levels: subjective, electroencephalographic, and cardiovascular levels. Furthermore, a comparison was made between three different age groups in order to verify the assertion of increasing effect of noise with increasing age.

Material and Methods

Experiment 1. 6 young adults of both sexes slept in the laboratory for 20 consecutive nights. Traffic noises with peak intensities of 45, 55, and 65 dB (A) were presented semi-randomly at a rate of 90 noises per hour for 15 consecutive nights (4th to 18th nights).

---

[2] This work has been supported by a grant from the Ministère Français de la Culture et de l'Environnement. Convention de Recherche no 76–22.
[3] For references see compound list on page 232.

Experiment 2. Three groups of 8 subjects each slept in the laboratory for 2 non-disturbed and 2 disturbed nights. These three groups were constituted respectively by 8 children (6–12 years old), 8 young adults (19–28 years old), and 8 elderly people (56–66 years old). Peak intensities of traffic noises were of 40, 50, and 60 dB (A) during the first disturbed night and of 45, 55, and 65 dB (A) during the second disturbed night. As for experiment 1, the traffic noises were presented semi-randomly throughout the night at a rate of 90 noises per hour. Sleep recording and sleep scoring were made according to *Rechtschaffen and Kales* [16] manual. The heart beat intervals as well as the beat by beat finger pulse amplitudes were calculated on-line by a computer. Then the heart rate response (HRR) and the finger pulse response (FPR) were computed for each noise taken individually and their amplitudes were calculated. The HRR amplitude was defined as the difference (in beats/min) between the acceleratory and the deceleratory components of HRR. The FPR amplitude was defined as the percent decrease of the smallest finger pulse amplitude contemporary to the noise, compared to the finger pulse amplitude just preceding the noise onset. In a following step, average HRR amplitude and average FPR amplitude were calculated for each noise intensity within each sleep stage and for all night. Every morning after awakening, the subjects had to answer a sleep questionnaire.

Dependent and independent Student's t-tests were used and $p < 0.05$ or better were considered as significant.

Results

The amplitude of HRR and FPR depended upon the peak intensity of the noise and the sleep stage. Thus, the amplitudes of cardiovascular responses increased as the noise peak intensity increased, and they were significantly smaller in SWS than in stage 2 and REM sleep. No habituation of the cardiovascular responses was found from the beginning to the end of the disturbed nights. Results published elsewhere [14] showed that, in the 15-night exposure to traffic noises, the subjective complaints ceased after a few nights and that there was a clear adaptation process to the noisy environment when considering the sleep stage criteria. However, when comparing the average amplitudes of the cardiovascular responses to the loudest noise, 65 dB (A) peak intensity, between the first two and the last two disturbed nights a significant reduction was observed only for the amplitude of HRR in stage 2 ($p < 0.05$).

Therefore, there was almost no habituation of the cardiovascular responses to the noise over the 15 disturbed nights.

The magnitude of the cardiovascular responses depended upon the age of the subjects (fig. 1). For example, the average HRR to the 65 dB (A) noise was significantly ($p < 0.01$) larger in children and in young adults than in the oldest group for each sleep stage category. The all-night and the SWS average FPR to the same noise were significantly ($p < 0.05$) larger in

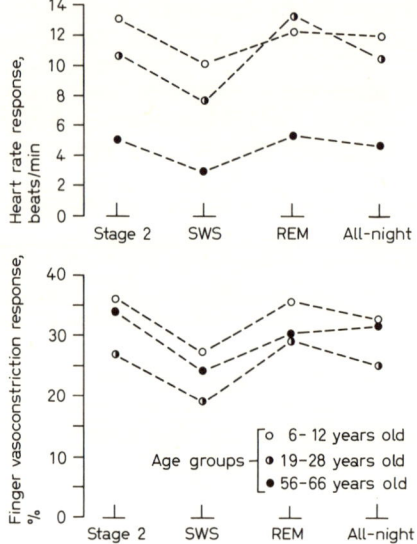

Fig. 1. Heart rate and finger vasoconstriction responses to the 65 dB (A) peak intensity noise during experiment 2. The average amplitudes of the cardiovascular responses were calculated for each sleep stage category in each age group.

children than in young adults, while of about the same magnitude as those observed in the oldest subjects.

Discussion

The most important finding of this study was the dissociation that appeared between a progressive disappearance of subjective complaints and of some of the sleep EEG disturbances due to the noise, and the maintenance of obvious cardiovascular responses that hardly habituated over 15 consecutive disturbed nights. These results suggest that the cardiovascular level, in spite of its unconscious nature, must always be taken into account in the study of the effects of noise on sleep. Previous findings of *Lukas and Kryter* [13] on the higher EEG reactivity of oldest subjects were confirmed [14]. However, the cardiovascular reactivity to noise appeared somewhat different. HRR amplitudes were almost of the same value in children and in young adults, but much higher than in oldest subjects. The lower cardiac reactivity observed in the latest group was certainly not due to aging-related hearing loss, because the amplitude of FPR in this group was comparable to that seen in the youngest subjects. These results indicate that protection against noise should be extended to all age categories, and that the common assertion that children are less disturbed by noise during sleep than adults should be reconsidered.

Noise and Sleep in the Home: General Methodology

B. Griefahn, Eckhard Gros, H. Kauth, Düsseldorf, FRG

Because of the fact that sleep disturbances are more and more related to environmental noise, it may be possible to prevent a considerable number of sleep disturbances by lowering the noise level and by finding upper limits above which nightly occurring noises due to road traffic should not be allowed. Reliable results cannot be obtained only by experimental work in the laboratory; they require field studies. The great number of variables influencing the sleep under normal conditions determines the number of recordings.

Therefore, the CEC has initiated and financed a study in order to find out the relation between long-term exposure to usual road traffic noise and physiological variables during sleep and subjective response and performance the following day.

General Design

The investigation was designed as a pilot study and was carried out in France, in the Federal Republic of Germany, in the Netherlands, and in the United Kingdom. Altogether about 1,000 nights have been recorded by the 4 teams. 63 male and female subjects took part in the study. Their age varied from 18 to 65. All subjects lived in streets with high traffic density.

The subjects were healthy, had normal hearing, and did not take drugs. Shiftworkers were excluded.

Dependent on the individual experimental design of each team 12–23 nights have been registered for each subject continuously or discontinuously at home under 'normal' conditions. Due to the relative small number of subjects sleep disorders cannot be clearly related to noise. Consequently, all subjects were recorded under both noisy and quiet conditions, meaning that registrations were done under 'normal' and experimental conditions as well. The experimental condition consisted of opening the windows (noisier than normal) or of wearing earplugs, fitting double glazing or moving to another bedroom (quieter than normal). The difference between both situations noise and quiet was required to be 10 dB (A) either for the equivalent noise level or for L_1. The majority of data were collected using an A-B-A design, but several variations were examined too.

Recordings and Evaluation

During each night EEG, EOG, and the equivalent A-weighted sound level were continuously recorded. EMG was recorded by the French and by the Dutch, EKG by all teams but the German, signalled awakening by the British, body motility by the Dutch and the German and respiration only by the Dutch team. In order to relate variations in performance and mood to different nightly occurring sound levels, several questionnaires and performance tests were administered by the 4 teams. A small sleep questionnaire was applied by all teams, mood and well-being by day was recorded in the Netherlands. The German team applied the four-choice-reaction-time test in the morning and in the evening for 10 min; all the other teams applied the simple-reaction-time test. In addition to this performance test the British team also recorded the four-choice-serial-reaction time, the auditory short-term memory and the Wilkinson vigilance test. Though a detailed analysis has yet to be completed, several data had been sufficiently evaluated. Where possible, the results of the different teams have been combined and they are presented in the subsequent papers together with individual team results.

Noise and Sleep in the Home: Effects on Sleep Stages

A. A. Jurriëns, Delft, The Netherlands

Introduction

In four countries possible influences of traffic noise on sleep and subsequent well-being have been examined with people at home, living in a noisy environment. The general methodology of this study is presented by *Griefahn et al.* [this volume] and consisted of a common base upon which variations were possible. Here it is recalled that 63 subjects slept both in relatively noisy and in relatively quiet conditions. For 45 subjects the normal condition was noisy, for 18 subjects it was quiet. The experimental condition has been realized by fitting double glazing (UK and NL), by using earplugs (FRG), by moving subjects to a quieter bedroom (FR) and by opening windows (FRG and NL).

All teams recorded continuously EEG and EOG during each night, and sleep stage scoring has been agreed upon as a common analysis. Other analyses on EEG have been done or will be done by some teams (see *Vallet et al.* and *Wilkinson* [this volume]). In this paper combined results on sleep stages are presented.

Method

Sleep stage scoring has been performed visually (FR) and with two different automatic systems (UK and NL/FRG). The three methods turned out to be equivalent [9]. For each night the total sleep time (TST), the sleep latency (SL), the REM latency (RL), the amounts of sleep stages, both in minutes and in percentage TST, and the barycentres of stage 4 and of stage REM have been calculated. Then for each subject values for each night of the variables mentioned have been averaged per sound condition, and differences between these condition averages have been determined. Results are presented here as numbers of subjects with a positive and with a negative difference between condition averages, ignoring essentially zero differences.

Results

Table I gives an overview of results thus obtained. The clearest trends suggest that during noise the total sleep time and the time in stage REM decrease, whereas the amount of stage W increases. But also a differentiation has been made between the two groups with different normal conditions, and normal and experimental condition have been compared (for the group with normal condition quiet this means an inversion of + and −). When in the latter comparison a trend becomes more obvious or appears, such a trend is probably related to the experiment itself, while on the other hand diminution or disappearance of a trend points in the direction of noise influence. Table II shows that both possibilities occur, for TST and time in REM, respectively. The change in stage W was about the same in both comparisons. A change in TST also effects certain sleep stage scores. When no other influences are present, with more TST one should usually expect more time in stage 1+2, the same amount of stage 3+4, more time in stage REM, barycentre 4 more negative and barycentre REM more positive. With less TST just the reverse. In table III for the combined results a differentiation is made between subjects with more TST and with less TST, in terms of noisy versus quiet and of normal versus experimental. Noticeable deviations from the expectations mentioned occur for stage 3+4 (a slight decreasing trend

Table I. Numbers of subjects with a positive and with a negative difference between averages of the conditions noisy and quiet

	FRG		FR		NL		UK		T	
	+	−	+	−	+	−	+	−	+	−
TST	8	12	8	11	5	7	5	6	26	36
SL	15	5	12	8	4	8	4	7	35	28
RL	10	10	14	4	5	7	3	8	32	29
1+2	9	11	12	8	6	6	4	7	31	32
1+2, %	11	9	10	6	4	8	6	5	31	28
3+4	10	10	10	10	9	2	4	7	33	29
3+4, %	9	10	8	9	7	5	7	3	31	27
REM	7	13	4	14	3	9	9	2	23	38
REM, %	8	12	5	14	6	6	8	3	27	35
W	9	11	13	5	9	3	4	7	35	26
W, %	8	12	14	5	9	2	4	7	35	26
BAR 4	11	9	9	10	2	2	6	3	28	24
BARREM	10	10	10	9	1	3	5	4	26	26

Table II. Influence of normal condition on some results of table I

	Noisy – quiet						Normal – experimental	
	N		Q		N or Q			
	+	–	+	–	+	–	+	–
TST								
FRG	1*	9	7	3	8	12	4*	16
FR	8	11	–	–	8	11	8	11
NL	1	3	4	4	5	7	5	7
UK	5	6	–	–	5	6	5	6
Total	15*	29	11	7	26	36	22*	40
REM								
FRG	3	7	4	6	7	13	9	11
FR	4*	14	–	–	4*	14	4*	14
NL	2	2	1	7	3	9	9	3
UK	9	2	–	–	9	2	9	2
Total	18	25	5	13	23	38	31	30

* Significant distribution (sign test).

Table III. Influence of changes in TST on some results of table I, also taking into account differences in normal condition

	Noisy – quiet				Normal – experimental			
	more		less		more		less	
	+	–	+	–	+	–	+	–
1+2	23*	3	7*	29	17*	5	5*	35
3+4	14	12	18	17	9	13	17	22
REM	12	13	11*	24	14	7	17	22
BAR 4	7	12	21	11	6	11	22	12
BARREM	11	7	14	19	12	4	17	18

* Significant distribution (sign test).

during the experimental condition) and for stage REM (increasing numbers of subjects with less time in REM, comparing noisy and quiet).

Discussion

Despite results in the opposite direction of one team, the combined results presented here show a trend of less time in stage REM during the noisy condition. And although better balanced subgroups would have been more profitable, this trend cannot be explained by different normal conditions, as is probably the case with the simultaneous decrease in TST. A first inspection showed that other experimental differences are not distributed over the results in such a way that they might explain this finding about REM. On the other hand, an examination of the influence of a change in TST underlines the tendency of REM to decrease during noise. From the combined results on sleep stage REM it is therefore concluded that sleeping in noisy circumstances tends to reduce REM sleep.

Sleep 1980. 5th Eur. Congr. Sleep Res., Amsterdam 1980, pp. 220–224
(Karger, Basel 1981)

Noise and Sleep at Home: Stage Changes and Arousals [4, 5]

M. Vallet, J. M. Gagneux, V. Blanchet, Bron, France

Noise, as other external stimulations, provokes on the one hand structural changes in the different sleep stages and on the other hand there are temporary reactions to a single noise event. These short effects are at least one epoch long and present three steps in intensity: arousals, sleep stage changes as defined by *Lukas* [11], and transient reaction as described by *Muzet et al.* [15]. One considers the total number of arousals per night, natural as well as caused by noise, for 44 subjects recorded with the following experimental design: noise-quiet, 6–11 nights per condition, or noise-quiet-noise, 3–6 nights per condition, i.e. a total of 700 nights.

[4] This work has been supported by a grant from the Commission of European Communities, Brusselles.
[5] For references see compound list on page 232.

Table IV. Increasing number of arousals in quiet condition

Subjects and group	TST, min	W, min
8/20 subjects 1st group	↗ 25 ↗ 33 ↗ 15 ↗ 12 = 0 ↘ 11 ↗ 52 ↗ 20	↗ 4 ↗ 1 ↗ 6 ↗ 27 23 ↗ 33 ↗ 58 ↘ 4
7/11 subjects 2nd group	↗ 13 ↗ 50 ↗ 56 ↘ 24 ↘ 3 ↘ 8 ↗ 26	↗ 8 ↗ 6 ↗ 30 ↗ 1 ↗ 8 ↗ 5 ↗ 20
1/5 subjects 3rd group	↗ 41	↘ 4
6/10 subjects 4th group	↗ 30 ↗ 52 ↗ 13 ↗ 35 ↗ 45 ↗ 8	↗ 8 ↗ 1 ↗ 1 = 0 ↗ 23 ↗ 1
	17 ↗ 4 ↘ TST ↗ 26	W ↗ 10

Results

22 subjects show a decreasing number of arousals in quiet condition and 22 subjects provide an increase (table IV). In the first case the arousal number decreases whilst the total sleep time (TST) is stable: TST, –7 min, i.e. –1.5% in quiet conditions; W (total time of awakening), –10 min, i.e. –35%.

For the same duration of sleep in noisy and quiet conditions a better sleep quality appears in quiet conditions with 10 min reduction of the awakening period. This is a significant result. In the second case we observe that the increasing number of arousals in quiet conditions is connected with a higher TST: TST, +26 min, i.e. +6% in quiet conditions; W, +10 min, i.e. +35%. The mean duration of awakenings increases, but the subject response is variable: sometimes W grows, sometimes it is stable or decreasing (cf. table IV).

One cannot conclude that arousals are more numerous in quiet conditions. However, one observes, for some subjects, a strong increase of the arousals number in quiet conditions, especially in the first nights that one can attribute to the situation changes, the noisy conditions being usual. Everything taken into account, the noise provokes brief EEG sleep arousals that occur during all categories of stages: light sleep, delta and REM, their number decreases in quiet conditions for a majority of the subject sample.

Subjective Evaluation

After each night was recorded, subjects replied to a short questionnaire about the quality of their sleep. We have investigated in the data of the French team to know whether there exists a relationship between physiological values of arousals (sleep latency, number of arousals, duration of arousals) and the subjective assessments (of the same aspects) reported in the questionnaire of the morning. There is no significant relationship between these two categories of data concerning the number of arousals. Further analysis should be carried out, taking into account the duration of each EEG arousal, as in previous studies by *Baekeland and Hoy* [1] and *Schneider* [17]. A significant product moment correlation exists between the effective time of awakening and the time estimated by the questionnaire ($r = 0.33$; $n = 202$). The data seem to limit the use of survey by questionnaires to assess the effect of noise on sleep, or any other isolated physical factor (there is no parallel between physiological reactions and their subjective evaluation generally speaking), especially when questions concern physiological events.

Temporal Evolution

In the experience at home, and for the observed numbers of noise (300–1500), the traffic provokes in mean value about 30 short disturbances caused by noise after a 5-year exposure to noise; these are generally sleep stage changes, and sometimes awakening. These effects contribute to interrupt continuously the sleep of people living near motorways which is unacceptable physiologically and psychologically.

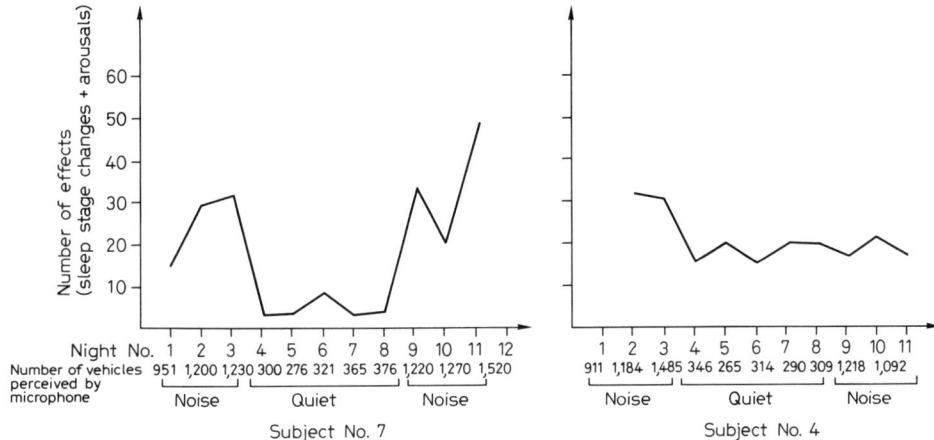

Fig. 2. Immediate effects of the noise peak event and change with the noise environment.

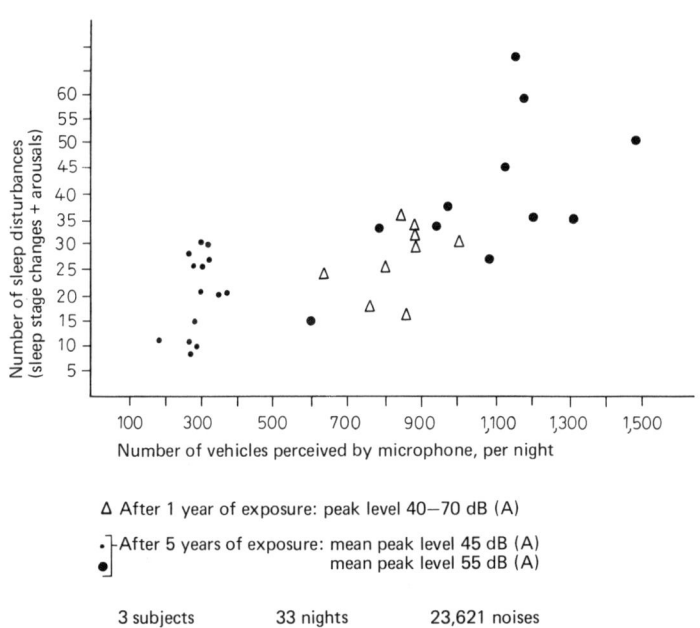

△ After 1 year of exposure: peak level 40–70 dB (A)

• ⎱ After 5 years of exposure: mean peak level 45 dB (A)
• ⎰   mean peak level 55 dB (A)

3 subjects   33 nights   23,621 noises

Fig. 3. Evolution of the number of immediate effects due to the noise.

We observe large differences between the subjects: for some of them sleeping in a quiet bedroom provokes a large decrease in the number of sleep disturbances (fig. 2, subject 7). For others, the change in noise environment does not bring about any improvement (fig. 2, subject 4).

We propose to conclude that for people like subject no. 7, the sleep environment could be largely improved by double glazing; indeed they have an interesting margin of habituation. For other subjects the 10 dB(A) decrease of the noise peak levels does not change anything.

We have analyzed the temporal evolution of the arousals and of the stage change frequency with some subjects, already recorded in 1974 in a previous experiment, and exposed to noise during this period. The number of these temporary reactions increases slightly, maybe with the growing traffic. It changes from 24 reactions per night in 1974 to 35 reactions in 1978, with 300 vehicles more per night; the noise peak values vary from .45 to 80 dB(A) for the 2 periods. Contrary to the results obtained in laboratory on short periods, we observe an increase of the numbers of EEG reactions with the exposure time (fig. 3).

These data, provided by the same subjects, show an evolution between 1974 and 1978. If we consider now 3 degrees of sleep disturbance: transient reaction, sleep stage changes, and arousals, we observe that the percentage of arousals and sleep stage changes increases whilst the percentage of transient reactions decreases with the length of exposure to noise (fig. 4).

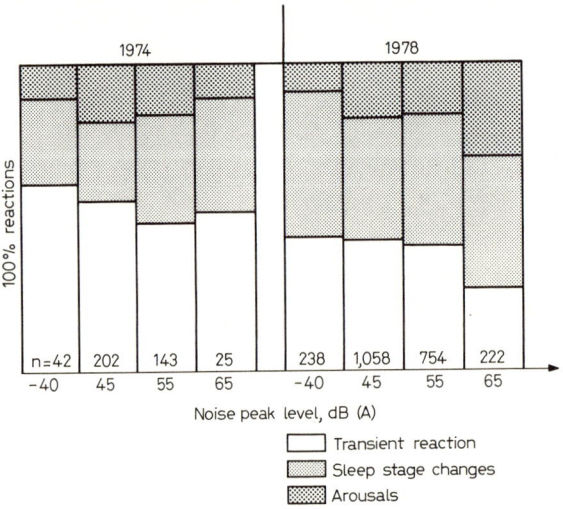

Fig. 4. Effects of exposure time to noise on sleep in 1974 and 1978.

Effects of Traffic Noise upon Sleep in the Home

Subjective Report, EEG, and Performance the Next Day [6]

R. T. Wilkinson, Cambridge, England

This report forms part of a joint project, supported by the Commission of the European Communities (CEC) to assess sleep in the home as a function of traffic noise. Four teams have taken part from laboratories in Britain, France, Germany and the Netherlands. Part of this paper will be concerned with presenting the results obtained from pooling data from the teams taking part.

Three possible methods of assessing the quality of sleep will be considered. The first is the one which is used most in practice: To ask people to say how well they have slept. The question is usually presented in a questionnaire and is included among a number of others relating also to the previous night's sleep. The questionnaire is given soon after waking the following morning. The answers are made using a visual analogue scale, typically a 10-cm line with extreme values, e.g. 'best I can remember', 'worst I can remember' defining each end of the line. The subject puts a mark at an appropriate place along the line to indicate his estimate of the quality of the night's sleep. The score is expressed in centimetres from the left hand side, in other words the better the sleep the higher the score.

The common method of the project was to modify the level of traffic noise, either by double glazing the bedroom windows, opening windows, or having subjects wear ear plugs or sleep in a back rather than a front bedroom. Measurements were then taken for periods in the normal and in the modified condition in an A-B-A design. Noisy and quiet nights were then compared.

Let us take for each subject of each team the response to the common question, 'How well did you sleep last night?', averaged for all his (or her) 'Noisy' nights, and the same over all his 'quiet' nights. Over all teams there were 62 subjects for whom this analysis could be made. The average score

[6] For references see compound list on page 232.

out of 10 (best sleep possible) was 6.8 for noisy and 7.5 for quiet conditions. The difference between the two, though small in magnitude, is highly significant by a Wilcoxon test ($p<0.001$). In none of the individual teams was the corresponding comparison significant, so here we see the advantage of combining data in a multi-team project. Thus, over all subjects there is good evidence of a modest improvement in subjective sleep quality as a result of attenuating traffic noise by about 10 dB (A) Leq in the bedroom.

If we are concerned with the restorative value of sleep, and its ability to prevent fatigue and drowsiness the next day, a purely subjective impression of the previous night's sleep may not be the most appropriate index. Are we aware during the night of the moment-to-moment restorative quality of our sleep? And even if this is so, can the sleeping brain adequately store the information for retrieval the next morning as an accurate index of how good a job the night's sleep had done in restoring us?

A good measure of restorative function of sleep may be our performance the next day, providing the task is one whose sensitivity to the effects of reduced sleep has been verified. Such a test is the unprepared simple reaction time. Three teams used the portable cassette recording form of this test, produced in this laboratory and now available commercially. This test has been shown [4] to be significantly impaired by one night's loss of sleep. It has the further advantage of showing little improvement with practice after the first 20 min work. In the present project, therefore, the test was administered to each subject, each morning for 10 min. The subject watches a window for the appearance of a *000* display which immediately starts counting up in milliseconds. The subject has to stop it as soon as he can by pressing a button. When he does so the arrested number displays his reaction time for 1.5 s before going out, thus indicating the start of another waiting period for the same stimulus to occur again, and so on. The inter-stimulus waiting period varies randomly between 1 and 10 s, so the test is more one of concentration than of sheer response speed.

The three teams were able to test 27 subjects in all, with each subject receiving at least 10 tests, some on days following noisy nights, others following quiet nights. Analysing as before, the overall average reaction time was 285 msec for noisy nights and 273 ms for quiet ones. Thus, performance was better after a quiet night than a noisy one, and of the 27 subjects from three teams 22 showed this effect. This is significant on sign test alone ($p<0.05$). A Wilcoxon test in the noise-quiet differences for each subject gives a T of 69 ($p<0.005$). Thus, for the group as a whole reducing the level of traffic noise during the night appeared to improve unprepared simple reaction time performance the next day. It follows that under the prevailing levels of traffic noise sleep was less restorative than normal.

A third way in which the quality of sleep might be assessed is by physiological recordings taken during sleep, particularly the EEG. Even when automatically scored by computer an element of visual inspection and subjective judgement must arise with these assessments. A more objective alternative is provided by analysing the EEG on the basis of frequency. The results of the British team using this method will now be reported briefly.

The details of the method are as follows: We recorded the EEG during the sleep of 11 people living beside noisy, suburban, arterial roads. Noise conditions were: 1st week, normal noise; 2nd and 3rd weeks, quiet (double-glazing in bedroom); 4th and 5th weeks, back to normal noise (double-glazing removed). Recording was on the weekdays of the 1st, 3rd and 5th weeks. EEG sleep records were analysed on a phase-lock loop system for the presence of alpha, spindle, delta frequencies and (using a band-pass, level-detecting system) rapid eye movements.

The average minutes of EEG delta activity over all subjects and over all noisy nights was 81.1; for quiet nights the average was 88.8. The difference between the two was significant ($p<0.05$) using a Wilcoxon test. Corresponding averages for alpha, spindle and REM activity showed no signifi-

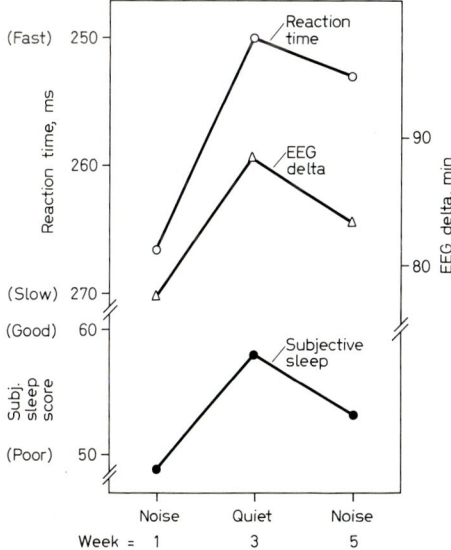

Fig. 5. Scores on the three possible sleep measures, reaction time, subjective report, and EEG delta activity during weeks of 'noise' and 'quiet'.

cant difference between noisy and quiet nights. This is the result we would expect on the twin assumption that presence of delta activity indicates sound sleep and a higher level of traffic noise will tend to disturb sleep.

We can now, again with the data of the British team only, look at the results of the three possible measures of sleep quality together, as they varied across the first week of normal noise, the third of double-glazed 'quiet', and fifth back at the normal noise level. Figure 5 shows the levels of subjective sleep questionnaire score (How well did you sleep?), unprepared simple reaction time, and minutes of EEG delta activity per night, averaged over the nights of the weeks concerned. Clearly in these data there is an indication of some agreement between the three scores in showing an improvement in sleep quality when traffic noise is reduced and when the data are collapsed over both nights and subjects. That this agreement is not so apparent from the point of view of the whole four-team project may be instructive, particularly as much of the disagreement is in the sleep stage scoring, which is more visually oriented and subjective, even when computer analysis is involved. The joint four-team nature of this project may therefore have been valuable not only in revealing positive effects of traffic noise on performance and subjective report, but also in providing documentary evidence of the lack of uniformity possible in results derived from sleep stage scoring from one laboratory to the next.

Sleep 1980. 5th Eur. Congr. Sleep Res., Amsterdam 1980, pp. 228–231
(Karger, Basel 1981)

Noise and Sleep in the Home: Effect on Heart Rate[7]

W.F. Hofman, A. Kumar, R. van Diest, Amsterdam, The Netherlands

In a recent review by *Kryter* [8] on the effect of noise on health, it was suggested that cardiovascular responses to noise show habituation. The results of our study are in contradiction with this. They are, however, in agreement with the results of the study of *Muzet* [14] who found that cardiovascular responses to artificial noise stimuli do not show any habituation.

[7] For references see compound list on page 232.

Methods

12 subjects participated in this study for 20 nights according to the design: 10 nights with traffic noise (windows open)–10 nights quiet (double glazing), 10 nights quiet–10 nights noise, 5 nights noise–10 nights quiet–5 nights noise, 5 nights quiet–10 nights noise–5 nights quiet.

EEG, ECG and dB(A) level of noise were recorded in the homes of the subjects according to the general procedure described earlier. For each night the time interval between two R peaks of the ECG was used to calculate for each minute the number of beats per minute and the coefficient of variability of the number of beats per minute. For each minute also L_{eq} and the standard deviation of noise were calculated. Correlations between the noise parameters and the ECG parameters were calculated for the whole night.

Results

In table V the significant correlations for the first 8 subjects are summarized. No difference could be found between the noise and the quiet con-

Table V. Number of significant product moment correlations between two noise variables (L_{eq}, standard deviation of noise) and two ECG parameters (beats/min, coefficient of variability of beats/min)

Subject	Condition	n	L_{eq}				Standard deviation noise			
			beats/min		coeff. var., beats/min		beats/min		coeff.var., beats/min	
			+	−	+	−	+	−	+	−
P.G.	N	9	5	1	4	1	2	1	5	0
	Q	10	9	1	9	1	4	1	9	0
W.G.	N	8	7	0	8	0	7	0	7	0
	Q	6	6	0	4	1	6	0	6	0
H.K.	N	8	6	1	2	2	1	6	3	0
	Q	8	8	0	6	0	8	0	8	0
L.K.	N	6	3	0	2	0	0	2	0	0
	Q	3	2	0	3	0	3	0	3	0
C.H.	N	3	2	1	1	0	1	2	1	2
	Q	8	4	2	7	0	7	0	7	0
C.V.	N	9	7	0	8	0	4	1	8	0
	Q	8	8	0	6	0	8	0	8	0
H.V.	N	4	3	0	3	0	2	0	4	0
	Q	4	3	1	4	0	3	0	3	0
P.V.	N	8	7	0	7	0	1	0	6	0
	Q	6	5	0	5	0	3	0	3	0

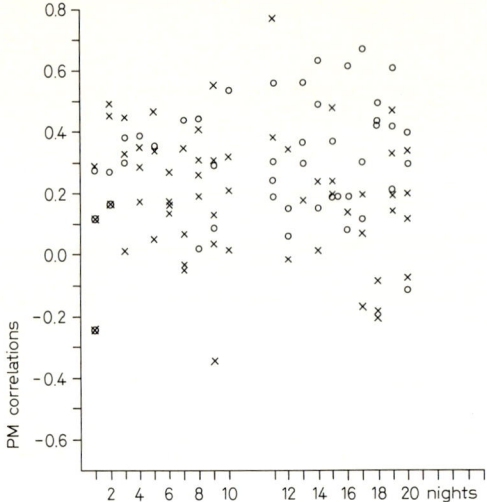

Fig. 6. Correlations of L_{eq} with the number of beats/min over 20 nights. O = Quiet; × = noise.

dition. The correlations in the quiet condition were as high or even a little bit higher than in the noise condition.

As each subject participated in this study for at least 20 nights, we could see if any habituation of the cardiac response to noise occured within this period. In figure 6 the correlations of L_{eq} with the number of beats per minute of 8 subjects are plotted for all subsequent nights.

Linear regression, calculated over the 20 nights resulted in a statistically not significant correlation of 0.072, which means that there was no variance presence in the correlations, that could be attributed to the order of the nights. In short: no habituation occurred within the 20 nights.

In the United Kingdom a similar analysis was done with 11 subjects, 12 nights each with a N-Q-N design. Their results were in agreement with the results of this study: (1) a higher level of L_{eq} and a higher standard deviation of noise per minute was correlated with a higher amount of beats per minute and higher coefficient of variability per minute; (2) no habituation of the cardiac response to noise could be found over the 13 nights. Looking at the correlations of noise with heart rate in the different sleep stages showed no less response of heart rate to noise during stage 3+4 together than during stage 1+2 together. The percentages of significant positive correlations between L_{eq} and number of beats per minute in the different sleep stages are as follows: stage 1+2, 59%; stage 3+4, 76%; stage REM, 44%.

Discussion

From the result that no difference could be found between the noise and the quiet condition it had to be concluded that the 'quiet' condition was not really quiet, as the traffic was not stopped. Normal insulation by means of double glazing lowers the overall dB(A) level, but is not able to protect against the lower frequencies. This could mean that not only the level of the noise, but also the noise transients disturb the cardiac activity.

It was found that EEG responses and subjective feelings do habituate after some days of exposure to noise during sleep [2, 14]. The cardiovascular system, however, does not seem to habituate at all, neither to artificial noise nor to traffic noise.

Though the arousal threshold to tonic sounds during stage 3+4 is known to be higher than during stage 1±2, yet in our study cardiac disturbance by noise was found to be equally prominent in both slow-wave and fast-wave sleep.

Sleep 1980. 5th Eur. Congr. Sleep Res., Amsterdam 1980, pp. 231–232
(Karger, Basel 1981)

Comments on Traffic Noise and Sleep[8]

Harold L. Williams, Oklahoma City, Okla., USA

The diverse results reported here suggest that a perspective other than 'noisy vs quiet' might be useful. That dimension may not denote the same noise parameters between settings. Another approach would examine discrete responses and physiological profiles *across* 'noisy' and 'quiet' conditions. The literature provides hints on what to look for [see 20 for a review].

[8] For references see compound list on page 232.

Stimulus Effects. Although intense stimuli with fast rise times do disturb sleep, such effects are not simple. *Schieber* in *Metz's* laboratory found that low density traffic at about 61 dB was more disruptive of sleep than high-density traffic at about 70 dB. With this in mind, a comparison of the noise profiles in these four settings would be useful.

Response Systems. Analysis of discrete responses, of the hierarchy of sensitivity of various systems and of time of night effects should be continued and extended. In addition one should examine the ultradian rhythms of sleep, particularly the REM and delta cycles. Thus, variability of REM periodicity is a sensitive index for certain chronic conditions and may covary with waking performance.

Individual Differences. Responsiveness to noise during sleep varies with age, sex, psychopathology and physical condition. Data on available individual differences should be compared across settings and future studies should be extended to special groups such as the aged, the physically ill, the neurotic and depressed. They are probably most vulnerable to chronic noise exposure.

Conclusion

Though extensive analyses of these data should continue, one notes that some outcomes already reported are so varied from group to group as to defy interpretation. Are these due to differences in design, methods of noise reduction, measurement styles, samples or what? Obviously future research will benefit from firm standardization across laboratories on all variables that can reasonably be controlled.

List of References of Workshop E

1   Baekeland, F.; Hoy, P.: Reported vs recorded sleep characteristics. Archs gen. Psychiat. *24:* 548–551 (1971).
2   Diana, I.P.F. de; Kumar, A.; Hofman, W.F.: Sleep quality and human performance; the specific sleep quality scale. Sleep Res. *5:* 128 (1976).
3   Firth, H.: Habituation during sleep. Psychophysiology *10:* 43–51 (1973).

4   Glenville, M.; Broughton, R.; Wing, A.M.; Wilkinson, R.T.: Effects of sleep deprivation on short performance measures compared to the Wilkinson auditory vigilance task. Sleep *1:* 169–176 (1978).
5   Griefahn, B.: Effects of sonic booms on finger pulse amplitudes during sleep. Int. Arch. occup. Environm. Hlth *36:* 57–66 (1975).
6   Johnson, L.C.; Lubin, A.: The orienting reflex during waking and sleeping. Electroencephal. clin. Neurophysiol. *22:* 11–21 (1967).
7   Johnson, L.C.; Townsend, R.E.; Naitoh, P.; Muzet, A.: Prolonged exposure to noise as a sleep problem, publ. No. 550/9–73/008, pp. 559–574 (US Environmental Protection Agency, 1973).
8   Kryter, K.D.: Physiological acoustics and health. J. acoust. Soc. Am. *68:* 10–14 (1980).
9   Kumar, A.; Campbell, K.; Hofman, W.; Vallet, M.; Jurriëns, A.A.; Van Diest, R.: Evaluation and validation of automatic and visual methods of sleep stage classification of human sleep recordings done at home. Proc. 5th Eur. Congr. of Sleep Res., Amsterdam 1980, p. 78.
10 Ludlow, J.E.; Morgan, P.A.: Behavioral awakening and subjective reactions to indoor sonic booms. J. Sound Vib. *25:* 479–495 (1972).
11 Lukas, J.: Predicting the response to noise during sleep. Proc. Int. Congr. on Noise as a Public Health Problem, Dubrovnik, publ. No. 550/9– 73008, (US Environmental Protection Agency, 1973).
12 Lukas, J.S.; Dobbs, M.E.: Effects of aircraft noises on the sleep of women. Report No. CR-2041 (NASA, Washington 1972).
13 Lukas, J.S.; Kryter, K.D.: Awakening effects of simulated sonic booms and subsonic aircraft noise on six subjects 7 to 72 years of age, report No. CR-1599, (NASA, Washington 1970).
14 Muzet, A.: Modifications végétatives entraînées par le bruit au cours du sommeil, (Rapport de Recherche No. 76.22, CEB, Strasbourg 1980).
15 Muzet, A.; Shnieber, J.P. et al.: Relationship between subjective and physiological assessments of noise disturbed sleep. Proc. Int. Congr. on Noise as a Public Health Problem, Dubrovnik, publ. No. 550/9 73008, (US Environmental Protection Agency, 1973).
16 Rechtschaffen, A.; Kales, A.(eds.): A manual for standardized terminology, techniques and scoring system for sleep stages of human subjects (US Government Printing Office, Washington 1968).
17 Schneider, N.: Evaluation subjective du sommeil pertubé par le bruit; thèse CNRS, CEB, Strasbourg (1973).
18 Thiessen, G.J.: Disturbance of sleep by noise. J. acoust. Soc. Am. *64:* 216–222 (1978).
19 Townsend, R.E.; Johnson, L.C.; Muzet, A.: Effects of long-term exposure to tone pulse noise on human sleep. Psychophysiology *10:* 369–376 (1973).
20 Williams, H.L.: Effects of noise on sleep: a review. Proc. Int. Congr. on Noise as a Public Health Problem, Dubrovnik, publ. No. 550/9 73008 (US Environmental Protection Agency, 1973).

Addresses of Authors

Dr. B. Griefahn, Institut für Arbeitsmedizin (Prof. Dr. Dr. *G. Jansen*), Moorenstrasse 5, D-4000 Düsseldorf 1 (FRG)

Dr. W. F. Hofman, Psychophysiology Laboratory, University of Amsterdam, Eerste Constantijn Huygensstraat 20, 1054 BW Amsterdam (The Netherlands)

Dr. A. A. Jurriëns, TNO Research Institute of Environmental Hygiene, Delft (The Netherlands)

Dr. A. Kumar, Psychophysiology Laboratory, University of Amsterdam, Eerste Constantijn Huygensstraat 20, 1054 BW Amsterdam (The Netherlands)

Dr. A. Muzet, Centre d'Etudes Bioclimatiques du CNRS, F-67087 Strasbourg Cédex (France)

Dr. M. Vallet, Centre d'Evaluation et des Recherches des Nuisances et de l'Energie, Institut de Recherche des Transports, 109, Avenue S. Allende, F-69500 Bron (France)

Dr. R. T. Wilkinson, Medical Research Council, Applied Psychology Unit, Psychophysiology Section, 5 Shaftesbury Road, Cambridge (England)

Dr. H. L. Williams, Department of Psychiatry and Behavioral Sciences, University of Oklahoma, Health Sciences Centre, Oklahoma City, 73190 OK (USA). On sabbatical leave at Medical Research Council, Applied Psychology Unit, Psychophysiology Section, 5 Shaftesbury Road, Cambridge (England)

Part III. Special Lectures

On the Role of Active (REM) Sleep in Ontogenesis of the Central Nervous System

Michael A. Corner, Majid Mirmiran, Harry Bour

Netherlands Institute for Brain Research, Amsterdam, The Netherlands

Sleep in immature mammals is a state characterized by a high level of often intense, spontaneous motor activity. Since total sleep time too is extremely high in early life, 'active' sleep (AS) is the physiological state in which neonatal and late fetal animals spend the largest part of the time. These facts have led many developmental physiologists to propose that AS may play an important role in the maturation of the nervous system [e.g. 9, 15, 27]. The 'AS (REM sleep) hypothesis' is supported by observations of intense, generalized neuronal excitation throughout AS in adult mammals [31], and also in the embryonic chicken [25] during periods of spontaneous motility (which closely resembles sleep twitching).

Possible Mechanisms of AS Effects upon Brain Development

We have recently been able to verify that AS in infant mammals is indeed a highly excited state from the neuronal vantage point [6, 18]. Spontaneous action potentials appear abruptly in the occipital cortex around day 12 but are already well developed at day 9 in the pontine reticular formation (PRF). In both regions during AS the increase in firing rate above the QS level is 2–3 times, on the average, and is probably considerably more than that in some of the PRF units. Neuronal bioelectric discharges, generated endogenously by most CNS regions from early embryonic stages, have recently been implicated in selective 'stabilization' at the neuromuscular junction [8] – this process could conceivably be occurring at central synapses as well. If so, AS excitations would constitute an ideal epigenetic factor in 'genetic readout' mechanisms of this type [15].

An interesting possibility, based upon isolated cortex experiments in vivo, is that inhibitory synapses in particular require adequate stimulation for maintaining their effectiveness [29]. In the absence of such stimulation a state of generalized neural hyperexcitability results: this mechanism has also been proposed in order to explain physiological effects observed after REM sleep deprivation [33]. It must be emphasized here, however, that tissue culture experiments have placed constraints upon the extent to which neurogenesis may reasonably be expected to depend upon functional activation. Thus, total suppression of bioelectric excitation failed to prevent either the morphological or the electrophysiological differentiation of nerve cells, nor the formation of functional synaptic interconnections [10]. Even the numerical density of synapses formed in vitro has recently been shown to be unaffected by the blockage of all neuronal firing during the culture period [28]. Presumably it is the more subtle aspects of synaptogenesis (affecting the efficacy of transmission and/or spatial distribution) which are sensitive to stimulus levels during development.

It is unnecessary, of course, to suppose that spontaneous bioelectric activity could affect neurogenesis only via direct 'trophic' interactions among the neurons. For example, hormone release patterns are known to be subject to neuronal control, and there are several hormones which have demonstrated effects upon brain maturation [4]. It would in fact be surprising if AS proved *not* to be involved in this process. Still another useful working hypothesis is suggested by recent experiments on the role of monoaminergic neurotransmitters in mediating long-lasting changes in the organization of the cerebral cortex in response to visual stimulation [24]. Since these monoamine systems are intimately involved in sleep mechanisms, AS could interact with sensory stimulation to 'modulate' the degree to which the developing brain will be lastingly affected by experiences to which it is exposed [see ref. 7]. Pursuing this line of reasoning, we have recently been able to show that rats reared in an 'enriched' environment have more than normal amounts of AS throughout the period of extra sensory stimulation [19].

Still another recent suggestion is that AS might contribute to brain maturation in endothermic organisms by facilitating aerobic respiration, while simultaneously reducing energy expenditure [17]. Breathing in kittens is in fact accelerated during bursts of REMs, and the higher their REM activity the greater their resistance to the lethal effects of hypoxia. Finally, the possibility of a purely behavioral function for infantile REM sleep deserves to be mentioned. For example, we have noticed that neonatally AS-deprived rat pups often appeared to be neglected while the rest of the litter was retrieved and nursed [unpublished observations], as if the continual movements in normally sleeping pups served as a sign of life for the mother rat.

Experimental Studies on the Role of AS in Neural Development

The hypothesis that REM sleep is in some way needed for normal brain maturation has been subjected to several tests since it was first formulated [27]. Acute deprivation experiments have been carried out in infant or juvenile rats, cats, monkeys, and humans on the assumption that a developmental requirement for AS might express itself as a stronger tendency than in adult animals to compensate for AS losses ('rebound' effect). Quite the contrary was observed in all four species [2, 3, 5, 14] but this negative result can hardly be said to refute the hypothesis in question. Moreover, a rebound in PGO waves (with almost 100% recovery of the lost phasic activity) has since been reported in kittens following short-term manual deprivation of AS [2]. Further indirect support for the AS hypothesis was obtained in isolation-rearing experiments with rabbits: as predicted, the reduction in external stimulation was partially offset by increased AS time (i.e. retardation of the normal decline with age) during most of the sensory deprivation period [11]. In rats, the hypothesis has been tested by comparing the acute effects upon locomotion of AS deprivation at different ages [13, 23]. Conflicting results are on record, and the approach seems to be inherently unsuited for an unambiguous interpretation of the data obtained.

A variety of *chronic* AS deprivation experiments have also been carried out. Instrumental AS suppression (pedestal technique) applied continuously for *14* days to juvenile deermice resulted in a dramatic increase, as compared with several control groups, in locomotor activity which persisted for at least 3 weeks following cessation of the treatment [21]. Sleep deprivation for a few hours per day immediately following a socialization period in mice which were otherwise reared in isolation from 1 to 4 months caused severe physiological and behavioral abnormalities, which were similar to those seen in animals which had been reared without any social experience whatsoever [34]. Since AS is probably involved in the consolidation of learning which occurs immediately prior to sleep onset, this result is one of the most convincing on record in support of a developmental role for REM sleep. Further support for this hypothesis comes from experiments in which AS was strongly reduced throughout the first 3 weeks of life in rats and cats by means of daily α-methyldopa injections. When tested at 1 month of age, the drug-treated kittens showed clear signs of increased anxiety (less exploration, stronger escape tendencies than in the controls) on several kinds of behavior tests [30]. The rats were tested in adulthood only for locomotor activity, and the experimental group scored higher than did the controls [16]. More extensive follow-ups on these animals would have been most desirable, since the pharmacological approach really is a very promising one.

We therefore decided to use chlorimipramine (CPM) in order to chronically reduce AS in infant rats, based upon its demonstrated long-term potency and selectivity both in clinical practise and in our own pilot studies. This approach was found to be feasible from 1 week of age on, and treatment was continued up to the end of the third week of postnatal life [see ref. 20]. A profound reduction in AS could be achieved throughout this period, with only a gradual recovery in the week thereafter. The AS loss was mostly compensated for by increased time in QS, and the QS epochs lasted on the whole as long or longer than in the control animals. EEG amplitude measurements suggested that CPM-induced QS was relatively deep (prolonged high-voltage slow waves) while multiple unit firing rates tended, correspondingly, to be somewhat lower than they normally are during QS [18]. Also in the PRF, in 5 out of 6 preparations, action potential discharges during sleep became stabilized at or below the normal QS level for some hours following injection of CPM (fig. 1, 2); only upon the return of AS were epochs of relatively high frequency neuronal firing again observed. Since CPM-induced QS at the expense of AS thus appeared to be physiologically normal, a large-scale assessment of eventual sequelae of the treatment in infancy appeared to be justified.

In the absence of reliable guidelines for predicting the effects of infantile AS deprivation, it was decided to survey a wide range of behaviors after the animals reached maturity. Open-field behavior was abnormal in the males: CPM-treated rats showed low levels of exploration and rearing, so as to suggest heightened anxiety levels in this group. These same males also showed signs of anxiety in a passive-avoidance learning situation, being predisposed on the pre-shock trial to enter the dark box sooner than did the controls. Furthermore, although their learning efficiency proved to be normal, the experimental animals performed on the whole more rapidly in operant conditioning situations. As a final disturbance, they were severely deficient in the consummatory aspects of masculine sexual behavior (i.e. intromissions and ejaculations) although the motivational components (latency and frequency of mounting the test females) were undisturbed.

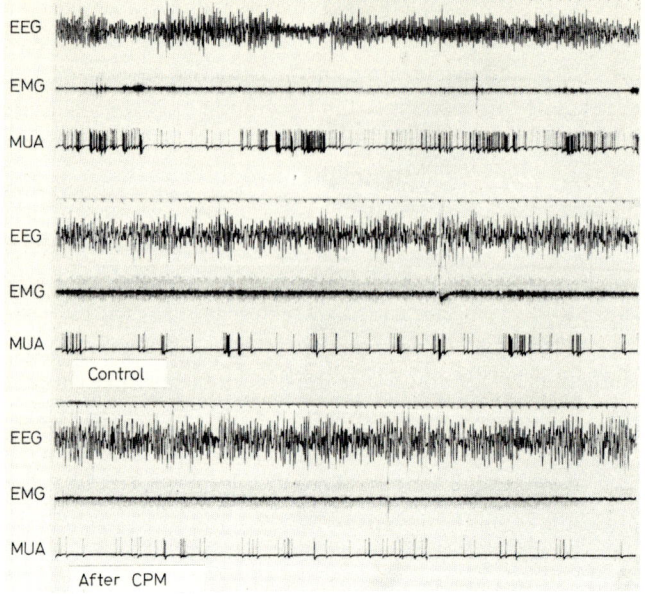

Fig. 1. Multiple-unit activity (MUA) recorded from the pontine reticular formation of a 17-day-old rat pup during *active* sleep (top), *quiet* sleep (middle) and 1–2 h after injection of CPM (bottom). The EEG was recorded from the dorsal hippocampus, the EMG from nuchal musculature. Time calibration is 1 s per division.

Some of these behavioral abnormalities could conceivably be secondary to *sleep* disturbances, since the experimental males – but not the females – showed a striking abnormality of AS. Not only was there twice as much AS as in the controls, mostly due to a higher frequency of AS epochs, but approximately half of the epochs were characterized by unusually frequent and intense muscular twitching [9, 20]. These abnormal epochs lasted much longer on the average than did seemingly normal epochs in the same animal, and they sometimes occurred directly from wakefulness (as in infant animals).

The brains in both male and female rats were analyzed at 15 months of age for possible permanent structural effects resulting from the chronic CPM treatment in infancy [9]. Experimental females proved to have significantly heavier brains than controls, which was attributable to an enlargement of the hippocampus and cerebral cortex. The experimental males, in contrast, showed enlargement only of the hypothalamus. DNA and protein

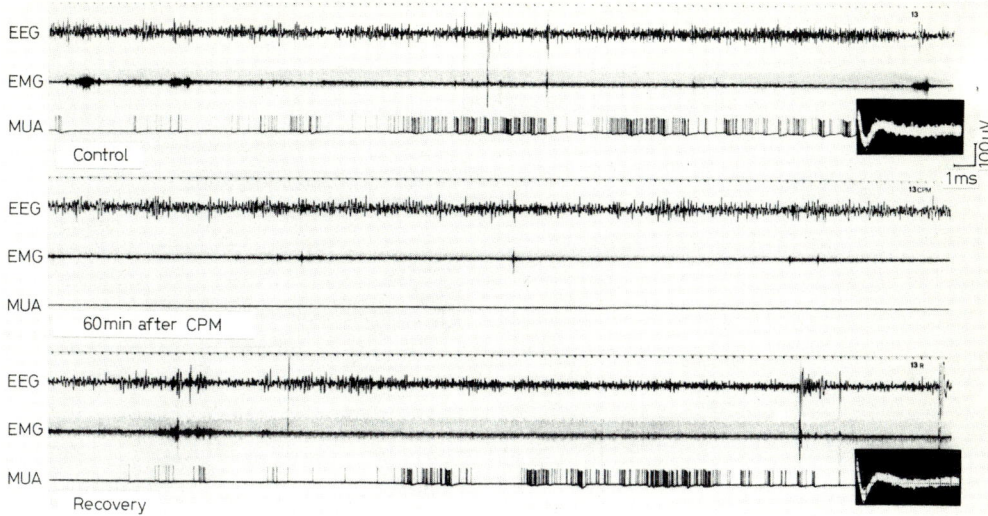

Fig. 2. Activity of a unit in a 13-day-old rat pup (see inserts: 10 superimposed oscilloscope sweeps) which displayed a high rate of firing only during active sleep (top), and which ceased firing altogether during CPM-induced quiet sleep (middle). The original discharge pattern reestablishes itself after recovery (bottom). Time calibration is 1 s per division.

measurements in the affected brain regions showed the expected increases in CPM treated animals, proportional to the observed weight increments. In addition, the cerebral cortex of the males showed an unexpected *decrease* in cell number from the control value. Since total protein was about the same as in control rats, many of the remaining cells must have enlarged considerably (by ca 20% on the average). The question of whether or not the abovementioned effects may be attributed to REM sleep loss suffered in early life must now be considered.

Strategies for the Analysis of AS Effects on Brain Development

Inherent in all experimental studies on ontogenetic mechanisms – regardless of whether the technique used is chemical, mechanical or electrical in nature – is the question of *specificity* of the effects produced. This means that side-effects of CPM must be considered as possible causal factors when interpreting the results of our AS-deprivation experiments. Given the general nature of this problem, along with apparently widespread misunderstanding of its implications [26], it may be useful to outline here some of the 'control' strategies which are at our disposal.

(1) Side-effects monitored during the treatment period (e.g. abnormal food intake or metabolism, motor activity, vigilance level, autonomic variables, etc.), and suspected of contributing to the brain/behavior syndrome which appears in later life, can be studied by using a method which produces such effects *without* affecting sleep. This approach would seem to be an inefficient one in most experimental situations, however. More promising is to combine CPM injections with some treatment which selectively neutralizes the AS-reducing action of CPM (eserine has been suggested as a possibility, but we have been unable to verify its usefulness for this purpose in the infant rat).

(2) AS suppression could be carried out using several fundamentally different procedures, on the assumption that eventual side-effects (if few in number) will in general differ according to the mechanism of intervention. *Instrumental* deprivation is therefore an especially attractive procedure, but it must first be established that this can be accomplished in neonatal animals over a long enough period of time. *Environmental* conditions such as low temperature or disturbing sounds, which are known to preferentially affect AS, ought also to be considered seriously. The importance of cholinergic systems in AS opens an alternative *pharmacological* deprivation possibility, by using selective antagonists such as hemicholinium or scopolamine – if a way can be found to prolong their action in rat pups when injected directly into the brain [unpublished observations]. Localized *brain inactivation* is still another methodological candidate, although the ineffectiveness of even large brainstem lesions which have been tried so far gives some grounds for pessimism [1].

An intriguing line of approach has been opened recently by the discovery that *antibodies* to brainstem perfusates can strongly and preferentially suppress AS for several hours, and that for an even longer period thereafter there is a complete absence of the usual phasic excitation during the AS epochs [12]. It may well prove possible, therefore, to eliminate the characteristic rise in neuronal firing during AS by means of one or two daily injections directly into the brain of developing rats (experiments in progress). The last deprivation technique which will be mentioned is the *genetic* approach in mice [32]; we are eagerly awaiting the successful breeding and testing of very highly AS deficient strains!

It should be added that, with all these approaches, the dose dependency of putative 'AS deprivation syndromes' might give valuable additional clues for assessing the true causality of AS in the developmental effect. For instance, augmentation of the dosage beyond that required for a complete neurophysiological suppression effect ought not to further increase the severity of the symptoms which appear later on in the experiment [e.g. 22].

(3) A different approach exploits the developmental dimension of the problem, either by extending the period of life during which AS is present in large amounts and intensity, or by identifying specific 'sensitive' periods for producing later effects by means of AS deprivation. Especially the latter approach would be very convincing indeed, in the event that a restricted age range during ontogeny could be found which is related to given abnormalities exhibited upon reaching maturity.

The various experimental possibilities have been outlined in some detail because it is by no means clear which of them will eventually prove to be the most useful. Furthermore, it may be doubted that any single 'control' procedure will yield a fully convincing answer, so that a judicious combination of the suggested approaches may be required if an acceptable degree of confidence in a given hypothesis is ever to be achieved. As a final remark, we would urge that the technique of monitoring neuronal discharge rates be adopted whenever possible in sleep-manipulation studies. Only in this way, it would seem, can the efficacy of the intervention be accurately evaluated and its causal relationship to subsequent changes in CNS function established in a quantitative fashion.

References

1 Adrien, J.: Ontogenesis of some sleep regulations: early postnatal development of the monoaminergic systems; in Corner, Baker, Van de Poll, Swaab, Uylings, Maturation of the nervous system. Prog. Brain Res., vol. 48, pp. 393–406 (Elsevier, Amsterdam 1978).
2 Adrien, J.: The nature of sleep in development; its monoaminergic regulation and possible functional influences. Sleep Res. *9* (in press).
3 Anders, T.F.; Roffwarg, H.F.: The effects of selective interruption and deprivation of sleep in the human newborn. Dev. Psychobiol. *6:* 77–91 (1973).
4 Balazs, R.: Hormonal influence of brain development. Biochem. Soc. Spec. Publ. *1:* 39–57 (1974).
5 Berger, R.; Meier, G.: The effects of selective deprivation of states of sleep in the developing monkey. Psychophysiology *2:* 354–371 (1966).
6 Bour, H.L.; Corner, M.A.: Bioelectric activity of brainstem reticular neurons during sleep and waking in developing rats (this volume).
7 Bradley, R.M.; Mistretta, C.M.: Effects of early sensory experience on brain and behavioral development; in Gottlieb, Early influences; studies in development of behavior and the nervous system, vol. 4, pp. 215–247 (Academic Press, New York 1978).

8   Changeux, J.P.; Mikoshiba, K.: Genetic and 'epigenetic' factors regulating synapse formation in vertebrate cerebellum and neuromuscular junction; in Corner, Baker, Van de Poll, Swaab, Uylings, Maturation of the nervous system. Prog. Brain Res., vol. 48, pp. 43–67 (Elsevier, Amsterdam 1978).
9   Corner, M.; Mirmiran, M.; Bour, H.; Boer, G.; Van de Poll, N.; Van Oyen, H.; Uylings, H.: Does rapid-eye-movement sleep play a role in brain development? in McConnell, Boer, Romijn, Van de Poll, Corner, Adaptive capabilities of the brain. Prog. Brain Res., vol. 53, pp. 347–356 (Elsevier, Amsterdam 1980).
10 Crain, S.M.; Bornstein, M.B.; Peterson, E.R.: Maturation of cultured embryonic CNS tissues during chronic exposure to agents which prevent bioelectric activity. Brain Res. 8: 363–372 (1968).
11 De Santis, D.; Waite, E.; Thoman, E.; Denenberg, V.: Effects of isolation rearing upon behavioral state organization and growth in the rabbit. Behav. Biol. 21: 273–285 (1977).
12 Drucker-Colin, R.; Tuena de Gómez-Puyon, M.; Del Carmen Gutiërrez, M.; Dreyfus-Cortés, G.: Immunological approach to the study of neurohumoral sleep factors: effects on REM sleep of antibodies to brainstem proteins. Expl. Neurol. (in press).
13 Hicks, R.A.; Pettey, B.; Okuda, A.; Thomsen, D.: The effects of REM sleep deprivation and age on locomotor activity in rats. Psychol. Rec. 29: 355–360 (1979).
14 Holdstock, T.L.; Verschoor, G.J.: Propensity for paradoxical sleep following deprivation in rats of different age groups. Psychon. Sci. 29: 39–40 (1972).
15 Jouvet, M.: Paradoxical sleep and the nature-nurture controversy; in Mc Connell, Boer, Romijn, Van de Poll, Corner, Adaptive capabilities of the brain. Prog. Brain Res., vol. 53, pp. 331–346 (Elsevier, Amsterdam 1980).
16 Juvancz, P.; Nowaczyk, T.: Effects of early postnatal α-methyl-L-dopa treatment on behavior in the rat. Psychopharmacology, Berlin 42: 95–97 (1975).
17 McGinty, D.J.: Rapid-eye-movement sleep and CNS excitability. Sleep Res. 9 (in press).
18 Mirmiran, M.; Corner, M.A.: Polyneuronal discharges in the cerebral cortex of developing rats during sleep (this volume).
19 Mirmiran, M.; Van den Dungen, H.: Influence of a complex ('enriched') environment on the sleep-waking patterns of developing rats (this volume).
20 Mirmiran, M.; Van de Poll, N.; Corner, M.; Van Oyen, H.; Bour, H.: Suppression of active sleep by chronic treatment with chlorimipramine during early postnatal development: effects upon adult sleep and behavior in the rat. Brain Res. 204: 129–146 (1981).
21 Mitler, M.: Some developmental observations on the effects of prolonged deprivation of low-voltage fast-wave sleep in the deermouse. Dev. Psychobiol. 4: 293–311 (1971).
22 Nelson, P.G.; Bergey, G.: Pharmacological and developmental studies on mammalian central neurons in cell culture; in Giacobini, Vernadakis, Shahar, Tissue culture in neurobiology, pp. 221–228 (Raven Press, New York 1980).
23 Ogilvie, R.D.; Broughton, R.J.: Sleep deprivation and measures of emotionality in rats. Psychophysiology 13: 249–260 (1976).
24 Pettigrew, J.D.: The paradox of the critical period for striate cortex; in Cotman, Neuronal plasticity, pp. 311–330 (Raven Press, New York 1978).
25 Provine, R.R.; Rogers, L.: Development of spinal cord bioelectric activity in spinal chick embryos, and its behavioral implications. J. Neurobiol. 8: 217–228 (1977).
26 Rechtschaffen, A.: The function of sleep: methodological issues; in Drucker-Colin, Shkurovich, Sterman, The function of sleep, pp. 1–18 (Academic Press, New York 1979).

27 Roffwarg, H.P.; Muzio, J.N.; Dement, W.C.: Ontogenetic development of the human sleep-dream cycle. Science *152:* 604–619 (1966).
28 Romijn, H.J.; Mud, M.; Habets, A.; Wolters, P.: A quantitative electronmicroscopic study on synapse formation in dissociated fetal rat cerebral cortex in vitro. Brain Res. (in press).
29 Rutledge, L.T.: Effects of cortical denervation and stimulation on axons, dendrites and synapses; in Cotman, Neuronal plasticity, pp. 273–289 (Raven Press, New York 1978).
30 Saucier, D.; Astic, L.: Effets de l'alpha-methyl-dopa sur le sommeil du chat nouveau-né: évolution comportementale au cours du 1er mois postnatal. Psychopharmacology, Berlin *42:* 299–303 (1975).
31 Steriade, M.; Hobson, J.A.: Neuronal activity during the sleep-waking cycle. Prog. Neurobiol. *6:* 155–376 (1976).
32 Valatx, J.-L.: Possible embryonic origin of sleep interstrain differences in the mouse; in Corner, Baker, Van de Poll, Swaab, Uylings, Maturation of the brain. Prog. Brain Res., vol. 48, pp. 385–391 (Elsevier, Amsterdam 1978).
33 Vogel, G.W.: A motivational function for REM sleep; in Drucker-Colin, Shkurovich, Sterman, The function of sleep, pp. 233–250 (1979).
34 Watson, F.M.C.; Henry, J.P.: Loss of socialized patterns of behavior in mouse colonies following daily sleep disturbance during maturation. Physiol. Behav. *18:* 119–123 (1977).

M.A. Corner, PhD, Netherlands Institute for Brain Research, 28 IJ Dijk,
1009 DB Amsterdam (The Netherlands)

Dreams and Dream Research

David Foulkes
   GMHI-Emory University, Atlanta, Ga., USA

It is an honor for me to address this distinguished audience. I am particularly gratified that I have been asked to speak in a general way about dreaming and dream research. I think that this is a time when some useful things can be said and need to be said about both the subject matter and the methods of dream psychology.

As has been indicated by *Arkin et al.* [1] in their recent book-length review, empirical dream psychology now seems to stand in a state of some greater than usual disarray. The psychophysiological research paradigm which initially was responsible for the formation of a systematic empirical science of dreams, and which has guided the course of this field for almost the past three decades, seems about to have run its course or, at the least, to have reached a state of greatly diminished returns. Naturally enough, this state of affairs has raised retrospective questions about what it is that we really have learned about dreams from psychophysiological research and prospective questions about what sorts of other research paradigms may prove useful in extending our current state of knowledge about dreams.

The psychophysiological paradigm for dream research followed from *Aserinsky and Kleitman's* [2] demonstration, in 1953, of apparent physiological correlates of dreaming. The potential of the paradigm must, at first, have seemed practically limitless. Judging by the appearance of eye movements and other physiological indicators, one could tell when dreaming actually was occurring. One could employ manipulative techniques to determine both the sources and the consequences of the dream process. One had an independent arena in which to test clinically based models of dreaming. By awakening human subjects as they dreamed, one could get detailed, and presumably minimally revised, accounts of the dream process itself, and by further investigation of the physiological accompaniments of that process, one could perhaps even begin to understand the reasons *why* dreams are as they are.

But, as we know, everything did not work out according to these early, exhilarating and, in retrospect, somewhat grandiose, plans. Research has cast doubt on whether there are in fact distinctive physiological correlates of dreaming and, more significantly, there has been reevaluation of whether such correlates ever could explain, and of how they could help us to understand, dreaming as a psychological event. In dream studies themselves, experimental findings have proliferated, but they have been subject to the usual problems of replication and interpretation and, more importantly, rarely seem to have built upon one another in the cumulative manner of research in other areas of biology or even psychology. Finally, diverse theories, models, and viewpoints continue to flourish, even as they did in dream psychology's pre-experimental past, and empirical studies often seem to be conducted more to illustrate a theoretical possibility than rigorously to test a precisely formulated hypothesis.

But it would be mistaken, I think, to take any or all of these situations as implying that there has been little real progress in our understanding of dreams in the past three decades. The reinterpretation of a field's breakthrough discovery, the seeming untidiness of its subsequent findings, the persistence of diverse theoretical viewpoints – all of these things can be signs of that healthy skepticism and pluralism which characterize any developing science at its best. In fact, I think, we have learned a great deal about dreams and dreaming in the past three decades, more than in any comparable period of history and more than we probably had much right to expect. It would be unfortunate if a momentary state of pessimism engendered by the waning utility of the psychophysiological research paradigm obscured the substantial advances that have been made under its aegis.

Many of these advances, not unsurprisingly, given the primitive state of empirical dream psychology 30 years ago, have been not so much at the level of observation as at that of conceptualization and of method. We have, for example, learned that dreaming is both a simpler and a more complex phenomenon than previously imagined. It is simpler because the typical REM sleep dream has been found to be less different from waking cognition than earlier characterizations suggested. It is more complex, because somewhat different forms of dreaming have been found to be associated with different stages of sleep, and this diversity of dream forms suggests that there will be no easy answers to questions about dreams in general – each answer will have to be qualified in terms of the kind of dream to which it best applies. These are major breakthroughs in our grasp of the subject matter of our discipline and, in each case, the finding on which the new conceptualization rests was directly dependent upon the use of a psychophysiological dream-monitoring strategy.

As method, then, psychophysiology has served us well. While concurrent physiological monitoring has not allowed us to collect dreams which, because they come from a uniform stage of sleep, are perfectly homogeneous, it surely has made our job easier by allowing us to sort dream reports into groups with more meaningful interconnection than otherwise would have been possible. Physiological dream monitoring also has enabled us to get closer to the dream phenomena which we want to study, and to know them with minimal contamination by what *Cohen* [4] calls the reconstructive and deductive operations of waking memory. Even in research whose explicit aim has been not merely to describe dreams but also to relate them to extrinsic psychological variables, it is easy to overlook the degree to which psychophysiological methods of dream collection have contributed to the demonstration and interpretation of such relationships. In my own research, described elsewhere in this volume, I was able to show that children's REM dream reports evidence a comprehensible developmental progression in the way in which dreams are constructed and that individual differences in REM-report phenomena are meaningfully related to children's waking cognitive status. Had I relied on spontaneously recalled home dreams, I doubt that these findings could have been demonstrated with equal clarity. Certainly their force as bearing on actual dream experience would greatly have been diminished. For those of us who are curious about the processes of dreaming and about what the mind actually does during sleep, there continues to be no serious alternative to the method of psychophysiological dream monitoring.

If we have been let down by the psychophysiological research paradigm, it has been by its promise of explanatory, as opposed to observational or methodological, value. I remember hearing, when I came into dream research, that definite mind-body parallels had been found in sleep of an order of strength then quite unattainable in wakefulness. Sleep was to be the frontier along which evidential barriers to resolution of the mind-body problem first would be penetrated. Physiological observation not only had shown us when dreams occurred, it also would show us why, and how, dreams occurred.

In retrospect, of course, all that sounds somewhat naive. But certainly a part of dream psychologists' interest in the past three decades in monitoring the substantial progress made in the neuroanatomical, neurophysiological, and neurobiochemical delineation of REM sleep systems lay in the hope that this work would help us to explain, or at least to understand, the nature of REM dreaming. Some daring souls even sought to show us how this might be accomplished. But it didn't work, it doesn't work, and it won't work, at least for the forseeable future. There is simply too large a gap between the level at which neurophysiologists, for instance, proceed – the firing of brain stem neurons – and the level at which dreams are generated – as outputs of complex cerebral and cognitive integration. We have no terribly compelling neural models of complex waking thought, and there is little reason to believe that we soon shall have any. The situation must be precisely identical for dreaming. The neurosciences simply are not yet ready to tackle our kind of problem, except in the most simplistic and least helpful of ways.

Closer to home, many dream psychologists, myself included, devoted much time and effort to looking, at the human level, for more intimate and fine-scaled correspondences between pre-awakening REM physiology and post-awakening REM reports. The search for such correspondences seems to have been doubly motivated: we, too, wanted to explain the course of dream scenarios on the basis of REM physiology, but also more modestly, we simply thought that it would be helpful to have externally observable indices of the course of dream experiencing. The explanatory motive probably was ill-conceived, but the indicational one remains plausible. Unfortunately, however, as has been amply documented by *Arkin et al.* [1], the empirical yield from this kind of research has proved to be meager: findings are difficult to replicate, and of such a low order of strength as not to be terribly useful. Perhaps we have chosen the wrong physiological variables, or perhaps we have been knocking our heads against the limits of the human subject's ability to attend to the infrastructure of the dream. Whatever the reasons, psychophysiological correlational research now appears to offer such a low rate of return for effort expended as not to be a wise place for dream psychology to continue to commit much of its limited resources.

Herein lies the current crisis of the psychophysiological research paradigm and, perforce, of empirical dream psychology. We still want to use psychophysiological recording techniques to collect dreams, but it no longer appears likely that psychophysiological correlational studies will enrich our understanding of what dreams are like or of how they are constructed. We can collect a lot of dreams, but why? To what end? What will we do with them? Increasingly, the answer of dream researchers' own practice seems to be that we shall address an eclectic set of problems from social psychology, personality psychology, and cognitive psychology with the novelty of our dream data. These forays no doubt have been valuable in their own right, and they have had the not inconsiderable additional value of alerting non-dream researchers to the potential relevance of dream data to their own concerns. But, taken together, they do not describe a very coherent kind of dream psychology. In retrospect, we can recognize that psychophysiology for many years gave dream psychology not only a set of methods and a set of problems but also a central focus, a way of thinking about what we were doing that made most research seem to be interrelated and part of some larger plan. But that center no longer holds.

However, if dream psychology is to make systematic and cumulative progress in the years to come, it will require some new central focus, some new way of conceptualizing the problem of dreaming such that what we dream psychologists do coheres and builds. We need something to replace the concept of the dream as a mind-brain frontier. We do not want, however, to return to the earlier concept of the dream as a symptom. Clinical psychiatry no doubt will continue to influence the work that we do, but the time is long past when it could lay claim to being the core of our discipline. Dream psychology has gone 'experimental' and there is no turning back. Its appetite has been whetted for more substantial and potentially falsifiable forms of truth than clinical psychiatry is equipped to provide. We cannot, of course, ignore the sometimes substantial contributions of our clinical forebears. But we must see them for what they are: they are more often hunches than evidence, more often sources of hypotheses than the means of their verification. The kind of reconceptualization which experimental dream psychology requires is one which aligns its research with that of today's empirical science, rather than one which ties its fortunes to outdated theories or inadequate methods.

As prognosticator, I think I know what that reconceptualization will be. As propagandist, I know what I want to argue that it should be. Stated quite simply, it is this: *dreaming is a mental process,* and it must be studied as we now study other mental processes. Whatever brain events *accompany* dreaming, whatever we may *read out* of the dream's content about the life situation of the dreamer, what the dream *is* is a mental act. This, I would argue, is the appropriate starting place for the next round of activity in empirical dream psychology.

How does this admittedly very elementary alignment of dreaming with other cognitive processes promise to advance the prospects of a scientific dream psychology? The answer to that, I think, is quite simple. Simultaneous with recent developments in neuroscience, equally great strides have been made in what now is beginning to be called 'cognitive science', a field embracing experimental cognitive psychology, developmental cognitive psychology and genetic epistemology, linguistics and psycholinguistics, neuropsychology, and the like. Experimentation, observation, and theory formation have been growing at a vast rate in these disciplines concerning how human beings acquire, store, reprocess, and utilize their knowledge. This burgeoning interest in waking mind functioning began at roughly the same time that dream psychology bet its fortune on neuroscience. In consequence, with a few notable exceptions, dream psychology has profited little from recent work in cognitive science, and cognitive science has devoted very little attention to dreaming. Collaborative possibilities between the two would seem to be enormous, and particularly salutary for a dream psychology that momentarily seems to have lost its bearings.

Effecting such cross-disciplinary collaboration will not be easy; it rarely is. There always are the problems of investigators from different backgrounds having different knowledge bases, different senses of what is appropriate methodology, and different theoretical orientations. But in the case of our collaboration with other cognitive-science disciplines, there will be additional handicaps. Some of these are of our own making. We are widely perceived by cognitive science as having a less than total commitment to the canons of empirical science, as being characterized by an inveterate tendency to go well beyond where our data stop to embroider them with allegedly 'deep' but generally untestable and often implausible impressionistic insights. The source of these alleged insights most often is felt to lie in our additional commitment to some form of psychoanalysis or other dynamic psychiatry.

These perceptions are damaging to our integration into cognitive science. They have a substantial basis in fact. During the psychophysiological phase of dream research, unsubstantiated and unwarranted clinical inference managed to live side by side with the measurement techniques of physiology and the research designs of experimental psychology. Psychophysiology gave us the data, and dynamic psychiatry told us what they meant. The temptation surely was great, once our REM dreams had been collected under the imprimatur of high science, to speculate in the loosest of ways about their Oedipal content or their archetypal images. I, for one, confess to having succumbed to this kind of temptation far too often.

What made possible the coexistence of psychophysiology and psychodynamics was the fact that psychophysiology was little concerned with meaning or mind-modeling, and willing to let us fill in that side of the equation in pretty much whatever way we wished. These conditions most clearly will *not* exist in a collaboration of dream psychology with cognitive science. The clipping of the sails of clinical pretension which accompanied our collaboration with psychophysiology was modest indeed in contrast to what will be required of us now. But, I would submit, this is precisely what our field itself requires if we are to make substantial further progress. This purgation must not entail such a mad rush to respectability that we disown portions of our clinical heritage which in fact are responsive to actual peculiarities of our subject matter, but it still will have to be considerable.

Not all of the handicaps to an effective collaboration with mainstream cognitive science will be of our own making. There are reasons other than our commitment to clinical methods and theory which have impeded, and will continue to impede, the consideration of dreams by cognitive scientists. Substantial progress has been made in recent years by experimental cognitive psychology; it recently has been identified by *Simon* [9] as one of the most productive and promising areas of behavioral science. Cognitive psychology probably is the cognitive-science discipline with most potential reason to be interested in dreaming. But the experimental paradigm under which it has progressed, and which it now, understandably, will want to preserve, seems largely incompatible with any consideration of dreaming. In this paradigm, experimentally controlled inputs are compared with carefully observed outputs in order most economically to model, psychologically, the information processing which mediated the person-environment exchange. The problem with dreaming, or with any spontaneous fantasy, is that the inputs to the process are largely memorial rather than sensory, and hence can be neither experimentally controlled nor even known with any great degree of certainty. Cognitive psychologists likely may believe that it is precisely by the sloughing off of problems inconsistent with their paradigm – such as dreaming – that they have been able to make the advances they have. Reluctance to reopen consideration of phenomena such as dreaming no doubt will be considerable, but I think that that resistance can, by dint of our own efforts, be overcome.

What we will need to do is to demonstrate that we can conform to the more general methodology of cognitive science, in particular to show that we can collect relevant observations with which to formulate, modify and, as the evidence demands it, discard, models of the mental processing of the sleeping mind. These models should articulate with observation and theory elsewhere in cognitive science, just as acceptable psychophysiological models of dreaming should conform to more general evidence and theory in the neurosciences. This means that there will be stringent limits on our propensity to smuggle concepts from clinical dream theory into our modeling. Perhaps we shall be able independently to establish the requirement for constructs with functions or mechanisms reminiscent of such clinically based concepts as censorship and repression, but such concepts cannot constitute initial assumptions of our modeling efforts.

I want to close my discussion by, at long last, being reasonably specific. I would like to suggest some of the kinds of observations we can collect and some of the kinds of models we can elaborate within a general cognitive-science framework. My list is by no means meant to be inclusive.

First, we need to address the question of the programs which guide the formation of the dream. Dreams generally are reasonably coherent narratives, stories whose elements are reasonably well interrelated. This fact suggests that there is an underlying narrative program for the dream, a story grammar at the level of the manifest dream which lends organization and even predictability to its moment to moment flow of imaged events. By examining dream-reports in which the intradream sequence is clear – note, here, our continuing dependence on psychophysiological dream collection – we can search for principles of the dream story's organization, much as developmental cognitive psychologists search for the story grammars of children's spontaneous waking narratives. Individual and group differences, as well as regularities across persons and groups, are obvious topics of interest, as also are variations by stage of sleep.

Next, we can attempt to approach another aspect of dream generation. While the narrative or story grammar I have just discussed will deal with the sequencing of *general categories* of dream events, such as the likelihood that a given class of interaction will be followed by another such class, we also need to better understand the bases on which *particular memorial elements* are recruited for dream usage and the ways in which they are transformed in their dream representation. By analogy to linguistics, we here are asking about lexical selection, or the particular meanings attaching in particular dreams to general structural elements in dream formation. This is, in a sense, the deeper, and certainly the more individualized, part of reconstructing how it is that dreams are put together. The research objectives here are quite similar to Freud's search for dream sources via free association and for dream transformations via a dream-work analysis. In a way that is not true of the story-grammar analysis, we here require the collaboration of the dreamer beyond the point of simply telling the dream. The dreamer also must suggest memorial sources of the dream and help us date and categorize these sources. I [5] have tried to show that reliable empirical means can be devised for modeling how memories are transformed in dreaming. I would not claim, and do not know, that the means I have suggested are the best ones available, but the very fact of their formulation does open the more general possibility of serious empirical research on how memories are recruited in, and how they are treated by, the dreaming process.

The ways in which these two levels of modeling dream generation can be shown to have to be dependent upon one another will be a particularly interesting feature of their analyses. Presumably the course of the dream at moment n+1 is some sort of joint function of the surface-structural regularities or sequential probabilities associated with the particular dream feature experienced at moment n and of the underlying memories responsible for and stimulated by that particular dream feature. It is not implausible to imagine that a distant but realizable goal of the integration of these two levels of dream modeling will be computer simulation of dreams. That is, given either general or idiosyncratic rules of narrative formation and of memory selection and transformation, and lexical entries from my own life experience, computer programs should be capable of being sharpened to such a point that *their* dreams are indistinguishable, for outside judges, at least, from my own.

Obviously, neither of the strategies I have discussed is going to be easily implemented. It is less the collection of relevant data that is at question than the apparently monumental task of organizing them. But the problems to which these strategies are addressed are the central ones of dream psychology, and if they had easy resolutions we no doubt already would have found them. With the helpful guidance of the not altogether unrelated mind-modeling already going on in the cognitive psychology of wakefulness, the time is ripe for tackling these problems head-on. We stand to lose little by making the effort, and it certainly seems preferable to continuing to dwell poetically on the 'mysteries' of dream formation.

A third research strategy is relatively more easily implemented, but nothing is truly easy in dream psychology. This strategy provides an experimental model of dream formation that *does* conform to the information-processing paradigm of cognitive psychology. It is, in fact, the old stimulus-incorporation paradigm of dream psychology, in which we try to influence dreams by controlled externally delivered stimulation during sleep, during, particularly, psychophysiologically monitored REM sleep. In employing this paradigm, however, we now will have to give more attention to stimulus-effects, rather than stimulus-incorporations – the stimulus may have influenced the dream even if it does not appear in it in a highly recognizable form – and we will have to be more concerned – along the model of *Berger's* [3] early study – with *how* the stimulus has been processed by the dreaming mind. The stimulus-incorporation paradigm is not important in the sense that it demonstrates the typical sources of dream imagery, because it no longer seems plausible to believe that dreaming often is or has to be initiated by aperiodic external stimuli. But the paradigm is a terribly important one because it may help us model the processing mechanisms of the dreaming mind, i.e. the processes which generally operate on memories rather than sense inputs, but which we experimentally have forced to operate on stimuli whose nature we can both identify and control.

A fourth research strategy bends toward the neurosciences. Neuropsychology sits astraddle the neuroscience-cognitive science boundary, in that it is concerned with the neural organization at a systemic level of higher human cognitive functions. Its case materials are patients with generally circumscribed neural lesions and cognitive deficits. We stand to gain much knowledge about the cognitive systems implicated in dream generation by coordinating psychophysiological dream monitoring techniques with more standard neuropsychological assessments, by implementing a neuropsychology of dreaming. Recently, for instance, we have studied two individuals [7, 8] with waking deficits not in visual imagination in general but in a particular kind of visual imagination: they could not mentally rotate visual images. They could not, for instance, solve problems in which test figures presented at different angles of rotation are to be judged as identical or non-identical to a standard figure. Associated with this defect in kinematic mental visualization was an either total or general failure to experience dream narratives in visual imagery. The subjects, one with Turner's syndrome and the other with extensive right-hemisphere damage, clearly dreamed: they simply did not see the dream events which they felt to be happening before their awakening. These observations are illustrative of the way in which special and unusual cases may illuminate the more general contribution of certain cognitive processing systems to dream formation.

It can be equally informative to begin with persons who make unusual claims about their dreamlife, and then to try to determine waking cognitive defects or anomalies to which these dream phenomena are related. We [6] found that two adolescent siblings with near-zero REM recall had specific cognitive deficits in skills assessed by block-design and embedded-figures tests. It would be interesting to track down adult subjects whose claims to be nondreamers *cannot* be contradicted by laboratory-awakening data, and to determine their cognitive profiles. Once again, the general point is that, in cognitive science as in personality study, the aberrant case can cast great light on the normal one.

Finally, the ontogeny of dream reporting on laboratory awakenings needs to be coordinated in a serious and disciplined way with children's accession to stages of waking cognitive development. My own research has made first steps in this direction, but its findings require replication and extension in studies that are more carefully focused on cognitive issues than was mine. The importance of research on dream ontogeny is, of course, at least twofold: we are interested in children in their own right and, as for waking thought, we stand to better appreciate what it is that we adults can do by understanding the developmental course which led us to these possibilities. At minimum, my experience suggests that meaningful laboratory dream research with children is possible. It further suggests that the course of dream development is determined and constrained by children's waking cognitive development. Preoperational children, for example, seem in their REM dreams to be unable to generate kinematic imagery, to effect definite self-representation, to symbolically reconstruct affect, and to decenter their thoughts from concurrently active stimuli such as their own body states. These features of early dreaming are mirrored, in each case, by concomitant waking observations of children's symbolic-representational behavior. My data further suggest that the cognitive constraints on what and how children can dream by no means lose their force in early childhood, but persist at least well into adolescence, where characteristic dream changes are strongly related to waking cognitive-representational variables and not at all related to children's particular psychosocial situations. More generally, my data suggest that the key to understanding developmental changes in dreaming lies in cognitive maturation rather than in personality development, that the contents of children's dreams better index their state of cognitive skill than their waking life behaviors or circumstances. This is so, apparently, precisely because dreams *are* mental acts, and because there are, particularly in early development, substantial gaps between what children do and their understanding of and ability to symbolically recreate what it is that they do.

These, then, are a few of the directions in which it seems to me that dream research might move to better achieve its own goals and to better effect collaboration with other disciplines studying high-level human mental processing. Central to all these suggested research plans is the conceptualization of dreaming as thinking. That conceptualization would have to be judged so elementary as hardly to merit notice were it not for the fact that its research implications have been so widely ignored or overlooked. But there are many such implications, and both theory and method increasingly are available to guide their implementation. I do not mean to argue that the kinds of research strategies I have discussed are the only ones that may be fruitful or that deserve serious attention. The unexpected way in which the *Aserinsky and Kleitman* [2] discovery burst upon the scene years ago reminds us that we never know whence to expect a critical finding or a productive method. Still, insofar as the future of a discipline can be foretold by reason and logic alone, it seems to me that now is a particularly auspicious time for the dream to establish its position as a legitimate and fruitful object of study by cognitive science. In a decade or two, someone else will no doubt come along to explain why *that* didn't work out, but in the meantime we can try to see if we can match the same kind of productivity as has characterized our immediate, psychophysiological, past.

References

1. Arkin, A.M.; Antrobus, J.S.; Ellman, S.J.: The mind in sleep: psychology and psychophysiology (Lawrence Erlbaum Associates, Hillsdale 1978).
2. Aserinsky, E.; Kleitman, N.: Regularly occurring periods of eye motility, and concomitant phenomena, during sleep. Science *118:* 273–274 (1953).
3. Berger, R.J.: Experimental modification of dream content by meaningful verbal stimuli. Br. J. Psychiat. *109:* 722–740 (1963).
4. Cohen, D.B.: Sleep and dreaming: origins, nature and functions (Pergamon Press, Oxford 1979).
5. Foulkes, D.: A grammar of dreams (Basic Books, New York 1978).
6. Foulkes, D.: Home and laboratory dreams: four empirical studies and a conceptual reevaluation. Sleep *2:* 233–251 (1979).
7. Kerr, N.H.; Foulkes, D.; Jurkovic, G.J.: Reported absence of visual dream imagery in a normally sighted subject with Turner's syndrome. J. ment. Imag. *2:* 247–264 (1978).
8. Kerr, N.H.; Foulkes, D.: Right hemispheric mediation of dream visualization: a case study. Cortex (in press).
9. Simon, H.: The behavioral and social sciences. Science *209:* 72–78 (1980).

D. Foulkes, PhD, GMHI-Emory University, Atlanta, GA 30306 (USA)

La narcolepsie

P. Passouant

Service de Physiopathologie des Maladies Nerveuses, Centre Gui de Chauliac, Cliniques St-Eloi, Montpellier, France

La narcolepsie ou maladie de Gélineau est le désordre du sommeil, ou plus exactement du cycle veille-sommeil, le plus curieux et le plus riche dans son expression clinique. Marquée par l'inversion des deux sommeils, avec production du sommeil paradoxal dès l'endormissement, c'est au cours de cette maladie que surviennent les aspects dissociés du sommeil dont la cataplexie et les hallucinations hypnagogiques sont les plus précis.

La narcolepsie décrite en 1880 par Gélineau a subi bien des vicissitudes sur une période de cent ans et son histoire est une excellente approche pour situer cette maladie [1].

Gélineau a donné le nom de narcolepsie à une affection traduite par de brusques accès de sommeil, associés à des chutes appelées «astasies». Par cette dénomination, l'accent était mis sur la brusquerie de l'accès. Mais si l'influence de l'émotion était retenue dans le déclenchement du sommeil, par contre les chutes étaient rapportées à ce dernier, rapprochées de celles «d'un ivrogne ou d'un enfant endormi». Le mérite de Gélineau est d'avoir donné une autonomie à cet état, de l'avoir différencié de l'épilepsie, et à cette époque la narcolepsie fut classée dans le groupe des névroses.

Si la dénomination de la narcolepsie remonte à 1880, sa description est antérieure. Une observation comparable fut rapportée en 1862 par le docteur *Caffé* de Paris [cité dans réf. 1] dans le *Journal des connaissances médicales et pharmaceutiques* et l'affection dénommée «maladie du sommeil» observation rapprochée par Gélineau de la sienne. Les accès de sommeil du sujet en question furent par la suite associés à des hallucinations ce qui entraîna une hospitalisation en milieu psychiatrique où il fut traité par le docteur *Semelaigne*. Ce dernier rejeta la diagnostic de maladie du sommeil, rapporta l'état à une épilepsie et répondit, non sans humour, au docteur Caffé: «il ne vous déplaira sans doute pas plus qu'à moi, mon cher confrère, de laisser aux nègres la maladie du sommeil, les blancs en ayant assez d'autres sans celle-là?»

Westphal [cité dans réf. 1] publia en 1877 l'observation détaillée d'un sujet atteint d'accès de sommeil à survenue brusque, associés à des épisodes de pertes de mouvements et de langage et signala de plus que la mère de ce malade était atteinte de la même affection, décrivant ainsi le premier cas familial de cet état. Hésitant sur la nature de cette affection, *Westphal* la rapprocha d'une «attaque épileptique».

Après cette première période, la narcolepsie reste une maladie rare et peu connue jusqu'à la deuxième décade du 20e siècle, époque de l'encéphalite épidémique, qui suscita un intérêt pour les troubles du sommeil. Les cas publiés deviennent de plus en plus nombreux. Les deux symptômes, accès de sommeil et cataplexie sont différenciés et leur association considérée comme nécessaire dès 1902, par *Lövenfeld* [cité dans réf. 1], pour le diagnostic de la narcolepsie. L'atonie cataplectique rapprochée de celle qui accompagne le sommeil est rapportée à une dissociation de la fonction hypnique: «Körperschlaf» des auteurs allemands ou «sommeil somatique» des auteurs français.

Les symptômes de la narcolepsie s'enrichissent. Les paralysies du sommeil décrites en 1876 par *Weir Mitchell* [cité dans réf. 1] chez des sujets normaux sont signalées par *Wilson et Lhermitte* [cités dans réf. 1] dès 1928 chez les narcoleptiques et une relation entre ces deux manifestations est retenue.

Les hallucinations hypnagogiques prédormitionnelles du sujet normal ont fait l'objet de nombreuses études de langue française au cours du siècle dernier. Leur fréquence fut signalée lors de l'encéphalite épidémique et leur relation avec une atteinte pédonculaire proche du noyau rouge retenue par *Van Bogaert* [cité dans réf. 1]. Décrites au cours de la narcolepsie dès 1924, *Schester* [cité dans réf. 1], souligna leur caractère pénible «flammes grimpant au mur, images de squelette, de tête de mort». Rapprochées du rêve, elles étaient interprétées par *Lhermitte* [cité dans réf. 1] comme «l'équivalent actif de la narcolepsie».

Si à cette époque qui va jusqu'en 1940 les symptômes de la narcolepsie sont mieux connus, par contre de nombreux auteurs considèrent la narcolepsie comme une affection secondaire à des causes les plus diverses et non comme une affection autonome. Pour *Wilson* [cité dans réf. 1] «il est non scientifique et réactionnaire de considérer un désordre du sommeil comme une maladie et en particulier la variété décrite comme narcolepsie ne saurait être un morbus sui generis.» La narcolepsie, au même titre que l'épilepsie, devient un symptôme de causes très variées: traumatiques, endocriniennes, épileptiques, circulatoires, tumorales, encéphalitiques.

De son côté, *Lhermitte* [cité dans réf. 1] n'hésite pas à écrire «j'ai le regret de ne pouvoir présenter ou de ne pouvoir vous relater l'observation d'un malade personnellement suivi et chez lequel les caractères de la maladie eurent permis le diagnostic de narcolepsie purement idiopathique c'est-à-dire dépourvus de faits concomitants pathologique ou viscéral.»

C'est vers 1940, que la narcolepsie est isolée à nouveau en tant que *maladie autonome* et séparée des hypersomnies idiopathiques, pour une grande part grâce aux travaux de la Mayo Clinic.

La tétrade narcoleptique est bien individualisée avec l'accès de sommeil, la cataplexie, les hallucinations hypnagogiques et les paralysies du sommeil. L'accès de sommeil est commun à tous les cas, associé à la cataplexie dans 64% des cas, aux hallucinations hypnagogiques dans 30% des cas, aux paralysies du sommeil dans 24% des cas. Ces résultats d'une étude de *Yoss et Daly* [cité dans réf. 1] datant de 1960 sont toujours d'actualité.

L'application de l'EEG à l'étude de la narcolepsie n'a pas été sans inconvénients. Des anomalies signalées par certains auteurs ont fait reprendre l'analogie ancienne de narcolepsie-épilepsie. La localisation d'anomalies à la région temporale a fait envisager une participation du rhinencéphale.

Toutefois, deux expressions du sommeil sont signalées dès cette époque. L'une correspond à celle d'un sommeil avec tous les stades A, B, C, D, E, l'autre est marquée par un tracé d'endormissement (stade A) persistant environ 30 min. sans que les autres signes EEG et en particulier les «spindles» se produisent. Ce deuxième type, qui très probablement correspondait à des périodes de REM, fut interprété en tant que périodes de somnolence.

La période actuelle est celle de la REM-narcolepsie. Après la découverte du REM, *Vogel* [cité dans réf. 1] en 1960 étudiant le sommeil d'un narcoleptique, constata un endormissement en sommeil paradoxal. Ce résultat singulier, car l'entrée dans le sommeil se fait toujours en sommeil lent, allait entraîner un renouveau dans les études de la narcolepsie, après confirmation par *Rechtschaffen et al.* [2], *Takahashi et al.* [3] pour le sommeil de nuit et par nous-mêmes [4] pour le sommeil de jour et de nuit.

Le bilan d'aujourd'hui ou narcolepsie 1980 mérite d'être situé à plusieurs points de vue: l'inversion des deux sommeils, les anomalies du cycle veille-sommeil, la physiopathologie et le traitement.

L'inversion des deux sommeils

L'endormissement en sommeil paradoxal, soit de jour, soit de nuit, est actuellement unanimement reconnu.

Le nombre des accès de sommeil est des plus variables selon les sujets. Exceptionnellement, il a été observé chez des narcoleptiques, examinés en télémétrie, donc libres de leur activité, une périodicité de 2 h des accès débutant en sommeil paradoxal. Une telle expression est facilitée en modifiant l'environnement et en faisant allonger le sujet toutes les 2 h sur un lit et dans l'obscurité [5].

Ainsi, au cours de la narcolepsie, pourrait être identifiée une véritable «horloge biologique» du sommeil paradoxal dont le rythme ultradien de 2 h observé normalement durant la nuit est identifié durant le jour. L'anomalie narcoleptique, avec endormissement en sommeil paradoxal n'est toutefois pas constante. Certains accès de sommeil sont précédés de sommeil lent, d'autres sont uniquement formés de ce sommeil.

Cette variation peut dépendre de plusieurs facteurs dont la conservation du rythme circadien du sommeil paradoxal. La facilitation de ce sommeil dans la matinée, son inhibition en fin d'après-midi, observées chez le sujet normal, sont trouvées chez le narcoleptique. Au cours d'une étude faite chez 46 narcoleptiques, les accès en sommeil lent ont été constatés en fin de journée, alors que les accès en sommeil paradoxal prédominaient dans la première partie du jour. Dans les deux cas, la brusquerie de l'accès était identique [6].

Une autre expression est possible, faite de l'association de mouvements oculaires rapides, d'un aspect EEG de sommeil paradoxal ou parfois sommeil lent et de conservation du tonus musculaire. Cet aspect dissocié, habituellement observé chez les sujets traités par la chlomipramine peut être trouvé spontanément [7].

En complément des accès de sommeil, d'expression polygraphique différente, la narcolepsie se traduit par des périodes de somnolence de durée variable mais qui peut atteindre plusieurs heures. C'est au cours de ces périodes que surviennent des automatismes moteurs, qui ne laissent pas de souvenir et qui peuvent donner le change avec un équivalent épileptique. Un parcours à pied, fait normalement en 10 min, demande plus d'une heure, une dactylo continue à taper son texte, mais commet de nombreuses erreurs. Un conducteur automobile peut conduire sa voiture et se retrouver à 50 km de sa résidence, devant une maison qui n'est pas la sienne.

Ces périodes correspondent au stade I du sommeil lent avec absence du rhythme alpha, avec parfois des mouvements oculaires lents. De courtes périodes de sommeil paradoxal ou «micro-REM sleep» ont été décrites par *Dement* [8].

L'analyse de l'excès de sommeil des narcoleptiques durant le jour montre que l'anomalie ne concerne pas uniquement le sommeil paradoxal mais aussi le sommeil lent. Une telle éventualité ne saurait surprendre car les deux sommeils ne peuvent être dissociés. Il apparaît ainsi que la narcolepsie mérite d'être située dans un sens dynamique ou plus exactement d'une compétition des deux sommeils avec une prévalence du sommeil paradoxal.

La facilitation du sommeil lent peut être majorée au cours de l'évolution de la maladie. Chez certains sujets, les accès de sommeil peuvent être uniquement en sommeil lent alors qu'à d'autres périodes, l'anomalie narcoleptique est précise. Chez d'autres, ayant présenté des attaques de cataplexie lors de l'adolescence, des examens faits à plusieurs années d'intervalle, ont montré des accès de sommeil de jour en sommeil lent et une organisation normale du sommeil de nuit.

Ces diverses constatations posent un problème nosologique et les frontières de la narcolepsie peuvent être difficiles à délimiter, soit avec une hypersomnie idiopathique, soit avec d'autres hypersomnies par suite de la possibilité d'apnées ou de modifications des conduites alimentaires chez certains narcoleptiques.

L'inversion des deux sommeils permet une meilleure compréhension des aspects dissociés du sommeil et de la richesse de l'activité onirique du narcoleptique.

La cataplexie et les paralysies du sommeil qui en sont rapprochées ont été rapportées à un sommeil paradoxal dissocié, l'atonie musculaire étant comparable à celle qui survient durant ce sommeil. Cette interprétation retrouve partiellement celle qui avait été antérieurement donnée de «sommeil somatique».

Les hallucinations hypnagogiques sont liées à l'activité imaginaire du sommeil paradoxal.

Ces hallucinations peuvent intéresser tous les secteurs sensoriels. Elles sont tantôt uniquement visuelles ou auditives, tantôt polymorphes. Le narcoleptique est très souvent impliqué par l'imaginaire qu'il perçoit et toute une gamme de réactions est possible. La charge émotionnelle va de l'inquiétude à l'épouvante et les hallucinations terrifiantes sont plus fréquentes avant le sommeil de nuit. C'est ainsi qu'un de nos sujets entendait chaque soir des gens qui rôdaient autour de lui, avec intention de le tuer et l'heure du coucher était devenue pour cet homme une véritable épreuve.

Certes, la prise de conscience des caractères hallucinatoires est variable selon les sujets et les périodes de la maladie. Toutefois, même lorsque le sujet reste lucide et prétend ne pas dormir, sa conscience vigile est différente de celle de la veille. Ainsi s'exprimait un de nos narcoleptiques: «je suis éveillé puisque je sais où je suis et ce que je pense, mais mon esprit est incapable de s'écarter de la vision hallucinatoire qui m'absorbe.»

Il est à remarquer que les hallucinations peuvent être arrêtées par des moyens simples. C'est ainsi que le sujet qui présentait des hallucinations terrifiantes au moment du coucher arrivait à les empêcher par la présence d'une jeune enfant couchée dans son berceau.

La relation de l'hallucination hypnagogique et du rêve est certaine chez le narcoleptique et la séparation peut en être difficile.

Les narcoleptiques rêvent beaucoup et la richesse de leur activité onirique peut être à l'origine de réactions comportementales. La qualité du rêve peut être différente selon le nycthémère et celle qui accompagne le rêve du début de nuit est souvent associée à une importante angoisse. Des images souvent terrifiantes à type d'incendie, de bombardements, d'animaux fantastiques, sont alors relatées. Une véritable agitation peut accompagner de tels rêves et le sujet se débat contre des ennemis imaginaires. Une telle situation est proche du comportement du sommeil paradoxal observé par *Jouvet et Delorme* [18] chez des animaux après lésion caudale du locus coeruleus.

Si la richesse des rêves est liée à l'endormissement en sommeil paradoxal, elle l'est aussi par le nombre de rêves qui surviennent au cours d'une période de ce sommeil. Pour approcher ce problème, des éveils répétés ont été provoqués durant une phase de sommeil paradoxal. A la différence du sujet normal, le narcoleptique retrouve immédiatement ce sommeil. Durant une période de sommeil paradoxal de 30 min, 14 éveils provoqués ont été suivis d'un retour immédiat à ce type de sommeil et le sujet a, chaque fois, raconté un rêve différent [9].

Anomalies du cycle veille-sommeil

Il est fréquent de constater, au cours de la narcolepsie, une mauvaise qualité du sommeil de nuit. Les enregistrements polygraphiques ont en effet précisé la diminution ou l'absence du sommeil lent profond, les mouvements nombreux et les éveils fréquents de durée parfois prolongée. Ainsi les accès de sommeil et les périodes de somnolence de jour s'opposent à l'insomnie de nuit, aussi avons-nous proposé la dénomination de *dysomnie* pour différencier la narcolepsie des autres hypersomnies [4].

Il apparaît que l'organisation temporelle de la veille et du sommeil, acquise avec l'âge et dépendant de synchroniseurs en rapport avec la vie familiale, professionnelle et culturelle, est à des degrés divers, modifiée. Situer la narcolepsie par rapport à une modification des rythmes biologiques, n'est que pour le moment partielle, mais certaines données sont en faveur de cette orientation.

Le début de la narcolepsie peut être associé à une perturbation du rythme veille-sommeil. C'est ainsi qu'au cours d'une étude faite sur 50 narcoleptiques, chez 12 d'entre eux les premiers symptômes de la maladie ont suivi une modification du cycle veille-sommeil provoquée par le travail posté, ou un raccourcissement de la durée du sommeil dû à diverses circonstances.

Il peut exister, dans certains cas, une désorganisation complète du cycle veille-sommeil.

Au cours de privations sensorielles, le sujet étant couché dans l'obscurité durant 24 h, il a été constaté une importante augmentation de la durée totale du sommeil: 15–18 h pour 5–6 h de sommeil paradoxal avec des cycles raccourcis de 45 à 100 min. [5].

Un tel changement peut survenir spontanément lors des périodes d'aggravation de la maladie. La narcolepsie qui débute habituellement à la puberté est marquée par des périodes d'évolution avec une somnolence presque continue associée aux signes auxiliaires.

L'étude polygraphique a alors précisé une durée de sommeil très prolongée de 15–18 h au cours du nycthémère, avec une augmentation homogène pour les deux sommeils. Ces aggravations, de cause inconnue, surviennent en général quelques années après le début de la maladie. Leur durée est variable de quelques mois à quelques années.

L'aspect polyphasique de la veille et du sommeil est alors comparable à celui que l'on observe chez le jeune enfant et ces résultats nous ont fait soulever l'hypothèse d'une involution des états de vigilance chez le narcoleptique [10].

Ces périodes d'aggravation peuvent se traduire uniquement par des attaques de cataplexie. C'est ainsi que chez un sujet nous avons pu enregistrer 80 attaques de cataplexie durant le jour. Cet «état de mal cataplectique» a cédé en 48 h avec 50 mg de chlomipramine.

Parmi les autres anomalies des rythmes biologiques observées au cours de la narcolepsie, celles qui intéressent les sécrétions hypophysaires sont les plus précises.

Il est actuellement reconnu que la sécrétion de GH est augmentée au début du sommeil de nuit, lors de la première période de sommeil lent profond. Cette hypersécrétion liée au sommeil est retrouvée lors d'une inversion avec production du sommeil durant le jour. D'un autre côté, la sécrétion du cortisol dépendant d'un rythme circadien est sans liaison avec le cycle veille-sommeil.

Il a été montré [11–13] que le pic plasmatique de GH survenant au début du sommeil de nuit était dans la plupart des cas absent au cours de la narcolepsie. D'autre part, il a été indiqué [13, 14] que deux types de sécrétions pouvaient être observés au cours des 24 h, soit une très faible sécrétion, soit à l'inverse, une hypersécrétion pulsatile avec de nombreux pics de l'ordre de 4–18 ng/ml.

L'interprétation de ces résultats est difficile. Il est possible que l'absence de sécrétion de GH au début du sommeil de nuit soit en rapport avec une anomalie fonctionnelle hypothalamique, entraînant à la fois la production du sommeil paradoxal et l'abolition de sécrétion de GH. Quant aux deux expressions, il nous a paru qu'elles pouvaient avoir une relation avec l'aggravation de la maladie, l'hypersécrétion pulsatile avec de nombreux pics étant observée au cours des poussées évolutives de la narcolepsie.

Problèmes physiopathologiques

Diverses études, en particulier, celle de *Bruhova et Roth* [15] et de *Kessler et al.* [16] ont précisé le caractère familial de la narcolepsie. Ces études génétiques ont, d'autre part, montré que dans les familles de narcoleptiques il existait d'autres troubles du sommeil, soit paralysie isolée, soit hypersomnie idiopathique, comme si la narcolepsie était «l'expression finale de divers désordres du sommeil». Bien que ces résultats retrouvent la discussion nosologique que nous avons envisagée, il n'en reste pas moins que l'incidence familiale est en faveur d'une prédisposition individuelle et d'une origine biochimique de la maladie. Toutefois, les différentes recherches concernant les anomalies des métabolites des monoamines, soit dans les urines, soit dans le liquide céphalo-rachidien, n'ont jusqu'à maintenant pas apporté de résultats convaincants.

Sur le plan neurophysiologique, l'approche physiopathologique la plus précise concerne la cataplexie. Il est actuellement reconnu que l'atonie musculaire qui accompagne le sommeil paradoxal est due à un effet postsynaptique sur le motoneurone de la formation réticulaire inhibitrice bulbaire [17].

Le déclenchement de la cataplexie par une émotion peut trouver une interprétation dans les relations de cette formation avec le système limbique et la région orbitaire.

La production de narcolepsies réflexes a été plus difficile. Sur des préparations pontiques *Jouvet et Delorme* [18] ont obtenu par des stimulations proprioceptives ou nociceptives le sommeil paradoxal. Cette mise en jeu reste soumise à une phase réfractaire, alors qu'il est possible d'obtenir de façon itérative des épisodes d'atonie musculaire sans signe électrique ou oculaire du sommeil paradoxal.

Sur une préparation «encéphale isolé» concernant des chats privés antérieurement de sommeil paradoxal, *Dell et Puizillout* [19] ont provoqué par stimulation vago-aortique une narcolepsie réflexe. Le bombardement des chémorécepteurs entraîne une activation du noyau du faisceau solitaire, puis une déactivation du système réticulaire et provoquerait une accélération des processus biochimiques du REM.

Enfin *Mitler et Dement* [20] auraient provoqué le sommeil paradoxal par injection de carbachol dans la région pontine.

L'existence de narcolepsie-cataplexie chez certains animaux (caniches, chats, poneys) a donné un nouvel essor aux recherches expérimentales. Ces modèles étudiés, en particulier par le groupe de Stanford, paraissent pleins de promesses. Toutefois, pour le moment, les études anatomiques n'ont apporté aucune information et seules certaines études pharmacologiques peuvent être retenues.

Approche thérapeutique

Les psychostimulants ont été pendant longtemps le seul traitement de la narcolepsie en vue de contrôler les accès de sommeil et les périodes de somnolence. Après le citrate de caféine, proposé par *Gowers* [cité dans ref. 1] en 1907, le sulfate d'éphédrine puis le sulfate d'amphétamine ont été utilisés à partir de 1931.

Les amphétamines, et en particulier la forme lévogyre, contrôlent les anomalies de la vigilance dans environ 50% des cas et pour des doses faibles comprises entre 25 et 50 mg.

L'utilisation chronique de ces drogues n'est pas sans inconvénient. En complément des modifications de l'humeur, de la production de mouvements anormaux et de l'installation d'une hypertension artérielle, les amphétamines provoquent habituellement une tolérance et parfois une pharmaco-dépendance. Aussi le méthylphénidate, psychostimulant mieux toléré et sans effet sur la tension artérielle a-t-il été proposé par le groupe de Stanford.

Si les amphétamines ont un effet sur les accès de sommeil, leur action sur les signes auxiliaires est pratiquement nulle. Aussi une nouvelle orientation thérapeutique avec les tricycliques a-t-elle été proposée dès 1960.

Plusieurs tricycliques ont été utilisés: imipramine, deséthyl-imipramine, et surtout la chlomipramine dont l'effet sur la cataplexie est le plus précis. La dose de 50 mg est habituellement suffisante, mais la prise de ces médicaments peut entraîner certains inconvénients dont l'impuissance chez l'homme et une tolérance après 6 mois de traitement.

Les tricycliques diminuent ou suppriment le sommeil paradoxal dès le premier jour de leur prise. La durée de cet effet est limitée; toutefois la chlomipramine peut entraîner une suppression d'une durée d'un mois et par la suite le sommeil paradoxal reste à environ 50% du taux normal. Un tel effet peut être à l'origine de l'amélioration des signes auxiliaires de la narcolepsie, mais n'est pas accompagné d'une diminution des accès de sommeil. En complément, l'effet thymoleptique de ces médicaments n'est pas négligeable, car l'influence de l'émotion est importante dans la production des signes auxiliaires.

La troisième direction thérapeutique concerne le mauvais sommeil de nuit et un hypnotique, type benzodiazépine, est utile au coucher.

Le traitement de la narcolepsie reste actuellement symptomatique et les données neurochimiques sur les mécanismes du sommeil n'ont pas apporté les améliorations attendues.

D'après la théorie monoaminergique du sommeil, la sérotonine induit le sommeil lent et facilite le déclenchement du sommeil paradoxal. D'autre part, l'acétylcholine favorise la production du sommeil paradoxal. Les tricycliques et surtout la chlomipramine entraînent une augmentation de la sérotonine au niveau des récepteurs postsynaptiques mais provoquent une diminution du turnover de cette amine et une baisse des métabolites dont le 5-HIAA, dont l'influence dans la production du sommeil paradoxal a été retenue [21]. Les IMAO [22], dont l'action bien que différente peut être rapprochée de celle des tricycliques, sont difficiles à manier. La PCPA utilisée par *Wyatt et al.* [23] est sans effet, par contre, les antagonistes des récepteurs sérotoninergiques tel que le méthysergide [24] et le propranolol [25] diminueraient la cataplexie.

Bien qu'une influence cholinergique ait été indiquée dans la production du sommeil paradoxal et que la prostigmine facilite ce type de sommeil, l'atropine ou les drogues anticholinergiques sont pratiquement sans effet sur la narcolepsie.

Enfin, parmi les neuromédiateurs inhibiteurs, le GABA, dont le rôle dans les mécanismes du sommeil reste imprécis, a été utilisé.

L'hydroxabutyrate de sodium entraînerait une amélioration chez certains narcoleptiques, peut-être en favorisant le sommeil de nuit [26].

Le traitement de la narcolepsie est encore imparfait et cette maladie reste pour une part invalidante, avec de nombreuses complications qui jallonnent son histoire.

Conclusion

La narcolepsie est le désordre du sommeil le plus spectaculaire. Marquée par la compétitivité des deux sommeils avec priorité du sommeil paradoxal, elle l'est aussi par les aspects dissociés du sommeil. Elle est associée à une désorganisation du cycle veille-sommeil et à la liaison des sécrétions hypophysaires avec le sommeil.

Maladie à caractère héréditaire, elle est probablement dépendante d'une anomalie biochimique, et il est possible qu'une meilleure connaissance des mécanismes du sommeil permette demain un traitement mieux adapté.

Bibliographie

1 Passouant, P.: The history of narcolepsy; dans Guilleminault, Dement, Passouant, Adv. Sleep Res., vol. 3, Narcolepsy, pp. 3–14 (Spectrum, New York 1976).
2 Rechtschaffen, A.; Wolpert, W.; Dement, W.C.; Mitchell, S.; Fischer, C.: Nocturnal sleep of narcoleptics. Electroenceph. clin. Neurophysiol. *15:* 599–609 (1963).
3 Takahashi, Y.; Jimbo, M.: Polygraphic study of narcoleptic syndrome, with special reference to hypnagogic hallucinations and cataplexy. Folia psychiat. neurol. jap. suppl. 7, p. 343 (1963).
4 Passouant, P.; Schwab, R.; Cadilhac, J.; Baldy-Moulinier, M.: Narcolepsie-cataplexie. Revue neurol. *111:* 415–426 (1964).
5 Passouant, P.; Halberg, F.; Genicot, R.; Popoviciu, L.; Baldy-Moulinier, M.: La périodicité des accès narcoleptiques et le rythme ultradien du sommeil rapide. Revue neurol. *121:* 155–164 (1969).
6 Billiard, M.: The significance of SOREMPS in narcoleptics and in sleep apneic subjects. 3rd Int. Congr. Sleep Res. APSS, Tokyo 1979 (à paraître).
7 Passouant, P.; Cadilhac, J.; Billiard, M.; Besset, A.: La suppression du sommeil paradoxal par la chlomipramine. Thérapie *28:* 379–392 (1973).

8   Dement, W.C.: Daytime sleepiness and sleep 'attacks'; dans Guilleminault, Dement, Passouant, Adv. Sleep Res., vol. 3, Narcolepsy, pp. 17–42 (Spectrum, New York 1976).
9   Passouant, P.: Activité onirique et narcolepsie. J. Psychol. 2: 171–187 (1967).
10   Passouant, P.; Billiard, M.: The evolution of narcolepsy with age; dans Guilleminault, Dement, Passouant, Adv. Sleep Res., vol. 3, Narcolepsy, pp. 179–196 (Spectrum, New York 1976).
11   Takahashi, K.; Takahashi, Y.; Takahashi, S.; Honda, Y.: Growth hormone and cortisol secretion during nocturnal sleep in narcoleptics and in dogs; dans Hatotani, Psychoneuroendocrinology, pp. 67–76 (Karger, Basel 1974).
12   Higuchi, T.; Takahashi, Y.; Takahashi, K.; Niimi, Y.; Miyasita, A.: Twenty-four hours secretory patterns of growth hormone, prolactin and cortisol in narcolepsy. J. clin. Endocr. Metab. 49: 197–204 (1979).
13   Besset, A.; Bonardet, A.; Billiard, M.; Descomps, B.; Craste de Paulet, A.; Passouant, P.: Circadian patterns of growth hormone and cortisol secretions in narcoleptic patients. Chronobiologia 6: 19–31 (1979).
14   Weitzman, E.: Twenty-four hour neuroendocrine secretory patterns: observations on patients with narcolepsy; dans Guilleminault, Dement, Passouant, Adv. Sleep Res., vol. 3, Narcolepsy, pp. 521–542 (Spectrum, New York 1976).
15   Bruhova, S.; Roth, B.: Heredofamilial aspects of narcolepsy and hypersomnia. Archs suisses Neurol. Neurochir. Psychiat. 110: 45–54 (1972).
16   Kessler, S.; Guilleminault, Ch.; Dement, W.C.: A family study of 50 REM narcoleptics. Acta neur. scand. 50: 503–512 (1974).
17   Pompeiano, O.: The neurophysiological mechanisms of the postural and motor events during desynchronized sleep. Res. Publs Ass. Res. nerv. ment. Dis. 45: 351–423 (1967).
18   Jouvet, M.; Delorme, F.: Locus coeruleus et sommeil paradoxal. C. R. Soc. Biol. 159: 895–899 (1965).
19   Dell, P.; Puizillout, J.J.: Experimental reflex narcolepsy in the cat; dans Guilleminault, Dement, Passouant, Adv. Sleep Res., vol. 3, Narcolepsy, pp. 451–472 (Spectrum, New York 1976).
20   Mitler, M.; Dement, W.C.: Cataplectic-like behavior in cats after microinjections of carbachol in pontine reticular formation. Brain Res. 68: 335–343 (1974).
21   Buguet, A.; Petitjean, F.; Jouvet, M.: Suppression des pointes ponto-géniculo-occipitales du sommeil par lésions ou injections in situ de 6-hydroxydopamine au niveau du tegmentum pontique. C. r. Séanc. Soc. Biol. 164: 2293–2298 (1970).
22   Wyatt, R.; Fram, D.; Buchbinder, R.; Snyder, F.: Treatment of intractable narcolepsy with a monoamine oxidase inhibitor. New Engl. J. Med. 285: 987–991 (1971).
23   Wyatt, R.; Kupfler, D.; Scott, J.; Snyder, F.; Engleman, K.: Effects of para-chlorophenyl-alanine on sleep in man. Electroenceph. clin. Neurophysiol. 27: 529–532 (1969).
24   Wyler, A.; Wilkus, R.; Troupin, A.: Methysergide in the treatment of narcolepsy. Archs Neurol., 32: 265–268 (Chicago 1975).
25   Kales, A.; Soldatos, C.; Cadieux, R.; Bixler, E.O.; Tan, T.; Scharf, M.B.: Propranolol in the treatment of narcolepsy. Ann. int. Med. 91: 742 (1979).
26   Broughton, R.; Mamelak, M.: The treatment of narcolepsy-cataplexy with nocturnal gamma-hydroxybutyrate. Can. J. neurol. Sci. 6: 1–3 (1979).

Prof. P. Passouant, Service de Physiopathologie des Maladies Nerveuses,
Centre Gui de Chauliac, Cliniques St-Eloi, F-34059 Montpellier (France)

Part IV. Free Communications and Posters

A. Pharmacology and Endocrinology of Sleep

The Effect of Atropine on the Hypnogenic Action of Basal Forebrain Stimulation[1]

G. Benedek, F. Obál, Jr., F. Bari, F. Obál

Department of Physiology, University Medical School, Szeged, Hungary

The participation of cholinergic pathways in the basal forebrain hypnogenic mechanisms has been the subject of serious controversy [4–6]. The present experiments were carried out in order to obtain electrophysiological and behavioural data on the effect of a cholinergic antagonist, atropine, on the EEG synchronization and sleep induced by electrical stimulation of the basal forebrain in acute immobilized and chronically prepared cats, respectively. Since we had reported that high-frequency stimuli delivered into the cortical part of the olfactory tubercle were as effective as preoptic stimulation in eliciting cortical synchronization and sleep [1, 3, 7], the effects of the olfactory tubercle (TbOf) stimulations were tested in the present experiments.

Methods

20 cats were used for the acute experiments. The animals were operated on under ether anaesthesia, the anaesthesia was then discontinued, and the cats were immobilized and artificially respirated [1]. In 6 experiments 'cerveau isolé' preparations were used. Recording electrodes were placed over the anterior and posterior sigmoid gyri and over the median and lateral suprasylvian gyri. Atropine sulphate was administered intravenously in a dose of 1 mg/kg three times during the experiments, at intervals of 30–40 min. Before atropine treatment and after each dose, the effect of high-frequency TbOf stimulations was repeatedly tested. The stimulation parameters were: 100 Hz, 0.5 ms pulse duration, 250 µA. A stimulus train lasted for 10–30 s.

---

[1] Supported by the Scientific Research Council, Ministry of Health, Hungary (4-05-0303-0422/0) and MTA-OM-MÉM-EÜM 70.211/79.

In the chronic experiments 5 cats were used. Recording electrodes were implanted over the anterior and posterior sigmoid gyri, and the median and lateral suprasylvian gyri, into the dorsal hippocampus, orbita and neck muscles. 2 cats were injected with 10 mg/kg atropine sulphate, while 3 cats received the drug in a dose of 1 mg/kg intraperitoneally. After the treatment the cats were replaced into their sound-attenuated home-cage and the TbOf stimulations were carried out at random intervals during an 8-h observation period. Bilateral stimulations were used. The animals' behaviour was monitored on a TV screen. The stimulation parameters were: 200 Hz, 0.2 ms pulse duration, 50–400 μA. A stimulus train lasted for 2 min.

The electrode placements were confirmed via Klüver-Barrera stained histological sections.

Results

Atropine exerted a pronounced synchronizing effect on the spontaneous EEG activity in both acute immobilized and chronically prepared cats. Primarily the number of delta waves increased.

In the acute animals, high-frequency stimulation in the TbOf elicited cortical synchronization consisting of slow waves and spindles. The atropine treatment facilitated the synchronizing response; the number of spindles increased significantly during the stimulations [2]. After the third dose of the drug, the stimulations sometimes induced convulsive activity. Similar findings were obtained in the 'cerveau isolé' preparations.

The atropine-treated freely moving cats were restless; they circled and moved incessantly. The EEG activity was continuously synchronized with a predominance of slow waves. The stimulations failed to inhibit on-going behaviour: most of the stimulations had no apparent effects (table I). In contrast to the behavioural findings, the stimulations were still able to increase cortical synchronization: spindle activity appeared and the amplitude of the slow waves increased. The animals treated with low and large doses of atropine responded in the same way (table I).

Table I. Percentages of EEG and behavioural effects of TbOf stimulations in cats treated with 1 mg/kg (3 cats, 72 stimulations) and 10 mg/kg (2 cats, 61 stimulations) atropine

EEG effects	1 mg/kg	10 mg/kg	Behavioural effects	1 mg/kg	10 mg/kg
Desynchronization	5.6	0	activation	16.7	18.0
No effect	26.4	29.5	no effect	63.9	68.9
Synchronization	68.1	70.5	inhibition	19.4	13.1

Discussion

Hernandez-Peon [4] stressed the importance of acetylcholine as a transmitter substance mediating the sleep-inducing effect of the basal forebrain. Other reports, however, failed to support the cholinergic hypothesis [5, 6]. The present results suggest that the EEG and behavioural effects of TbOf stimulation are brought about by atropine-resistant ascending pathways and atropine-sensitive descending ones, respectively. Acetylcholine might be implicated in the behavioural effects of the basal forebrain hypnogenic area, but the mechanism of the cortical synchronization elicited by TbOf stimulation seems to be non-cholinergic, or at least of non-muscarinergic type.

References

1. Benedek, G.; Obál, F., Jr.; Szekeres, L.; Obál F.: Arch. ital. Biol. *117:* 164–185 (1979).
2. Benedek, G.; Obál, F., Jr.; Réti, G.; Obál, F.: Acta physiol. hung. *55:* 41– 49 (1980).
3. Benedek, G.; Obál, F., Jr.; Rubicsek, G.; Obál, F.: Behav. Brain Res. (accepted, 1980).
4. Hernandez-Peon, R.: In Akert, Bally, Schadé, Sleep mechanisms. Progress in brain research, vol. 18, pp. 96–117 (Elsevier, Amsterdam 1965).
5. MacPhail, E.M.; Miller, N.E.: J. comp. physiol. Psychol. *65:* 499–503 (1968).
6. Myers, R.D.: Can. J. Psychol. *18:* 6–14 (1964).
7. Obál, F., Jr.; Benedek, G.; Réti, G.; Obál, F.: Expl. Neurol. *69:* 202–208 (1980).

G. Benedek, MD, Department of Physiology, University Medical School, Dóm tér 10, H-6720 Szeged (Hungary).

Increased Slow-Wave Sleep in Man after Several Serotonin Antagonists

René Spiegel

Pharmaceutical Division, Clinical Research, Sandoz Ltd., Basel, Switzerland

Although numerous sleep studies with serotonin (5-HT) precursors, blockers and synthesis inhibitors have been performed, it is still unclear how sleep in man is regulated by the serotoninergic system. Hypnotic effects of L-tryptophan and 5-HT are controversial, and the influence of parachlorophenylalanine on sleep parameters was investigated in patients who are not representative for a normal population [1, 11]. In experiments with methysergide high and repeated doses were used which may have influenced sleep in an unspecific way [4].

Here I would like to report new results obtained with eight different 5-HT antagonists in man. All compounds (table I) are potent peripheral 5-HT antagonists as shown in a variety of test systems; some of them also block 'central' 5-HT effects such as tremor and head twitches in rats.

Materials and Methods

Sleep studies were carried out in healthy male volunteers aged 20–30 years. Each study included 6–12 subjects who slept for 5 consecutive nights in a sleep laboratory. The first 2 nights were used for adaptation and on the following nights 2 different doses of the experimental compound or placebo were administered under double-blind conditions 30 min before recordings were started. Doses applied were chosen from clinical experience with the compounds so as not to produce undue side-effects. Treatments were arranged according to Latin squares so that sequential effects were equalled out. Polygraphic sleep recordings lasted 7.5 hours and were visually assessed on the basis of the *Rechtschaffen and Kales* [7] criteria.

Results (cf. table II)

Rapid eye movement (REM) sleep was significantly reduced after mianserin, pizotifen and GC 46 Organon, an effect similar to that observed after numerous tricyclic antidepressants. This action was not observed after

FQ 27-096, FU 29-245 or PZ 100-862 – compounds which also lack clinical antidepressant activity. One of the ergolene derivatives, CF 25-397, also reduced REM sleep, whereas methysergide was ineffective.

Slow wave sleep (SWS) was increased after several compounds: this effect was dose-dependent and significant after FQ 27-096 and less pronounced after FU 29-245; after GC 46 Organon SWS was also increased, although not significantly, due to the small number of subjects. Increased SWS was observed in some, but not all, subjects after 10 mg mianserin; a recent study [8] has reported SWS increases of almost 50% after 30 mg mianserin in healthy subjects. Pizotifen produced a small, but significant, increase of SWS in the lower dose only, and neither methysergide nor CF 25-397 or PZ 100-862 had significant effects on SWS.

Table I. Compounds investigated

Methysergide	ergolene derivative [cf. 3]
CF 25-397	ergolene derivative [cf. 10]
Mianserin	tetracyclic [cf. 9]
GC 46 Organon	tetracyclic related to mianserin
Pizotifen	tricyclic [cf. 2]
FQ 27-096	investigational piperidyl derivative
FU 29-245	investigational piperidyl derivative
PZ 100-862	investigational butyl-thiazol derivative

Table II. Actions on sleep of eight 5-HT antagonists in healthy male subjects. Statistical comparisons with ANOVA and subsequent pair comparisons after Tukey: (*) ≙ $p<0.10$; * ≙ $p<0.05$; ** ≙ $p<0.01$

Compound (number of subjects)	Dose in mg p.o.		SWS: median % change		REM: median % change	
	lower dose	higher dose	lower dose	higher dose	lower dose	higher dose
Methysergide (12)	1	2	± 0.0	+ 1.0	+ 1.0	− 11.0
CF 25–397 (12)	1.25	2.5	− 5.5	− 14.5	− 9.5	− 33.0*
Mianserin (6)	5	10	− 6.0	+ 7.5	− 19.5*	− 36.0**
GC 46 Organon (6)	10	20	+ 37.0	+ 48.0(*)	− 29.0**	− 30.5**
Pizotifen (8)	2.5	10	+ 13.0*	− 2.0	− 17.0(*)	− 24.0*
FQ 27–096 (9)	100	200	+ 46.0(*)	+ 71.0*	− 6.0	+ 4.0
FU 29–245 (8)	50	100	+ 10.0	+ 21.0(*)	+ 2.0	− 8.5
PZ 100–862 (6)	25	50	+ 0.5	+ 1.5	+ 20.0	+ 1.5

The compounds investigated differ widely with regard to their chemical structures and pharmacological activities other than 5-HT antagonism. As they were part of different drug development programs at different times in different laboratories, methods used for establishing 5-HT antagonism were not identical. In order to have a common basis, all compounds were subsequently compared with regard to their affinities for 5-HT receptors and α-adrenoceptors using spiral strips from canine vessels. 5-HT antagonism was determined on basilar arteries by measuring drug concentrations necessary for reducing by 50% the maximum contractile responses to 5-HT (pD'_2 values) [5]. α-Adrenoceptor blockage was determined on femoral veins by calculating drug concentrations necessary for reducing contractile responses to a double dose of noradrenaline to that of a single one (pA_2 values) [6].

All compounds were confirmed to be 5-HT antagonists and all displayed α-adrenoceptor blocking properties as well; however, for antagonism of responses to noradrenaline 10–100 times higher concentrations were necessary than those required for 5-HT antagonism. In addition to their blocking properties some of the compounds also showed intrinsic agonistic activity when used at higher concentrations than those necessary for antagonism. Thus, pizotifen, CF 25-397, methysergide and PZ 100-862 induced increases in tone of basilar arteries, and mianserin, pizotifen, GC 46 Organon and CF 25-397 displayed some venoconstrictor activity [*Müller-Schweinitzer,* unpublished results].

Discussion

Comparison of the effects observed during sleep with the results obtained from the in vitro studies revealed that neither 5-HT blockage nor α-adrenoceptor antagonism alone were correlated with the effects observed during sleep, but that the presence or absence of intrinsic activity on basilar arteries (probably serotoninergic) and intrinsic activity on femoral veins (probably α-adrenergic) appeared to be crucial for the actions of the compounds on sleep: 1. 5-HT antagonists without intrinsic activity on basilar arteries increased SWS in man, while compounds with intrinsic activity on basilar arteries had no such effect, 2. 5-HT antagonists with intrinsic activity on femoral veins induced REM sleep suppression, while compounds without intrinsic activity on femoral veins had no such effect.

These results, which will have to be confirmed in studies with further 5-HT antagonists, are in contrast with previous studies with compounds which putatively interfere with central serotoninergic neurotransmission. It will now be necessary to study the effects of the present compounds on central 5-HT turnover and to perform receptor-binding studies on brain tissue. Such investigations are in progress at the present time.

References

1 Chernik, D.A.; Ramsey, T.A.; Mendels, J.: Br. J. Psychiat. *122:* 191–197 (1973).
2 Dixon, A.K.; Hill, R.C.; Römer, D.; Scholtysik, G.: Arzneimittel-Forsch. *27:* 1968–1979 (1977).
3 Fanchamps, A.; Döpfner, W.; Weidmann, H.; Cerletti, A.: Schweiz. med. Wschr. *90:* 1040–1046 (1960).
4 Mendelson, W.B.; Reichman, J.; Othmer, E.: Biol. Psychiat. *10:* 459–464 (1975).
5 Müller-Schweinitzer, E.: Arch. Pharmacol. *292:* 113–118 (1976).
6 Müller-Schweinitzer, E.; Stürmer, E.: Br. J. Pharmacol. *51:* 441–446 (1974).
7 Rechtschaffen, A.; Kales, A.: NIH publ. No. 204 (US Government Printing Office, Washington 1968).
8 Tormey, W.P.; Buckley, M.P.; O'Kelly, D.A.; Conboy, J.; Pinder, R.M.; Darragh, A.: Curr. med. Res. Opin. *6:* 456–460 (1980).
9 Vargaftig, B.B.; Coignet, J.L.; de Vos, C.J.; Grijsen, H.; Bonta, I.L.: Eur. J. Pharmacol. *16:* 336–346 (1971).
10 Vigouret, J.M.; Bürki, H.R.; Jaton, A.L.; Züger, P.E.; Löw, D.M.: Pharmacology *16:* 156–173 (1978).
11 Wyatt, R.J.; Engelman, K.; Kupfer, D.M.; Scott, J.; Sjoerdsma, A.; Snyder, F.: Electroenceph. clin. Neurophysiol. *27:* 529–532 (1969).

R. Spiegel, PhD, Pharmaceutical Division, Clinical Research, Sandoz Ltd., CH-4002 Basel (Switzerland)

On the Antagonistic Effects of Pimozide and Domperidone on Apomorphine-Disturbed Sleep-Wakefulness in Dogs

A. Wauquier, W. A. E. Van Den Broeck, C. J. E. Niemegeers

Department of Pharmacology, Janssen Pharmaceutica, Beerse, Belgium

Apomorphine, a dopamine (DA) agonist, is known to induce emesis, agitation and stereotypy in several species including human. It suppresses slow wave sleep (SWS) and REM sleep (RS) in rats and human, though small doses have been reported to induce sleep [1, 3].

Domperidone, a peripheral DA antagonist, is the prototype of a new chemical class of compounds with pronounced gastrokinetic properties [6]. In the dog, it has been described to possess potent anti-emetic effects [5]. Pimozide, the prototype of the diphenylbutylpiperidines [2], is a very specific central DA antagonist [4]. The present study in dogs examines the effects on the sleep-wakefulness of a subemetic, emetic and stereotypogenic dose of apomorphine and of an anti-emetic dose of domperidone and pimozide. Furthermore, the effects of domperidone and pimozide were studied on the apomorphine-induced disturbances of sleep-wakefulness.

Materials and Methods

5 adult male beagle dogs were stereotaxically implanted with cortical and subcortical electrodes [7] and after recovery adapted to a cage in a continuously illuminated room for repetitive 16-h sleep-wakefulness recordings from 3 a.m. to 7 p.m. The first 3 h were recorded on paper; the whole 16-h period was analyzed by the computer.

The following derivations were recorded on an Elema-Schönander mingograph: frontal-occipital cortex (left and right), frontal-frontal cortex, temporal-temporal cortex, occipital-occipital cortex, basolateral amygdala, dorsal hippocampus, lateral geniculate, pontine reticular formation, EMG and EOG. Other channels served to indicate time triggers. The paper speed used was 15 mm/s, occasionally 30 mm/s. Filters and time constants were for the EEG respectively 30 Hz and 0.3 s, for the EMG respectively 700 Hz and 0.015 s.

Visual- and computer-based analysis was done in 30 s epochs [7]. These epochs were classified in 5 stages: wakefulness, transitional stage, light SWS, deep SWS, and RS. On-line analysis was carried out with a PDP 11/10 computer.

Power spectral analysis, using Fast Fourier Transformation is done on the frontal-occipital derivation each 30-s epoch for 25.6 s (2,048 data points) which leaves 4.4 s for computations. The power contained in the frequency bands, delta (0.5–3.5 Hz), theta (3.5–7.5 Hz), alpha (7.5–13.5 Hz), and beta (13.5–25.0 Hz) is calculated. The power in the theta band is also calculated for the hippocampal derivation. The EMG and EOG mean amplitude is taken and a special algorithm detects spindle activity. These data were analyzed on-line (real time). Off-line automatic stage classification was done using a minimal distance approach.

After stabilization of the sleep-wakefulness patterns drug experiments were started. At least 1 week separated each treatment. Apomorphine was given s.c. at the doses 0.02 mg/kg (subemetic dose), 0.16 mg/kg (emetic dose) and 1.25 mg/kg (stereotypogenic dose). Pimozide, (0.063 mg/kg) and domperidone (0.16 mg/kg) were given orally. These doses represent 4 times the ED_{50} values against 0.31 mg/kg s.c. apomorphine-induced vomiting in dogs 4 h after oral administration of the compounds. Apomorphine was given immediately, and pimozide and domperidone 2 h before recording.

Results

Because of the relatively short duration of action of apomorphine, no dose-relationship was seen following analysis of the total 16-h period. However, a clear dose-relationship was observed during the first 4 h of the night (table I). At 0.02 mg/kg RS was significantly decreased, and at 0.16 and 1.25 mg/kg wakefulness (W) was increased and SWS and RS were decreased. Pimozide at 0.063 mg/kg and domperidone at 0.16 mg/kg did not

Table I. Mean (± SEM) percentages vs control during the first 4 h of the 16-h sleep recording after different doses of apomorphine (APOM) and the combination with 0.16 mg/kg of domperidone (DOMP) or 0.063 mg/kg of pimozide (PIM)

		Apomorphine, mg/kg		
		0.02	0.16	1.25
Wakefulness	APOM	115 (± 20.0)	194 (± 25.9)*	234 (± 38.7)*
	APOM + DOMP[1]	101 (± 13.9)	205 (± 35.7)*	233 (± 23.7)*
	APOM + PIM	113 (± 17.9)	129 (± 20.8)	186 (± 15.6)*
Slow wave sleep	APOM	99.5 (± 7.4)	39.6 (± 14.0)*	27.8 (± 10.5)*
	APOM + DOMP[1]	89.7 (± 7.0)	38.6 (± 11.2)*	26.8 (± 8.0)*
	APOM + PIM	101 (± 9.3)	105 (± 17.8)	58.0 (± 7.7)*
REM sleep	APOM	43.1 (± 10.0)*	12.7 (± 4.5)*	1.4 (± 0.4)*
	APOM + DOMP[1]	63.0 (± 11.9)*	22.8 (± 5.5)*	4.5 (± 4.5)*
	APOM + PIM	58.4 (± 18.2)*	48.0 (± 18.2)*	17.2 (± 5.4)*

*$p < 0.05$, Wilcoxon matched-pairs signed-ranks test
[1] The effects of apomorphine and apomorphine + domperidone were not significantly different from each other.

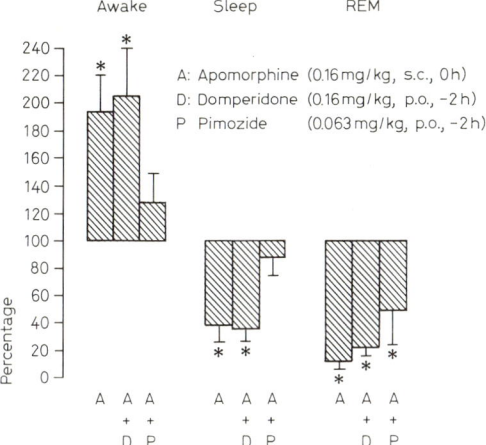

Fig. 1. Mean (± SEM) percentage of control after apomorphine or the combination of apomorphine with domperidone or pimozide at the indicated doses and times.

significantly change the sleep-wakefulness pattern. In the antagonistic studies, domperidone did not significantly antagonize the disturbed sleep-wakefulness pattern caused by the different doses of apomorphine (table I).

Pimozide (table I) partially antagonized the RS reduction caused by 0.02 mg/kg of apomorphine. It completely antagonized the increased W and decreased SWS caused by 0.16 mg/kg of apomorphine and partially the decreased RS. Pimozide partially antagonized the increased W, the decreased SWS and the decreased RS caused by 1.25 mg/kg of apomorphine. The effects of pimozide on the intermediate dose of apomorphine are illustrated in figure 1.

Discussion

In dogs, 0.02 (subemetic), 0.16 (emetic) and 1.25 (stereotypogenic) mg/kg of apomorphine caused a dose-related increase in W and a decrease in SWS and RS. Even the subemetic dose 0.02 mg/kg induced a significant suppression of RS. The disturbances of the sleep-wakefulness pattern induced by apomorphine were not affected by the peripheral DA antagonist domperidone; although domperidone antagonized completely the emetic effect induced by 0.16 mg/kg of apomorphine. Thus, the disturbed sleep-wakefulness pattern is centrally mediated and obviously at 0.16 mg/kg of apomorphine not related to the emetic properties of the compound.

Reduction of RS is the first effect seen with apomorphine and this appears more difficult to normalize with pimozide. Even when pimozide normalized SWS, RS was still suppressed. This suggests that the effect of apomorphine on RS is not simply the consequence of a reduced SWS. In as far as the effects caused by apomorphine are due to its DA agonistic properties, the experiments suggest that DA plays a particular role in the physiology of RS [8].

References

1. Corsini, G.V.; Del Zompo, M.; Mansoni, S.; Cianchetti, C.; Mangoni, A.; Gessa, G.L.: In non-striatal dopaminergic neurons, pp. 645–648 (Raven Press; New York 1977).
2. Janssen, P.A.J.; Niemegeers, C.J.E.; Schellekens, K.H.L.; Dresse, A.; Lenaerts, F.M.; Pinchard, A.; Schaper, W.K.A.; Van Nueten, J.M.; Verbruggen, F.J.: Arzneimittel-Forsch. *18:* 261–279 (1968).
3. Mereu, G.P.; Scannati, E.; Paglietti, P.; Pellegrini, O.; Quarantotti, B.; Chessa, P.; Di Chiara, G.; Gessa, G.L.: Electroenceph. clin. Neurophysiol. *46:* 214–219 (1979).
4. Niemegeers, C.J.E.; Janssen, P.A.J.: Life Sci. *24:* 2201–2216 (1979).
5. Niemegeers, C.J.E.; Schellekens, K.H.L.; Janssen, P.A.J.: Archs int. Pharmacodyn. Thér. *244:* 130–140 (1980).
6. Reyntjens, A.J.; Niemegeers, C.J.E.; Van Nueten, J.M.; Laduron, P.; Heykants, J.; Schellekens, K.H.L.; Marsboom, R.; Jageneau, A.; Broeckaert, A.; Janssen, P.A.J.: Arzneimittel-Forsch. *28:* 1194–1196 (1978).
7. Wauquier, A.; Verheyen, J.L.; Van Den Broeck, W.A.E.; Janssen, P.A.J.: Electroenceph. clin. Neurophysiol. *46:* 33–48 (1979).
8. Wauquier, A.; Van Den Broeck, W.A.E.; Janssen, P.A.J.: Life Sci. *27:* 1469–1475 (1980).

A. Wauquier, PhD, Department of Pharmacology, Janssen Pharmaceutica, B-2340 Beerse (Belgium)

Sleep Induction and Central Effects of Some α-Adrenoceptor Agonists

H. Depoortere

Biology Department, EEG Group, Synthelabo (LERS), Paris, France

In a previous publication [2] we have shown the existence of a positive correlation between the ability of α-sympathomimetics to reduce the ponto-geniculo-occipital activity induced by reserpine (PGO-R) in the cat and the decrease of paradoxical sleep in the rat. Moreover, the sedative effect of clonidine is marked by the appearance of sleep spindles in the EEG taken from the rat neocortex [3]. The effect of clonidine on reducing the number of PGO-R spikes is antagonised by $α_2$-receptor (presynaptic)-blockers (e.g. piperoxane, yohimbine or phentolamine) but not by $α_1$-receptor (postsynaptic) blockers (e.g. prazosin, phenoxybenzamine) [3, 4].

The aim of the present work was to compare the central effect and the sedative potential of various new clonidine-like antihypertensive agents (guanfacine: BS 100141; aryl-tetrahydro-pyrrolo imidazoles: ICI 101-187, ICI 106-207 [1] *p*-aminoclonidine and *N*-allylclonidine: St 567).

Methods

The central α-receptor effects have been defined by the study of PGO-R spikes in the cat. The protocol of the PGO-R test has been described previously [2, 4]. Cumulative doses of the drugs tested (0.001–1 mg/kg i.v.) were administered at 30-min intervals, 4.5 h after reserpine (0.75 mg/kg i.p.) to curarised cats (at least 3 animals per drug) which were artificially ventilated and implanted with cortical and geniculate electrodes. For the interaction studies the test drugs were injected 30 min before cumulative doses of piperoxane (0.01–1 mg/kg i.v.). The sedative effect of the drugs was studied by visual and automatic analysis (spectral analysis: Berg-Fourier analyzer of OTE Biomedica) of the electrocorticogram (ECoG) in curarised artificially ventilated Sprague-Dawley rats, 200–250 g body weight. The reference electrode was screwed into the interparietal bone.

Results

In the PGO-R test in the cat, the ED_{50} (reduction of number of spikes by 50%, μg/kg i.v.) was as follows: clonidine = 1; *p*-aminoclonidine = 10; *N*-allylclonidine = 314; guanfacine = 24; ICI 101-187 = 2 and ICI 106-270 = 7. Piperoxane (0.1 to 1 mg/kg i.v.) antagonised the effects of clonidine (10 μg/kg i.v.), guanfacine (300 μg/kg i.v.), ICI 106-270 (100 μg/kg i.v.) and ICI 101-187 (10 μg/kg i.v.) which suppressed the PGO-R spikes by 80–95% over a 2-h period.

In the ECoG test in the rat, doses of 0.1 mg/kg i.p., 0.5 mg/kg p.o. or 1 μg/kg i.c.v. clonidine provoked sleep spindles (12 c/s) in the sensorimotor and a hypersynchronisation of the theta rythm in the visual cortex. *p*-Aminoclonidine and *N*-allylclonidine provoked clonidine-like sleep recordings at doses of 3 and 10 mg/kg i.p., respectively. However, clonidine and *p*-aminoclonidine (0.2–1 μg/kg) were equipotent after i.c.v. injection in inducing sleep spindles. Guanfacine was found to be sedative at 3 mg/kg i.p. but the sleep recordings observed were closer to that of a physiological sleep pattern. ICI 101-187 and ICI 106-270 produced very mild sedative effects in doses up to 10 mg/kg i.p. or after i.c.v. injections up to 30 μg/kg.

At doses of 1–3 mg/kg i.p., yohimbine and piperoxane blocked the effects of clonidine on the ECoG spindles whilst phentolamine was ineffective up to 3 mg/kg i.p. (fig. 1).

Discussion

The various new drugs believed to be antihypertensive through a clonidine-like mechanism reduced the PGO-R spikes and this action was probably mediated by stimulation of α-adrenoceptors since it was antagonised by piperoxane. The fact that high doses of certain clonidine-like derivatives were required to produce relevant spindle activity is probably due to their poor penetration into the central nervous system as indicated by the results obtained following small i.c.v. doses of these compounds.

Guanfacine appears to exhibit an ECoG profile which differs from that for clonidine since it has the characteristics of physiological sleep. It is of interest that the ICI compounds, despite their suggested clonidine-like antihypertensive activity, do not possess the sedative effect of clonidine.

Phentolamine was reported to antagonise the effect of clonidine on the PGO-R spikes [4] and on paradoxical sleep in the cat [5]. However, this α-adrenoceptor antagonist failed to prevent the clonidine-induced sleep observed in rats (present results) and cats [6].

These results appear to indicate a dichotomy between the receptor sites

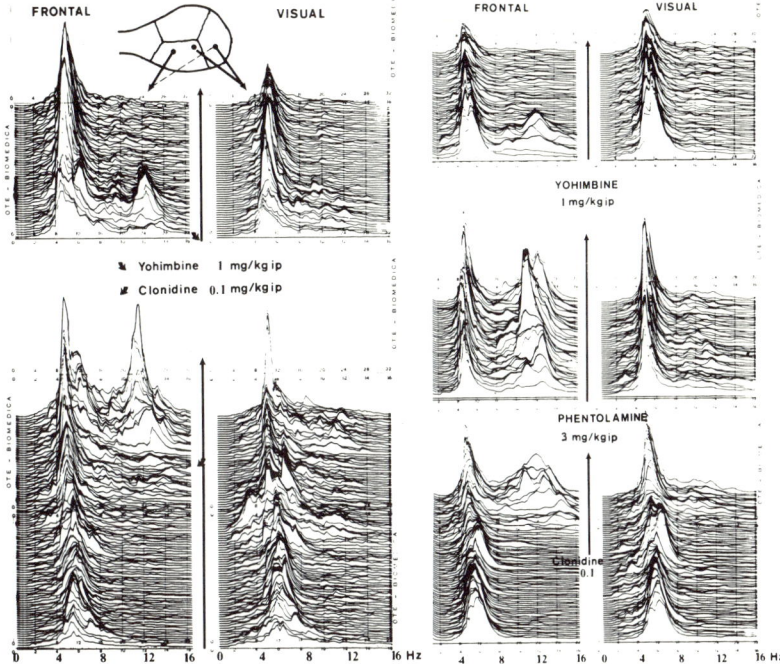

Fig. 1. Sequential spectral analysis (30-s periods) of the electrocorticogram before and after clonidine treatment and the interaction with yohimbine and phentolamine.

mediating both sedative and antihypertensive effects of the α-adrenoceptor agents studied. It is currently believed that clonidine decreases blood pressure mainly by stimulation of central α_2-adrenoceptors. However, the present study using several antihypertensive compounds, possessing a mechanism of action similar to that of clonidine, tends to exclude the direct implication of α-adrenoceptors with respect to the observed sedative (ECoG) activity. This is supported by the inability of the ICI compounds to induce sleep spindles together with the failure of phentolamine to antagonise the ECoG effects of clonidine.

References

1   Clough, D.P.; Hatton, R.; Pettinger, S.J.; Gillian Samuels, M.R.; Shaw, A.: Br. J. Pharmacol *62:* 385P–386P (1978).

2 Depoortere, H.; Honoré, L.; Jalfre, M.: 3rd Eur. Congr. Sleep Res., Montpellier 1976, pp. 358–361 (Karger, Basel 1977).
3 Depoortere, H.: Sleep Res. 7: 94 (1978).
4 Depoortere H.; Lloyd, K.G.: Adv. Biosci. 18: 173–178 (1979).
5 Leppävuori, A.; Putkonen, P.T.S.: Brain Res. 193: 95–115 (1980).
6 Tran Quang Lco; Tsoucaris-Kupfer, D.; Bogaievsky, Y.; Bogaievsky, D.; Delbarre, B.; Schmitt, H.: J. Pharmacol. 5: 51–62 (1974).

H. Depoortere, MD, Département de Biologie, Groupe EEG, Synthelabo (LERS),
31, avenue P.V. Couturier, F-92220 Bagneux (France)

α-Adrenergic Function and Sleep in Kittens

M. V. J. Miettinen

Institute of Physiology, University of Helsinki, Helsinki, Finland

Monoaminergic systems are postulated as having important function in regulation of sleep in adult cats [2, 3]. In kittens monoaminergic influences on sleep have been investigated by means of chemical and electrolytical lesions [1] which have minor effects in sleep regulation during the first postnatal weeks. In this study the effects of systemic injections of α-adrenergic agonists and antagonists on sleep during the early postnatal life have been investigated.

Kittens were studied at the ages of 2 weeks and 1 month. Electrodes for chronical recording were implanted under either Nembutal® or Phentanyl® and Ketalar® anaesthesia. 2–4 steel screws were used for frontal and occipital EEG recording and two silver wires were implanted into the neck muscles for EMG. Eye movements were recorded with silver wires, placed subcutaneously medially and laterally to the left eye. The animals were allowed to recover for at least 24 h.

The kittens were separated from their mother and litter mates during recording in silent dimly lighted boxes ($60 \times 60 \times 60$ cm) from 8–10 a.m. to 4–6 p.m. The drugs or saline for controls were given in single i.p. injection about 5 min before the recording started. Results were scored visually in 1-min epochs as: (A) active state (Q) quiet state, including drowsiness and quiet sleep, and (P) active sleep. Coded data was statistically analysed by computer and the significance of drug effects evaluated with t-test. In controls the distribution of vigilance states was similar to that described in other studies [4, 5].

The $α_2$-agonist, clonidine (CLO 10 μg/kg) at 2 weeks reduced P% from 43.1 in controls to 11.5 ($p<0.001$). The effect was most marked in the first 4-h period (1.0%, $p<0.001$), but was still significant during the second 4-h

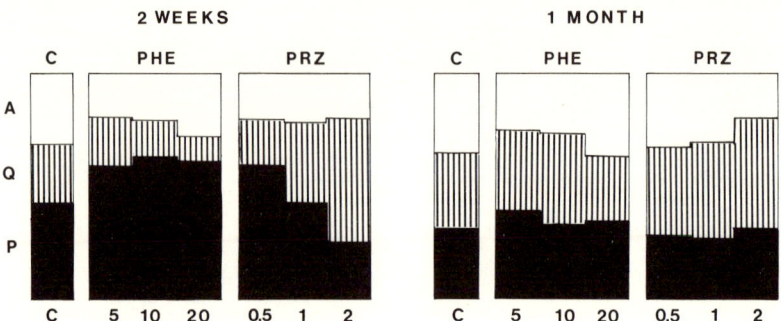

Fig. 1. Dose-dependent effects of phentolamine (PHE) and prazosin (PRZ) in kittens of 2 weeks and 1 month. Doses in mg/kg. C = control; A = active waking; Q = quiet state; P = active sleep.

period (21.7%, p<0.001). The effect was due to reduction in P episode count and to shortening of episode length. The mean episode length in control was 5.2 min, and after CLO 2.7 (p<0.001). This was also most significant during the first 4-h period (1.3, p<0.001, second 4-h period 2.7, p<0.01).

The effect was initially similar at 1 month's age, but did not last as long as at the age of 2 weeks. Control P% was 30.1 and in the CLO group it was 14.9 (p<0.001). In the first 4-h period the P% after CLO was 4.4 (p<0.001), but during the latter 4-h period there was no significant difference although the CLO effect was still obvious. The effect on P% was due to the same changes as in the 2-week-old group. The effect could be antagonised either by yohimbine 2 mg/kg (α_2-antagonist), or by phentolamine 10 mg/kg (PHE, α_2- and α_1-antagonist). The preferential α_1-antagonist prazosin 1 mg/kg (PRZ) potentiated the effect of CLO, and the P% was reduced to 10.7 at 2 weeks and to 8.4 at 1 month's age.

PHE alone dose-dependently (5, 10, and 20 mg/kg) increased P at the age of 2 weeks. The effect was significant (p<0.02) at the dose of 20 mg/kg. This latter dose also reduced the quiet state and produced active waking. At the age of 1 month only 5 mg/kg PHE significantly increased P. The effect was significant (p<0.02) during the first 4-h period. In both age groups, all doses favour the transition from A to P and long P episodes (fig. 1).

PRZ 0.5 mg/kg increased P at 2 weeks. The effect was most significant in the first 4-h period (p<0.001). PRZ 1 mg/kg had little effect on P%, and 2 mg/kg reduced P% and increased Q. When aged 1 month, PRZ did not affect the P% at any doses used (0.5, 1, and 2 mg/kg), but it dose-dependently reduced A and increased Q (fig. 1). In both age groups PRZ in all doses lengthened P episodes, and dose-dependently impaired the triggering of P. PRZ favoured the A–Q–P transition and reduced the narcoleptic tendency to A–P transition, typical for immature animals.

This study shows that effects on sleep can be elicited with pharmacological manipulation of the adrenergic system earlier than with lesion techniques [1]. The results indicated that moderate noradrenergic function favours active sleep in kittens like it favours paradoxical sleep in adults [6], as P was most significantly suppressed by the combination of CLO and PRZ, but enhanced by moderate doses of both PHE and PRZ. At 1 month's age, however, the occurrence of P could not be increased as in the younger group or in adults either by PHE or by PRZ. This may reflect dynamic changes in monoaminergic balance and adrenoceptor sensitivity during the developmental stage.

References

1 Adrien, J.: Prog. Brain Res. *48:* 393–403 (1978).
2 Jouvet, M.: Science *163:* 32–41 (1969).
3 Jouvet, M.: Ergebn. Physiol. *64:* 165–307 (1972).
4 Jouvet-Mounier, D.; Astic, L.; Lacote, D.: Dev. Psychobiol. *2:* 216–239 (1969).
5 McGinty, D.J.; Stevenson, M.; Hoppenbrouwers, T.; Harper, R.M.; Sterman, M.B.; Hodgman, J.: Dev. Psychobiol. *10:* 455–469 (1977).
6 Leppävuori, A.; Putkonen, P.T.S.: Brain Res. *193:* 95–115 (1980).

M.V.J. Miettinen, MD, Institute of Physiology, Siltavuorenpenger 20 J,
SF-00170 Helsinki 17 (Finland)

Effects of a Noradrenergic Agonist on Sleep in the Rat[1]

Laurence Lanfumey, Joëlle Adrien

Inserm U3. 91, Boulevard de l'Hopital, Paris, France

The noradrenergic system has been implicated in the regulation of paradoxical sleep (PS) [4]. Pharmacological investigations of this regulation were generally performed at the presynaptic level [2, 5]. In the present study, we have investigated the same phenomenon at the postsynaptic level. We have analyzed the sleep patterns after the injection of *L*-isoproterenol (an agonist of the β-adrenergic receptors) in animals pretreated by a β-antagonist, *DL*-propranolol, which is known to induce a PS insomnia [1]. This experiment was performed on two groups of adult rats: control animals, and a group which underwent previously a neonatal lesion of the locus ceruleus, resulting in the total disappearance of the noradrenergic presynaptic cerulean system [6].

Methods

Under ether anaesthesia, adult rats (Shermann, 4 controls and 3 neonatally lesioned) were implanted for standard polygraphic sleep monitoring. In addition, a stainless steel cannula (0.9 mm external diameter) was positioned in the lateral ventricle to allow intraventricular injection of *L*-isoproterenol. After recovery from the surgery and habituation to the recording conditions, the animals were submitted to the following pharmacological protocol: A dose of 10 mg/kg of *DL*-propranolol was injected intraperitoneally at 1130 hours. This treatment was shown previously to induce a PS insomnia during 4 h [1]. 30 min later, i.e. at 1200 hours, different doses of *L*-isoproterenol solved in 10 µl of serum saline were slowly perfused into the ventricle through the cannula. Each animal received 0–10 µg of *L*-isoproterenol according to this protocol. 1–3 recovery days were allowed between two pharmacological tests.

Results

In both groups, the injection of *L*-isoproterenol enhanced PS amounts during 4 h: the insomnia induced by propranolol was reversed by the

---

[1] This work was supported in part by INSERM (CRL 80-60-16) and DGRST (79-7-1238).

Table I. PS restoration (min) after the perfusion of different doses of *L*-isoproterenol following a single *DL*-propranolol injection (10 mg/kg). The recovery is expressed in minutes of PS during 4 h following the injection of *L*-isoproterenol (from 1200 to 1600 hours)

	L-isoproterenol (i. ve.)			
	0µg	2µg	5µg	10µg
Control	0	1.9	9.1	12.8
Lesioned	0	10.8	13.5	19.2

β-agonist, isoproterenol, in a dose-dependent fashion. As shown in table I, in the control group PS amounts were significantly increased after a dose of 5 µg of isoproterenol. In the lesioned group, however, a 2 µg dose was effective in restoring comparable PS values. For this dose, PS amounts were 5-fold higher in the lesioned animals as compared to the controls, whereas for larger doses the differences between the two groups persisted but were less important.

Discussion

The present work demonstrates that the β-agonist isoproterenol can reverse the PS insomnia produced by the antagonist propranolol. Whereas the blocking action of propranolol upon the postsynaptic receptors induced a PS deficit [1], the agonist reversed this blockade and consequently restored PS. Moreover, this effect was enhanced in the early-lesioned group as compared to the controls. This result is to be related to the biochemical supersensitivity phenomenon evidenced after neonatal noradrenergic lesions [3]: in this respect, the increased susceptibility of PS in the lesioned animals demonstrated by the present work could represent the same supersensitivity at a behavioural level.

References

1 Adrien, J.; Lanfumey, L.: Sleep Res. *8:* 74 (197).
2 Hartmann, E.: in Lipton, DiMascio, Killam, Psychopharmacology: a generation of progress, pp. 711–728 (Raven Press, New York 1978).
3 Jonsson, G.; Hallman, H.: Neurosci. Lett. *9:* 27–32 (1978).
4 Jouvet, M.: Science *163:* 32–41 (1972).
5 Kafi, S.; Gaillard, J.M.: Waking Sleeping *4:* 131–137 (1980).
6 Lanfumey, L.; Arluison, M.; Adrien, J.: Sleep Res. (in press 1980).

L. Lanfumey, MD, INSERM U3, 91, Boulevard de l'Hôpital, F-75013 Paris 1 (France)

Pre- and Postsynaptic Effect of Yohimbine on Rat Paradoxical Sleep

S. Kafi, J.-M. Gaillard

Psychiatric Clinic of the University of Geneva, Clinique de Bel-Air
(Director: Prof. *R. Tissot*), Chêne-Bourg, Switzerland

In previous studies, we have shown in the rat as well as in man that a low dose of chlorpromazine (CPZ) acting preferentially on presynaptic receptors increases the production of paradoxical sleep (PS), whereas the depression of PS observed under higher doses is probably related to a predominant postsynaptic blockade [1, 2]. It has been recently reported that the affinity of yohimbine (YOH) was about 100 times greater for presynaptic than for postsynaptic receptors [4]. However, these data were obtained using peripheral models and very little is known concerning the central effect of this drug. In rats, YOH administered at low doses has excitatory effects, but at higher doses induces a central depression [5]. We have been interested in studying the effect of YOH on the production of PS in the rat.

Methods

Male albino wistar rats weighing 250–300 g were used. They were implanted under pentobarbital anaesthesia (55 mg/kg) with four cortical electrodes for EEG recording, two for eye movements and two for electromyogram. After 15 days for recovery and habituation to experimental conditions, they were recorded for 12 h in the light period (light/dark schedule 12/12, with the light period between 7 a.m. and 7 p.m.). Drugs were given intraperitoneally after 1 h of control recording. The drugs studied included YOH (30 and 3,000 µg/kg), α-MPT, an inhibitor of tyrosine hydroxylase (150 mg/kg), the combination of 150 mg/kg of α-MPT and YOH (30 and 3,000 µg/kg) 30 min later, and the appropriate controls with saline. The drug-induced hypothermia was prevented by keeping the animals in a 27 °C environmental temperature. Polygraphic recordings were visually scored into four stages: waking (W), slow-wave sleep (SWS), intermediate stage (I), and PS.

Results and Conclusions

The effects of various doses of YOH alone on sleep stages have been already described earlier [3]. Briefly, YOH given intraperitoneally at the low dose of 30 µg/kg enhanced the production of PS in the 11-hour EEG record-

Table I. Effect of YOH alone and combined with α-MPT on the states of vigilance of the rat. Results are expressed in minutes ($\bar{X} \pm SD$)

Treatment	Dosage mg/kg	n	W	TS	SWS	I	PS
Control	equiv.	9	216 ± 33	434 ± 31	344 ± 30	25 ± 7	65 ± 9
α-MPT	150	5	178 ± 6	468 ± 28	383 ± 36	24 ± 9	61 ± 17
YOH	0.03	2	217 ± 62	443 ± 62	352 ± 50	18 ± 2	73 ± 8
YOH	3	3	334 ± 71	326 ± 71	250 ± 41	24 ± 11	52 ± 19
α-MPT + YOH	150, 0.03	3	225 ± 37	435 ± 36	355 ± 22	21 ± 4	58 ± 25
α-MPT + YOH	150, 3	3	162 ± 68	498 ± 68	460 ± 97	13 ± 12	25 ± 17

ing (table I). The higher dose of 3,000 µg/kg depressed PS slightly during the first 200 min of sleep. The administration of 150 mg/kg of α-MPT, ineffective by itself, 0.5 h before a small dose of 30 µg/kg of YOH prevented the increase of PS seen after YOH alone. The production of PS after this combination was similar to that of α-MPT 150 mg/kg alone. In contrast the combined treatment with α-MPT and YOH 3,000 µg/kg resulted in a very marked decrease of PS.

The enhanced production of PS after low dose of YOH can be attributed to a preferential blockade of presynaptic NA receptors, whereas much higher doses are necessary to block the postsynaptic receptors. The results of the experiments presented here show that as well as CPZ, YOH enhances the production of PS at low doses. In the same experimental conditions a small dose of CPZ after a small dose of α-MPT markedly depressed PS in the rat. The major difference between these two drugs at low doses with α-MPT probably resides in the fact that YOH acts on presynaptic receptors only, while low doses of CPZ also have a postsynaptic effect. The much higher dose of YOH than of CPZ necessary to depress the production of PS may be related to the large interval between the doses of YOH active on presynaptic and postsynaptic receptors.

References

1 Gaillard, J.-M.; Kafi, S.: Eur. J. clin. Pharmacol. *15:* 83–89 (1979).
2 Kafi, S.; Gaillard, J.-M.: Eur. J. Pharmacol. *49:* 251–257 (1978).
3 Kafi, S.; Gaillard, J.-M.: Waking Sleeping *4:* 131–137 (1980).
4 Starke, K.; Borowski, E.; Endo, T.: Eur. J. Pharmacol. *34:* 385–388 (1975).
5 Zebrowska-Lupina, I.; Kleinrok, Z.: Psychopharmacologia *33:* 267–275 (1973).

Dr. J.-M. Gaillard, Psychiatric Clinic of the University of Geneva, Clinique de Bel-Air, CH-1225 Chêne-Bourg (Switzerland)

α_2-Adrenoceptive Inhibition of Avian Paradoxical Sleep

M. Helojoki, P. T. S. Putkonen

Department of Physiology, University of Helsinki, Helsinki, Finland

Previous studies from our laboratory have shown that synthetic α_2-receptor stimulants inhibit paradoxical sleep (PS) in cats. This can be counteracted by α_2-blockade but not by preferential α_1- or β-antagonists [7]. Our failure to replicate this antagonism in rats [6] indicated species differences and suggested a comparative approach. Chickens were interesting because of their phylogenetic remoteness and a wealth of former pharmacologic studies on sympatichomimetic sedation and its antagonism in this preparation [e.g. 2, 3]. In our first experiments [8] with young chicks the α_2-agonist clonidine (CL) suppressed PS and, as in cats, this effect was antagonized by phentolamine, which has about equal α_1- and α_2- blocking potency. The present study was designed to determine, in birds, the importance of the receptor subtypes using two antagonists, yohimbine (YO) and prazosine (PR) representing the extremes in relative selectivity to α_2- and α_1-receptors [4].

Methods

5 adult White Leghorn cocks were prepared for sleep polygraphy under pentobarbital sodium anaesthesia. Two pairs of gold-plated screw electrodes were implanted into the skull around the orbits and over hyperstriatal brain areas to record eye movements (EOG) and cerebral activity (EEG). For EMG two wires were inserted into neck muscles.

During recovery and between experiments the birds lived in regular 12/12 h light/dark conditions. Polygraphic 6 h records were obtained simultaneously from 3 birds in isolated dark cages beginning at lights-off of their circadian cycle. These were preceded by two injections into the pectoralis muscles at 10-min intervals, the last at the start of the recordings. Each bird received: (1) saline (NaCl)×2 for control; (2) NaCl+Cl 0.04 mg/kg; (3) YO 2 mg/kg+CL 0.04 mg/kg, and (4) PR 5 mg/kg+CL 0.04 mg/kg. The order of these treatments varied from bird to bird. A rest and washout period of at least 2 days separated the drug tests.

The records were visually classified into 3 stages of vigilance: AW = Active waking or arousal was characterized by low, mixed frequency EEG, tonic and phasic EMG activity, and frequent eye movements and blinks. SS = Synchronized sleep, characterized by high amplitude, slow EEG, moderate to low EMG without phasic activity, and regular blinks but no eye movements in the EOG. PS = Paradoxical sleep intervened as brief (4–8 s), often repetitive periods in the mids of the SS-stage. During these EEG was desynchronized, EMG tone either lowered or unchanged from a previously low SS level. Short bursts of rapid eye movements and suppression of rhythmic blinking characterized the EOG.

The durations of successive vigilance periods were measured using a time scale of 1 s (10 mm with our paper speed). The coded data were fed to a computer for statistical processing.

Results

CL induced a relatively short lasting period of hypersynchrony and decrease of AW, which was about 55% below control during the first hour (table I). This effect was almost completely antagonized by YO but not by PR. CL had a more sustained inhibitory effect on PS, which led to a 74% relative decrement. YO antagonized this effect correcting the deficit to statistically non significant difference of 15% below control, whereas PR pretreatment increased the relative PS decrement to 83%.

Discussion

The sedative effect of CL and its antagonism by YO are in accordance with previous studies on behavioural sleep and sympathomimetics in chickens [2]. The lack of antagonism by the α_1-antagonist PR fits the pharmacological pattern emerging from previous studies, indicating that sed-

Table I. States of vigilance ($\pm$ SD) in percent of recording time with PS percent per sleep

Drugs	mg/kg	1 h			6 h			
		AW	SS	PS	AW	SS	PS	PS/SL
Control (NaCl)	–	50.3 ± 15.2	48.9 ± 14.9	0.8 ± 0.5	31.1 ± 8.6	65.1 ± 7.8	3.8 ± 1.4	5.4 ± 1.6
Clonidine	0.04	22.7** ± 6.4	77.3** ± 6.4	0	22.6 ± 2.6	76.4 ± 2.7	1.0** ± 0.6	1.3** ± 0.7
Clonidine Prazosin	0.04 5.0	22.1** ± 9.7	77.9** ± 9.7	0	23.8 ± 9.7	75.6 ± 9.8	0.6** ± 0.3	0.9*** ± 0.5
Clonidine Yohimbine	0.04 5.0	49.2 ± 8.2	50.8 ± 8.2	0	23.0 ± 4.9	73.7 ± 5.0	3.2 ± 0.8	3.9 ± 1.6

**p < 0.01; ***p < 0.001 (t-test).

ation depends on activation of α_2-receptors. In the cat CL increased the drowsy waking EEG pattern but not genuine slow wave sleep. In birds the EEG spindles of sleep are lacking and we could not define reliable criteria to dichotomize the SS stage, although an intermittent slow wave EEG has been classified as drowsiness in pigeons [9]. In any case CL did not increase intermittent synchrony, but induced a monotonous hypersynchronized slow pattern not readily distinguishable from deep SS. In comparison with our earlier series in chicks, adult birds were more sensitive to CL, as a tenfold dose (0.4 mg/kg) was required to produce a roughly comparable inhibition of PS at the age of 18–28 days [8].

One of the best known central effects of CL is to inhibit transmitter release in NA neurons, probably by activation of presynaptic and somatodendritic α_2-adrenoceptors engaged in negative feedback control of NA release [4]. The α_1-adrenoceptors appear to mediate postsynaptic, mainly excitatory effects – see *Leppävuori and Putkonen* [5] for detailed discussion and references. Against this background the present results with CL and selective α_1- and α_2-antagonists indicate that a critical level of postsynaptic NA-ergic activity appears to be required, in the chicken as in the cat, for the realization of PS. Similar conclusions were drawn by *Clarenbach and Cramer* [1], showing that PS is increased by injections of NA to chemically sympathectomized chicks. Furthermore, a positive correlation between EEG arousal and postsynaptic α_1-adrenoceptive activation would appear also valid for the avian brain.

References

1 Clarenbach, P.; Cramer, H.: Brain Res. *43:* 695–699 (1972).
2 Delbarre, B.; Schmitt, H.: Eur. J. Pharmacol. *13:* 356–363 (1971).
3 Fuegner, A.; Hoefke, W.: Arzneimittel-Forsch. *21:* 1243–1247 (1971).
4 Langer, S.Z.: The release of cathecolamines from adrenergic neurons, pp. 59–85 (Pergamon Press, Oxford 1979).
5 Leppävuori, A.; Putkonen, P.T.S.: Brain Res. *193:* 95–115 (1980).
6 Mäkelä, J.; Putkonen, P.T.S.: Acta physiol. scand. suppl. *473:* 62 (1979).
7 Putkonen, P.T.S.: Sleep research, pp. 19–34 (MTP Press, Lancaster 1979).
8 Putkonen, P.T.S.; Helojoki, M.: Acta physiol. scand. suppl. *473:* 61 (1979).
9 Van Twyver, H.; Allison, T.: Exp. Neurol. *35:* 138–153 (1972).

M. Helojoki, B.A., Department of Physiology, University of Helsinki, SF-00170 Helsinki 17 (Finland)

Electroencephalographic Studies on Clonidine and Para-Aminoclonidine, a Potent Peripherally Acting Imidazoline

H. Depoortere, F. Lefèvre-Borg, I. Cavero

Biology Department, Synthelabo (LERS), Paris, France

p-Aminoclonidine (PAC) decreased blood pressure in anaesthetised rats when injected intracerebroventricularly (i.c.v.) but failed to produce such a response when administered intravenously (i.v.) [5]. However, like clonidine, the compound decreases heart rate after either route of administration. This bradycardia [2] was shown to be entirely dependent on an intact and operative sympathetic drive to the cardiac pacemaker and probably of peripheral origin [1]. Moreover, in the spontaneously hypertensive rat, clonidine and PAC (0.5 mg kg p.o.) produced antihypertensive effects [2].

The aim of this investigation was to obtain information on the effects of PAC on the ponto-geniculo-occipital activity induced by reserpine (PGO-R) in the cat and on the sedative action as assessed by the electrocorticogram changes (ECoG) in normotensive and spontaneously hypertensive rats (SHR).

Methods

The central α-adrenoceptor agonist effects of a compound are easily detected by studying the PGO-R spikes in the cat. Cats were given reserpine (0.75 mg/kg i.p.) and 3 h later immobilised with gallamine (20 mg i.v./h) artificially ventilated and implanted with cortical and geniculate electrodes. At the end of the surgical procedure and an appropriate stabilisation time, PAC and clonidine (0.001–1 mg/kg i.v.) were administered cumulatively, maintaining a 30-min interval between two successive doses. The sedative effects of these compounds were studied by visual and spectral analysis (Berg-Fourier analyzer, OTE Biomedica) of the ECoG in curarised (alcuronium: 1 mg i.p.) artificially ventilated normotensive male rats (Sprague-Dawley, 200–250 g) and spontaneously hypertensive rats (Okamoto strain, 350–380 g). The recording electrodes were implanted in the fronto-parietal and occipito-parietal regions. The reference electrode was inserted in the interparietal bone.

Results

A 50% reduction (ED_{50}) in PGO-R activity in the cat was observed after 1 µg/kg i.v. clonidine as previously reported by *Depoortere* et al. [3] PAC was approximately 10 times less potent than clonidine in producing this effect (ED_{50}: 10 µg/kg i.v.).

In the normotensive rat, PAC (1–10 mg/kg i.p. or i.v.) was also 10–30 times less effective than clonidine (0.1 mg/kg i.p.) in inducing sleep spindles (12 c/s) in the ECoG recorded from the fronto-parietal (sensorimotor) region (fig. 1). Furthermore, both imidazolines produced hypersynchronisation (theta waves) on the ECoG from the occipito-parietal (visual) region. Unlike clonidine, these effects of PAC on the ECoG were of short duration (fig. 1). A 0.5-mg/kg oral dose of clonidine induced sleep spindle activity in SHR. The same effects were observed using a dose of PAC 10 times greater than that of clonidine. In contrast, both imidazolines (0.2–1 µg/kg) possessed bioequivalent ECoG activity when applied i.c.v. (fig. 2).

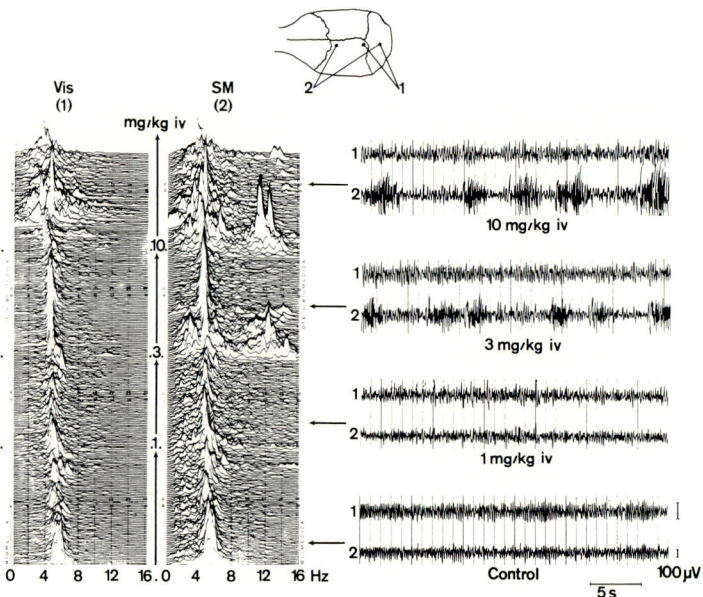

Fig. 1. Electroencephalographic recordings and sequential spectral analysis (30 s periods) of the sensorimotor and visual electrocorticogram before and after 3 doses of PAC in a curarised normotensive rat.

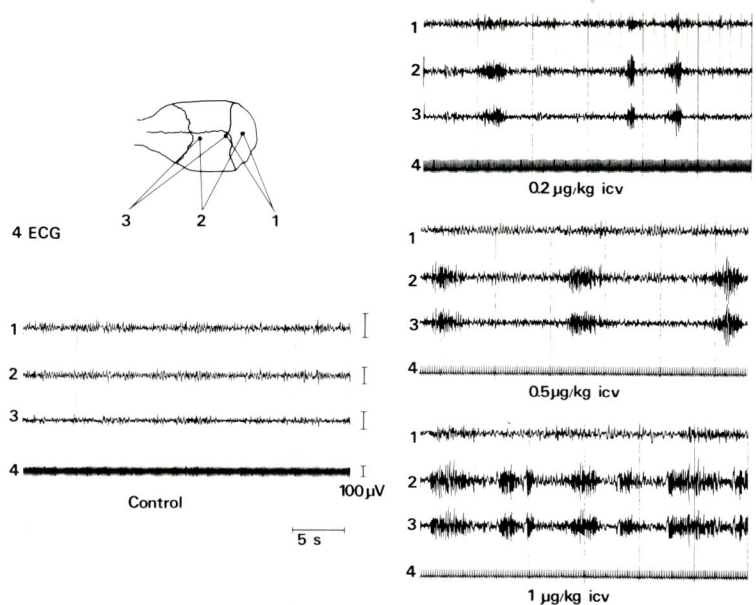

Fig. 2. Sensorimotor and visual electrocorticogram recording obtained before and after three different doses of PAC given i.c.v. to a curarised spontaneously hypertensive rat.

Discussion

These results indicate that PAC induces ECoG alterations in the rat which are characteristic of clonidine and may be related to the sedative effects of this imidazoline in man. PAC is less effective than clonidine when administered by the peripheral route (i.v., i.p. or p.o.). However, the two compounds become equipotent when injected into the cerebral ventricular space. Thus, it is reasonable to assume that PAC does not easily cross the blood-brain barrier both in rats and cats when given peripherally. The biological observation is confirmed by the physicochemical characteristics of PAC which render the passage of this compound through the blood-brain barrier extremely unlikely [5].

The fact that PAC and clonidine when given orally at 0.5 mg/kg [2] to the SHR produce an antihypertensive effect may suggest that PAC enters the central nervous system. This may occur through various possible mechanisms, for example the formation of a metabolite of PAC or alteration of the brain barrier due to the potent peripheral vasoconstrictor effect of PAC which occurs before the antihypertensive action. The present finding that PAC given p.o. at 0.5 mg/kg does not induce spindle activity in the SHR may indicate that the receptor mechanism mediating the latter effect and the antihypertensive action are probably not the same [4]. Finally, since PAC produced ECoG changes after 5 mg/kg p.o. in SHR, it is suggested that different biophase concentrations of this imidazoline may be required to elicit the antihypertensive and sedative effects.

References

1 Armstrong, J.M.; Cavero, I.; Lefèvre-Borg, F.: Br. J. Pharmacol. *70:* 169P (1980).
2 Cavero, I.; Depoortere, H.; Lefèvre-Borg, F.: Br. J. Pharmacol. *69:* 295P-296P (1980).
3 Depoortere, H.; Honoré, L.; Jalfre, M.: 3rd Eur. Congr. Sleep Res., Montpellier 1976, pp. 358–361 (Karger, Basel 1977).
4 Depoortere, H.: Sleep 1980. 5th Eur. Congr. Sleep Res., Amsterdam 1980, pp. 283–286 (this volume).
5 Rouot, B.; Leclerc, G.; Bieth, N.; Wermuth, C.G.; Schwartz, J.: C.r. hebd. Séanc. Acad. Sci., Paris *286:* 909–912 (1978).

H. Depoortere, M.D, Département de Biologie, Groupe EEG, Synthelabo (LERS), 31, avenue P.V. Couturier, F-92220 Bagneux (France)

Effect of DSIP on Diurnal and Nocturnal Sleep in Man

R. Blois, M. Monnier, R. Tissot, J.-M. Gaillard

Psychiatric Clinic of the University of Geneva, Clinique de Bel-Air,
(Director: Prof. *R. Tissot*) Chêne-Bourg, Switzerland

Delta sleep-inducing peptide (DSIP) has been shown to induce slow wave sleep in rabbits [3] and rats [2] and paradoxical sleep (PS) in cats [4]. This has led us to focus on the role of this peptide in man by studying its effects on diurnal alertness and nocturnal sleep.

Methods

Day polygraph recordings were realized in 10 subjects and the rating of stages was made by means of an automatic sleep analysis system [1], according to the Rechtschaffen and Kales' criteria. Subjects were administered i.v. 25 nmol/kg of DSIP or 10 ml saline; the recordings were balanced and separated by an interval of 8 days in order to avoid a possible prolonged effect of the peptide. The injection was made at 9 a.m., followed by a continuous polygraph recording of at least 220 min or more if the subject was still asleep.

24 night recordings were realized in 8 volunteers. After one habituation night, 3 consecutive nights were recorded as follows: the first one after i.v. 10 ml saline, the second after i.v. 30 nmol/kg DSIP, with no perfusion during the second night. After injection, light was turned off, and the awakening time left ad libitum. The recordings were analyzed as described above and stored in a data bank.

Results

In the day recordings, an increase of total sleep time was observed under DSIP (260 vs 230 min) concerning stages 3 and 4 and PS; however, none of these variations taken separately was significant. The increase of sleep time, especially marked during the last part of the recording, was significant from the 120th to the 160th min ($p<0.05$). A slight increase of waking stage was also observed in the first 65 min (NS).

From the dynamic point of view, appreciated by the calculation of the general trends of sleep stages, a greater production of total sleep and PS was noted under the peptide.

The study of the elementary activities showed that the density per min of spindles, theta, delta rhythms and K potentials was increased compared to the placebo condition; these increases were obvious from the 120th to the 180th min, particulary for theta rhythms (125th min: $p<0.01$; 135th min: $p<0.02$; 140th min: $p<0.05$).

The results of the night recordings did not show the same type of modifications. In fact, there was no difference between the placebo night and the DSIP night as regards total sleep and the different stages, either in time or as a percentage.

During the third night, a significant enhancement of total sleep occurred (513 min) as compared to the placebo night (478 min, $p<0.01$) and to the DSIP night (482 min, $p<0.05$); a significant increase of the second sleep cycle was also observed as compared to the first night (+27 min, $p<0.05$).

The general trend functions of stages indicated a lower waking stage production during the second and third nights. The production of other stages remained unmodified in any condition.

The study of the elementary activities did not show any important variation between the first 2 nights, but their distribution showed some variations. Under DSIP, a density peak occurred during the first hour, while this peak occurred during the third hour under placebo (delta, theta rhythms, K potentials).

Conclusions

This study yields a discrete but certain effect of DSIP, with different modifications in day and night recordings. An increase of day sleep, especially of PS, was found, as has been observed in cats. This effect also manifested itself in the elementary activities which are characteristic of the different sleep stages; moreover, an immediate and short-lasting activating effect seemed to exist. Night sleep physiological architecture was respected by DSIP which, however, reduced the production of waking.

The distribution of some elementary activities in the beginning of the night seemed to be characteristic of the effects of the peptide. The enhancement of total sleep in the next night could be related to a delayed effect, but the absence of injection during this night might be involved too.

References

1. Gaillard, J.-M.; Tissot R.: Comp. biomed. Res. *6:* 1–13 (1973).
2. Kafi, S.; Monnier, M.; Gaillard, J.-M.: Neurosci. Lett. *13:* 169–172 (1979).
3. Monnier, M.; Dudler, L.; Gächter, R.; Schönenberger, G.A.: Neurosci. Lett. *6:* 9–13 (1977).
4. Polc, P.; Schneeberger, J.; Häffely, W.: Neurosci. Lett. *9:* 33–36 (1978).

Dr. R. Blois, Psychiatric Clinic of the University of Geneva, Clinique de Bel-Air, CH-1225 Chêne-Bourg (Switzerland)

B. Neurophysiology and Biochemistry of Sleep

The Awakening of the Sleeping Ponto-Geniculo-Occipital Wave[1]

Robert M. Bowker

Marine Biomedical Institute, University of Texas Medical Branch, Galveston, Tex., USA

Ponto-geniculo-occipital (PGO) waves have long been regarded as fairly constant waveforms recorded only during paradoxical sleep (PS). The characteristics of PGO waves are dissimilar to those of eye movement potential (EMP) waves recorded during wakefulness and, as a result, they have been used to differentiate the sleeping state from that of wakefulness [2, 4]. The most notable differences are that: (a) PGO wave amplitudes are twice as large as those of EMPs; (b) cortical PGO wave amplitudes remain unaffected during darkness, while cortical EMPs are absent or greatly reduced in amplitude; (c) PGO wave relationships to eye movements are variable, but EMPs always follow them; (d) cortical PGO waves can possess initial negative deflections in contrast EMPs, and (e) PGO waves are more widely distributed on the visual cortex than are EMPs. The present series of studies will demonstrate that the characteristics of EMPs are dependent upon the excitability level of the brain rather than the environmental lighting conditions. Furthermore, during wakefulness with the presentation of novel stimuli, EMP characteristics are most like those of PGO waves recorded from the visual cortex and lateral geniculate body (LGN) during PS.

Cats were implanted for routine chronic recordings of the EEG, eye movements, and neck muscle activity. The PGO and EMP waves were recorded with bipolar electrodes implanted in the LGN and transcortically in the visual cortex (VC). Both sleeping and waking behaviours were observed and correlated with the recorded activities on polygraph paper. In wakefulness different stimuli were infrequently presented to the cat during varying illumination conditions.

---

[1] This work was supported by USPHS Grants NS08377 and NS13110 to *A. R. Morrison* and MH15767. Thanks to the assistance of Ms. *G. Mann* and Ms. *D. Pavlu.*

When first introduced into a well-lighted chamber, cats explored and gradually acclimated to the recording cage. During this time both head and eye movements were plentiful with each eye movement being accompanied by an EMP. Once acclimated the cat often assumed a partial sphinx posture with curled forepaws or casually groomed its face and paws. The EEG during this period of quiet wakefulness remained desynchronized. The EMPs recorded in the visual cortex and LGN at these times had an approximate mean amplitude of one third of the largest peak-to-peak amplitudes of PGO waves recorded during PS. Although an occasional large amplitude EMP wave could be recorded, most EMP amplitudes ranged from being absent in the baseline recording in which no detectable EMP was observed, even though eye movements were present, to 38% of the largest amplitude PGO waves. Disappearance of EMPs during well-lighted conditions occurred most often during visual scanning of the inside of the cage when the cat appeared relaxed and drowsy in association with eye movements.

With the presentation of a stimulus which elicited an orientation reaction, as defined by *Sokolov* [5], the amplitudes of the cortical and LGN EMPs increased to equal some of the largest PGO waves recorded during PS. The large amplitude EMPs gradually decreased in size after a period of several seconds, depending upon the degree of interest in stimulus that was exhibited by the cat. For example, large amplitude EMPs were recorded for

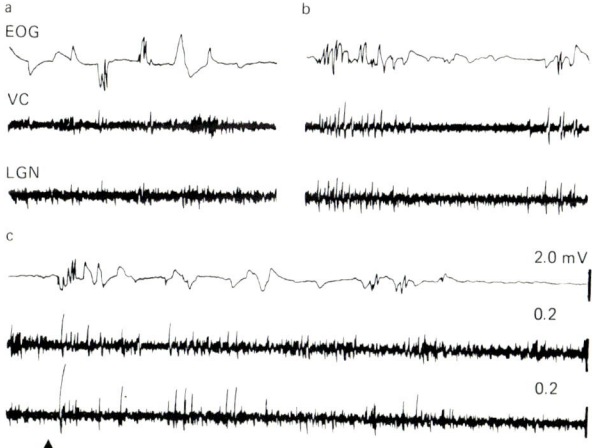

Fig. 1. Recording illustrating relative amplitudes of EMPs during quiet (a) and active (b) wakefulness and during orientation reaction (c) with presentation of noise (triangle). Time line: 1 s.

longer time periods (up to 1 min) whenever a rat or an insect was introduced into the cage in contrast to whenever an inanimate object, such as a pen, was presented. In figure 1c, a noise (triangle) was made by unlatching the cage door. Observe that an initial large amplitude EMP was recorded in the visual cortex and LGN. When the cat moved spontaneously within the cage, but without presentation of any apparent stimulus, EMP amplitudes had intermediate values between those observed during quiet wakefulness and orientation reactions (fig. 1b).

During darkness visual cortical EMPs became greatly reduced in amplitude or disappeared from the baseline recording. This observation is in agreement with other reports [2, 4]. However, if an external stimulus, such as a sound or olfactory odor, were presented to the cat, the cortical EMPs transiently reappeared for several seconds and then gradually decreased in amplitude. These findings indicate that cortical EMPs are not dependent upon different levels of background illumination, but appear to be related to changes in the level of alertness. In addition, during orientation reactions, the cortical EMP possessed an initial negative deflection and they became more widely distributed on the visual cortex. These characteristics of EMPs were similar to those of PGO waves recorded during PS.

These findings indicate that the characteristics of EMPs during orientation reactions elicited by the presentation of novel stimuli are similar to those of PGO waves during PS. Whereas PGO waves have long been differentiated from waking EMPs by certain criteria, the results of the present study indicate that these criteria are not valid indicators of the PS sleep state alone, but *they merely indicate that the brain is in a state of high excitability in both wakefulness and PS*. Furthermore, these findings support our original hypothesis that the PGO wave is merely an indicator of the activation of an alerting response network [1, 3].

References

1 Bowker, R.M.; Morrison, A.R.: Brain Res. *102:* 185–190 (1976).
2 Brooks, D.C.; Gershon, M.D.: Brain Res. *27:* 223–239 (1971).
3 Morrison, A.R.; Bowker, R.M.: Acta Neurobiol. exp. *35:* 821–840 (1975).
4 Sakai, K.; Cespuglio, R.: Electroenceph. clin. Neurophysiol. *41:* 37–48 (1976).
5 Sokolov, E.N.: A. Rev. Physiol. *25:* 545–580 (1965).

R.M. Bowker, PhD, Marine Biomedical Institute, University of Texas Medical Branch, Galveston, TX 77550 (USA)

The Role of Biological Rhythms versus Sleep-Wake Regulation in the Determination of Nocturnal Arterial Hypotension

F. Gnirss, D. Schneider-Helmert, J. Schenker

Research Department, Psychiatric Clinic, Königsfelden, Switzerland

According to a very old notion, nocturnal hypotension is the expression of an endogenous circadian biorhythm with lowest blood pressure (BP) in the hours after midnight. This assertion is to be examined with the alternative that hypotension is related to sleep per se or distinct sleep stages.

Although this problem was partially treated in several recent studies in the field of chronobiology, cardiology, and sleep research, a clear position is not yet established, primarily due to methodological inadequacies in previous investigations. To our knowledge, the following three essential requirements were not fulfilled: (1) Intra arterial continuous measurement of BP; (2) complete analysis of BP parallel to sleep EEG [cf. 1]; (3) more than 1 night recordings per subject for an analysis of eventual individual factors.

Our investigations were made in the habitual wake-sleep schedule of the subjects. Therefore, the study concerns the significance of circadian biorhythms for the natural situation.

Material and Methods

The material consists of 22 allnight polygraphic recordings in 6 healthy, normotensive volunteers, aged 22–49, 2 female and 4 male, 2–5 consecutive nights per subject. An intra arterial catheter was left in the brachial artery for the whole period of investigation. Parallel to continuous direct BP measurement, sleep EEG and different autonomic functions were recorded. TV monitoring with infrared light allowed visual observation and artifact detection. The BP recordings were completely processed with a computer. Means for 30-s epochs (identical time base for sleep stage scoring according to R.-K. criteria) were calculated for systolic, diastolic, mean BP, BP amplitude and heart rate [1]. Upon these reduced data nonparametric tests were applied to detect differences in the levels and time dependencies of BP.

Analyses and Results

BP at Sleep Onset

It was evident from the curves of continuous 30-s means of BP throughout the night [2] that BP falls considerably with sleep onset, and the general level of BP was indeed significantly lower during the sleep period (SP) than the evening waking level. An individual factor was significantly found to be involved in that a given subject tends to attain similar hypotensions from night to night [3]. With respect to the transition from evening waking BP to nocturnal sleep hypotension, a correlation analysis with the medians showed that the reduction of BP levels from waking to sleeping differs significantly between subjects, but is dependent upon evening resting levels within subjects [4]. BP measures of 22 night sleep onsets were submitted to a trend analysis for the period 'lights off to begin of SWS' with the following subperiods: 'lights off to sleep (stage 2) onset', 'sleep onset to begin SWS'. In addition, the 30 min after SWS onset were analyzed. The highest number of significant BP falls was found for 'lights off to begin SWS' (21 of 22), whereas the subperiods showed 11 and 14 significant falls, respectively. The period after SWS onset had 12 significant falls and 4 significant increases of BP trends [3].

It can be concluded from the analysis of BP changes with sleep onset that they are rather a part of a general vegetative regulation than a 'symptom' of cortical sleep onset. An individual factor plays a role for the amount of BP reduction and for the general nocturnal level.

BP in Different Sleep Stages

For a comparison of the general BP levels in the different sleep stages, the medians of the BP values within each sleep stage were considered representative, because time trends of BP within the sleep stages were present in the most cases (cf. below). No significant differences of BP levels were found between sleep stages. However, an analysis of successive differences of 30-s BP means revealed that the degree of BP variability within various sleep stages differs significantly, the variability being highest in S1 + SR and lowest in SWS. There was no time trend of the variabilities within a given sleep stage [2].

BP and Sleep Cycles

It follows from the above-reported analysis that the sleep cycles are represented in the BP tracings by alternating degrees of variability in REM sleep and NREM sleep. Beyond that we found no consistent relation of BP and sleep cycles, neither were there systematic BP level changes from cycle to cycle within the nights.

BP during the Whole Sleep Period

Time trends were analyzed on BP values within each sleep stage in order to separate the phasic component [2] from this analysis which is concerned with the time course of tonic BP. Run-tests revealed the existence of time trends in the most cases. If tonic BP follows a circadian biorhythm with its nadir after midnight, a wave trough should turn out as part of this curve. This was examined on the BP data of stage 2, because stage 2 is the best represented over the whole night and had significant time trends of BP in all recordings. But no consistent form was found, even not within individual subjects.

However, the well-known circadian curve of BP could be reproduced by taking a mean of all measurements within every hour of the night [3, 4]. Thus, this curve is a mathematical artifact, reflecting not more than the times of going to bed, sleeping, and getting up of the majority of the subjects under investigation.

Conclusions

Although time trends of BP appear frequently during night sleep they cannot be considered as a regular phenomenon following the rules of periodicity. But these would be the conditions for further holding the assumption that a circadian factor in the sense of a biorhythm determines the time and amount of nocturnal hypotension under natural night sleep conditions. However, the induction of the nocturnal hypotension is closely related to the process of falling asleep, obviously in the way of a complex regulation involving central and peripheral hemodynamic factors [3, 4], and as a part of the transition from ergotropy to trophotropy [*W. R. Hess*]. An individual factor is involved in determining the degree of hypotension. But the most important factor for the hypotension throughout the sleep period is sleep per se (as well as higher daytime BP is due to mental and physical activity), whereas the sleep stages differ from each other by the patterns of hemodynamic regulations.

References

1 Schenker, J., et al.: In Popoviciu, Sleep 1978, pp. 573–576 (Karger, Basel 1980).
2 Schneider-Helmert, D.; Schenker, J.; Gnirss, F.: In Popoviciu, Sleep 1978, pp. 531–534 (Karger, Basel 1980).
3 Schneider-Helmert, D.; Schenker, J.: Schweiz. med. Wschr. *110:* 563–570 (1980).
4 Schneider-Helmert, D.; Schenker, J.; Gnirss, F.: Sleep Res. *9* (in press, 1981).

F. Gnirss, MD, Research Department, Psychiatric Clinic, CH-5200 Königsfelden (Switzerland)

Fluctuation of REM Activity as Analyzed by REM Momentum

T. Kobayashi, Y. Saito, S. Endo, Y. Tsuji[a], H. Okuno[a]

Department of Psychophysiology, Psychiatric Research Institute of Tokyo, and
[a] Department of Electronic Engineering, Kogakuin University, Tokyo, Japan

It is well known that REM density provides us with some useful information for the study of drug effects and the diagnosis of depressed patients [2]. To obtain accurate information about REMs and to analyze the REM activity quantitatively, a new physical quantity called REM momentum (RM), which is a kinetic energy of eye movements, was introduced.

In order to examine whether there is some kind of regularity in the phasic event during REM sleep or not, the variation in time REM activity as the phasic event was analyzed by use of this new method.

Analysis Method and Material

In consideration of the physiological and dynamic characteristics of REMs, RM was defined as the product of the rotated angle (A) and the angular velocity (ω) of the eye ball [4]. That is, $RM = A \cdot \tan\Theta$ ($\omega = \tan\Theta$; Θ = rising angle of REMs wave form in EOG recordings). The new analysis method was applied to 133 REMP recordings in nocturnal sleep of 13 normal male young adults (ages 20–22).

Results

As shown from figure 1, REM momentum per time unit in REMPs of short duration, which appeared at the early part of nocturnal sleep, revealed a low value, whereas in REMPs of long duration, which appeared at the later part, RMs assumed high values. Therefore, this finding suggests that there is the physiological difference between the REM sleep appearing in the early part of nocturnal sleep and those in the later part.

To obtain a measure for the regularity in REM activity, we examined the relation between the duration of REMP and the relative position of peaks in the fluctuation patterns calculated by RM. A relative position of the peak is calculated for any of the durations of REMP as 100%. As shown

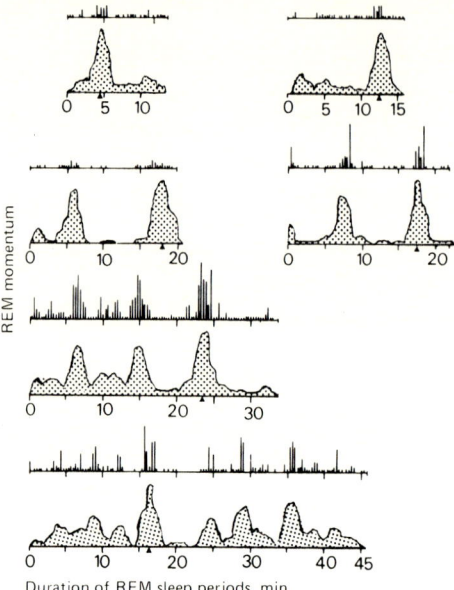

Fig. 1. The typical examples of REM activity during REM sleep as it appeared in nocturnal sleep. The upper bar graphs show the fluctuation pattern of RM calculated for every 20 s from REM sleep onset to the end. The lower graphs show the relative fluctuation pattern of RM. ▲ indicates maximum peak.

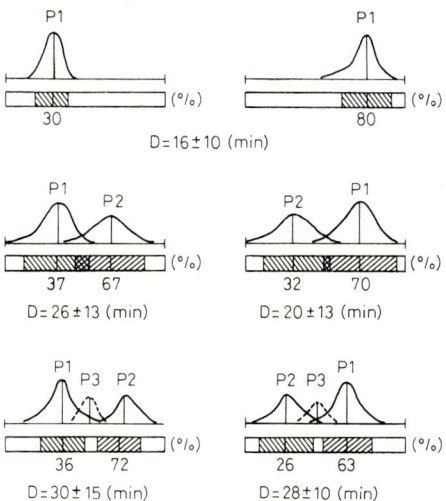

Fig. 2. Schematic representation of the relative position of peak in the fluctuation pattern of RM. The shaded area shows one standard deviation of the maximum peak position. P1 = highest peak; D = duration of REM sleep periods (min).

in figure 2, in REMPs of about 10 min duration, the maximum peak of the fluctuation pattern appears nearly at either 30 or 80% within the duration of REMP. In the REMPs of 20 min duration, there appeared two peaks; both appeared approximately at both 30 and 70%. In REMPs of over 30 min duration, there were more than three peaks; the maximum peak and the second one appeared at approximately 30 and 70%; the third one appeared in between the maximum peak and the second largest one; i.e. in almost all of REM sleep periods the maximum peaks in the fluctuation pattern of REM activity appeared either at 30 or 70% of their duration.

Discussion

Many studies on the regularity related to the fluctuation pattern of REM activity in REMP have been done [1, 5, 6]. Two viewpoints emerge; one is that there is some regularity in REM activity and the other is that REM activity changes irregularly. But the results obtained by conventional methods are not very reliable. Therefore, by the use of the new RM technique, the regularity related to the fluctuation pattern of REM activity was examined in this study. With this technique we recognized that the maximum peak of RM regularily appeared at a relatively fixed point in the course of the REM sleep period. This finding can be considered to mean either that the duration of REMP is predetermined and the fluctuation pattern of REM activity is decided upon secondarily, or that the fluctuation pattern is predetermined and the duration is fixed secondarily. In turn, there are as we have already reported, three kinds of periodicity components in REM activity, they are 5, 10, and 20 min [3]. From these two sets of regularities in REM activity, one can conclude that the fluctuation pattern of REM activity is preprogrammed rather than random.

References

1   Aserinsky, E.: Psychophysiology *8:* 361–375 (1971).
2   Foster, F.G.; Kuper, D.J.; Coble, P.; Mcpartland, R.J.: Arch. gen. Psychiat. *33:* 1119–1123 (1976).
3   Kobayashi, T.; Okuno, H.; Endo, S.; Saito, Y.: Sleep Res. *8:* 48 (1978).
4   Kobayashi, T.; Okuno, H.; Endo, S.; Tsuji, Y.: Jap. J. med. Electron. biol. Engng *17:* 222–229 (1979).
5   Lavie, P.: Electroenceph. clin. Neorophysiol. *46:* 683–688 (1979).
6   Salzaroulo, P.: Electroenceph. clin. Neorophysiol. *32:* 409–416 (1972).

T. Kobayashi, PhD, Department of Psychophysiology, Psychiatric Research Institute of Tokyo, 2–1–8 Kamikitazawa, Setagaya-ku, Tokyo (Japan)

Insomnia after Olfactory Tubercle Lesion in Cats[1]

F. Obál, Jr., G. Benedek, C. Józsa, F. Obál

Department of Physiology, University Medical School, Szeged, Hungary

Experiments with stimulation [8, 9] and lesions [4] revealed a powerful hypnogenic area in the preoptic region/diagonal band of Broca. Our own earlier results showed that the hypnogenic area extends more rostrally; stimulation of the olfactory tubercle (TbOf) resulted in cortical synchronization and increased the time spent in sleep in acute immobilized animals and chronically prepared cats, respectively [1, 6]. The aim of the present experiments was to study the effects of lesions in the TbOf on the sleep-wake activity.

Methods

6 male or female cats were used (LC 1–6). Recording electrodes were implanted over the anterior and posterior sigmoid gyri, and the median and lateral suprasylvian gyri, into the dorsal hippocampus, orbita and neck muscles. After a 1-month recovery period the cats were housed in a cage measuring $1.1 \times 0.7 \times 0.9$ m inside a sound-attenuated chamber equipped with low-level noise. The animals' behaviour was monitored on a TV screen. Standardized conditions were provided: a 10-h light/14-h dark cycle (with light on from 0800 to 1800 hours), with feeding at 1530, and animal care between 0800 and 0900. A 2-week habituation period was allowed. Prelesion recordings were obtained during 2–4 weeks, for 8 h a day (0900–1700). In 1 animal 24-h experiments were also carried out (LC 5). The EEG records were scored in 30-s periods according to *Ursin's* [10] criteria: wakefulness (W), light and deep slow wave sleep (LSWS and DSWS) and paradoxical sleep (PS) were distinguished. Postlesion recordings were obtained for 1–3 months in all but 1 cat, which died within 24 h (LC 4). The lesions were confirmed via Klüver-Barrera stained sections.

Bilateral TbOf lesions were produced by electrocoagulation (anodal current, 2 mA for 20 s). In 3 animals (LC 1–3) Teflon cannulae were implanted over the cortex according to the stereotaxic coordinates of the TbOfs. At the time of the lesion the animals were anaesthetized

[1] Supported by the Scientific Research Council, Ministry of Health, Hungary (4–05–0303–0422/0) and MTA-OM-MÉM-EÜM 70.211/79.

with Nembutal (30 mg/kg), and the coagulation electrodes were inserted into the TbOfs to make 4 coagulations at a distance of 1.5 mm from each other. In 3 animals (LC 4–6) 5 nicrothal wire electrodes were implanted into each TbOf in advance, and the lesions were produced without anaesthetics.

Results

Bilateral lesions in the TbOf resulted in a gradual increase in wakefulness and a decrease in sleep. Table I shows the average amounts of W and DSWS during the prelesion period, the maximum effect of the lesion, and the values obtained on the last day of the experiment. The cat which died within 24 h after the lesion was excluded from table I (LC 4). He was continuously awake during the 8-h observation period on the first day. In the other animals the amount of W increased to 80–90% during the second to third postlesion week and oscillated at this high level throughout the experiment. Though all sleep stages were affected, the reduction of DSWS was the most prominent (table I). To study the effect of the lesion on the diurnal distribution of sleep and wakefulness, 24-h recordings were obtained in LC 5 on a control day and 3 post-lesion days (table II). The increase of W was significant only during the light period; the high percentage of W during the night was hardly changed.

All animals showed a transient hyperactivity on the first postlesion week. The cats whose TbOfs were lesioned without anaesthetics walked con-

Table I. Percentages of wakefulness and deep slow wave sleep in 8-h recordings before TbOf lesion (mean ± SD), on the postlesion day of the maximum effect and on the last recording day. The numbers in parentheses indicate the day of the recording

Cat	Control	Max. effect	Last day
Wakefulness			
LC 1	44.7 ± 4.2	90 (14)	72 (28)
LC 2	42.1 ± 3.6	84 (15)	80 (83)
LC 3	23.6 ± 3.7	83 (6)	57 (28)
LC 5	49.9 ± 3.0	79 (13)	65 (59)
LC 6	41.9 ± 3.5	87 (18)	74 (26)
Deep slow wave sleep			
LC 1	25.3 ± 3.2	2 (16)	6 (28)
LC 2	28.0 ± 3.6	2 (15)	6 (83)
LC 3	43.2 ± 4.6	5 (6)	23 (28)
LC 5	27.5 ± 2.1	10 (13)	19 (59)
LC 6	26.9 ± 3.0	3 (18)	11 (26)

Table II. Sleep-wake percentages during the light (10 h) and dark (14 h) phase of the day in LC5 on a control day and on the 3rd, 13th and 17th postlesion days

Stage	Light				Dark			
	control	3rd	13th	17th day	control	3rd	13th	17th
W	54	67	79	73	83	88	84	84
LSWS	10	9	4	4	5	2	4	5
DSWS	24	12	10	13	9	4	8	7
PS	12	12	7	10	3	6	4	4

tinuously on the first day. After the hyperactive period the animals were less restless, but they still preferred a sitting posture to lying down. When they were sleeping their posture was normal. The pleasure responses to handling were highly exaggerated. In 3 cats sudden attack was sometimes observed, but no real harm was done: the attacks ended with purring. The behaviours of following a moving object and orientation in a new environment also increased. Food intake decreased in all animals on the first week. Circling was noted in 1 cat. The animals' behaviour in the presence of another cat (of the same or opposite sex), a rat or a mouse did not change.

Histology showed bilateral injuries in the caudal third of the TbOfs. In the most effective cases the lateral part of the diagonal band and the edge of the piriform cortex were also involved.

Discussion

The experiments of *Nauta* [5], *Sterman and Clemente* [8, 9], *Bremer* [2], *Siegel and Wang* [7] and *Hernandez-Peon* [3] indicated the existence of a synchronizing/hypnogenic area in the basal forebrain. This area included the preoptic region/diagonal band of Broca, the orbital cortex and probably the piriform cortex. Our results suggest that the TbOf acts as part of the hypnogenic area too. The behavioural findings indicate that the TbOf may contribute to the regulation of inhibitory processes and animals' affectivity.

References

1. Benedek, G.; Obál, F., Jr.; Szekeres, L.; Obál, F.: Arch. ital. Biol. *117:* 164–185 (1979).
2. Bremer, F.: Arch. ital. Biol. *111:* 85–111 (1973).
3. Hernandez-Peon, R.: in Akert, Bally, Schadé, Sleep mechanisms, Progress in brain research, vol. 18, pp. 96–117 (Elsevier, Amsterdam 1965).
4. McGinty, D.J.; Sterman, M.B.: Science *160:* 1253–1255 (1968).

5   Nauta, W.J.H.: J. Neurophysiol. 9: 285–316 (1946).
6   Obál, F., Jr.; Benedek, G.; Réti, G.; Obál, F.: Expl. Neurol. 69: 202–208 (1980).
7   Siegel, J.; Wang, R.Y.: Expl. Neurol. 42: 28–50 (1974).
8   Sterman, M.B.; Clemente, C.D.: Expl. Neurol. 6: 91–102 (1962).
9   Sterman, M.B.; Clemente, C.D.: Expl. Neurol. 6: 103–117 (1962).
10 Ursin, R.: Brain Res. 11: 347–356 (1968).

F. Obál, Jr., Department of Physiology, University Medical School, Szeged, Dóm tér 10, H-6720 Szeged (Hungary)

Polyneuronal Discharges in the Cerebral Cortex of Developing Rats during Sleep

M. Mirmiran, M. A. Corner

Netherlands Institute for Brain Research, Amsterdam, The Netherlands

There is good reason to believe that REM or 'active' sleep (AS) may be important for mammalian brain development, by virtue of the generally high level of neuronal excitation which is characteristic for this state, and the large proportion of time which it occupies in infant animals [1]. In order to study this question, chronic deprivation experiments have been carried out in infant rats [2] using chlorimipramine (CLM), a drug which produces a profound reduction of AS in favor of quiet sleep (QS). Since this reduction was monitored using only conventional polygraphic criteria, it was of great interest to confirm on the cellular level that the brain was truly in a QS-like state following the injection of CLM. We also wanted to know exactly how active AS actually is from the neuronal point of view, during early postnatal ontogeny. The occipital cortex was of special interest because this structure is known to be very susceptible to functional influence during development. It is furthermore the area from which the EEG was recorded in our AS deprivation experiments.

Nichrome wires were implanted under chloral hydrate anesthesia into the neck muscles and over the dura for recording the EMG and EEG, respectively. Using a micromanipulator, a bundle of three steel microwires (each 40 μm in diameter) was lowered slowly into the brain through a small hole over the left occipital cortex. At ca 1 mm depth the penetration was terminated, and the electrode bundle cemented in place, as soon as frequent action potentials exceeding 100 μV were recorded from at least one of the wires.

The animals were studied on the day following operation, and were freely moving except for a light cable connecting them to amplifiers. EEG, EMG and multi-unit activities (MUA) were stored on magnetic tapes while

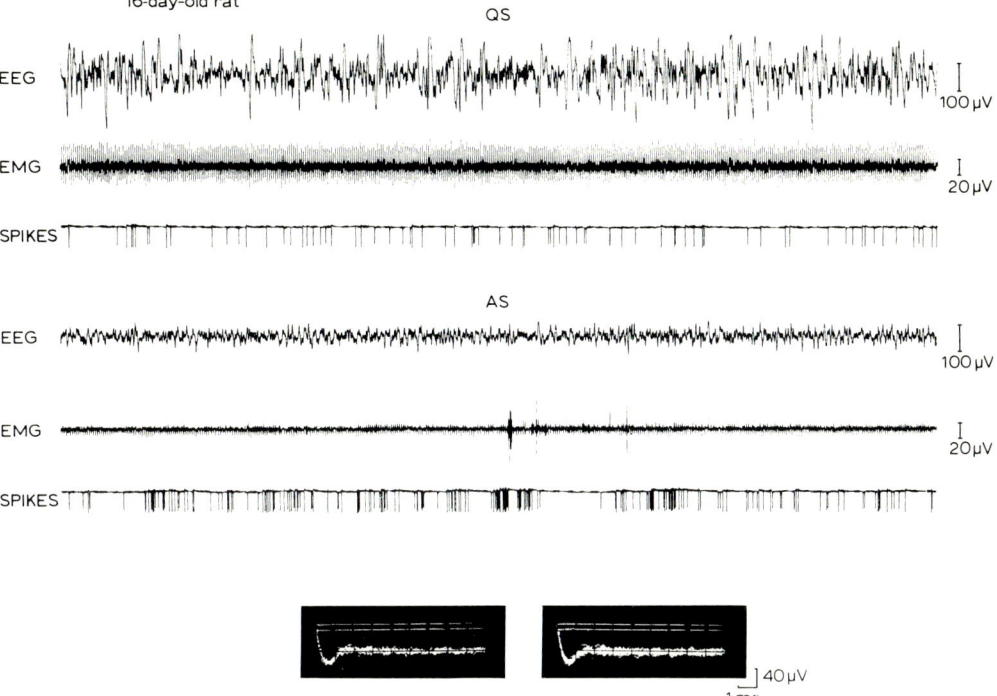

Fig. 1. Single unit activity during QS and AS (1 min stretch of each). Superimposed sweeps of the raw action potentials during the recording are shown in the inserts below.

the behavioral state was being observed continuously. The MUA was analyzed off-line by passing the signal through an amplitude 'window' discriminator, the threshold of which was set at twice the noise level, and from there into a digital integrator set to give a printed count every 2 s. In several cases where a single unit could be reliably discriminated from the others, it was analyzed separately. Mean rates of neuronal firing were calculated in each animal over several epochs of QS and AS, which states were identified on the basis of simultaneous polygraphic records (EEG and EMG) backed up by the direct behavioral observations.

Only sporadic neuronal firing, unrelated to the behavioral state, could be recorded prior to postnatal day 10 or 11. Thereafter the neurons discharged at a much higher rate, especially during AS (fig. 1). The QS pattern contained many isolated spikes and a few short bursts (up to 100–200 ms duration), in contrast to the frequent long bursts (500 ms to 1 s duration)

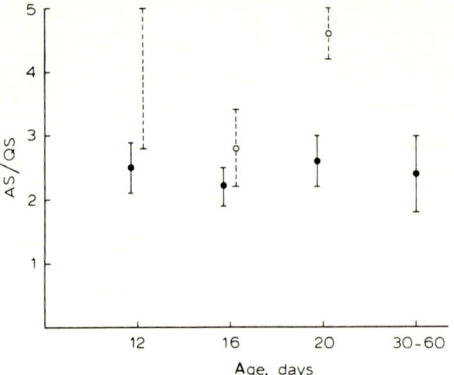

Fig. 2. The mean rate of polyneuronal discharges during AS over QS (AS/QS). Mean values (± SEM) over 5 animals in each developmental stage before (●) and after chlorimipramine (○).

and silent intervals (up to several seconds) seen during AS. Moreover, after CLM injection, the cortical firing consistently fell to a steady level during sleep which was at or even lower than the normal QS rate (fig. 2).

The present experiments therefore established that: (1) spontaneous neuronal activity emerges suddenly in the rat cortex at about the same age that QS can first be identified by means of the large amplitude EEG slow waves; (2) already at the onset, firing levels during the two phases of the sleep cycle are the same as in mature animals, and (3) chlorimipramine-induced quiet sleep is essentially normal even in terms of cortical neuronal activities.

References

1 Corner, A.A.; Mirmiran, M.; Bour, H.L.; Boer, G.; Van de Poll, N.; Van Oyen, H.; Uylings, H.: in McConnell et al., Adaptive capabilities of the brain. Progr. Brain Res., vol. 53 (Elsevier, Amsterdam, pp. 347–356, 1980).
2 Mirmiran, M.; Van de Poll, N.E.; Corner, M.A.; Van Oyen, H.G.; Bour, H.L.: Brain Res. *204:* 129–146 (1981).

M. Mirmiran, MD, Netherlands Institute for Brain Research, Ijdijk 28, NL-1095 KJ Amsterdam (The Netherlands)

Bioelectric Activity of Brainstem Reticular Neurons during Sleep and Waking in Free Moving Developing Rats

H. L. Bour, M. A. Corner

Netherlands Institute for Brain Research, Amsterdam, The Netherlands

In adult rats and cats, clusters of large neurons have been identified in the hindbrain reticular formation (RF) which sharply increase their rate of action-potential discharge during active wakefulness (AW) and sleep (AS), as compared with the rates observed during quiet sleep (QS) and wakefulness (QW). These neurons innervate extensive areas of the central nervous system, so that the excitation becomes widely distributed in the brain and spinal cord. They have therefore been implicated in the generation of certain types of movement during both waking and AS, as well as of the state of AS in its totality [1, 2].

AS plus AW together account for almost the entire behavioral repertoire of newly born rats, but during the next 3 weeks of life both the time spent in AS and the intensity of motor activity during this state fall off steeply, in favor of QS and QW. The large RF neurons are morphologically well differentiated already at birth, and thus would in principle be able to participate in the generation of AS at immature stages of development. However, nothing is known about their physiological activity in early life. The purpose of the present study, therefore, is to examine the temporal relationships between RF neural discharge levels and the shifts in physiological state, as they occur during postnatal development in the albino rat.

18 pups, varying in age from 9 to 39 days, were implanted with electrodes for recording hippocampal EEG and nuchal EMG in order to identify the physiological state. Spontaneous multi-unit bioelectric activity (MUA) was recorded by means of a bundle of 4 microwires (insulated steel, 25 or 40 µm in diameter) placed stereotaxically in the pontine RF. This method is presumed to preferentially pick up signals from larger neurons; spikes of relatively high amplitude (≥ 100 µV) were indeed found frequently in the

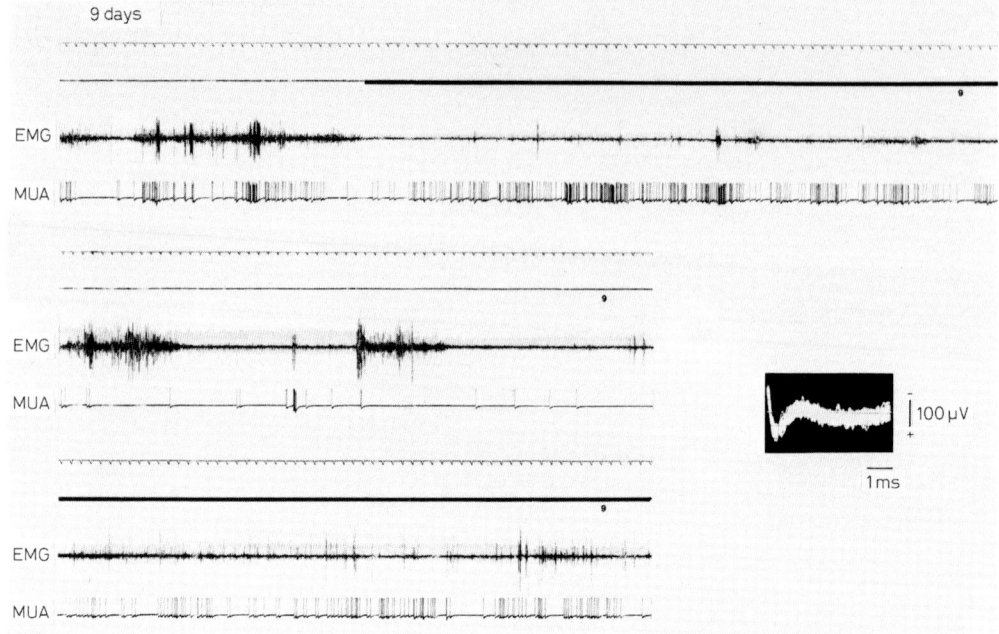

Fig. 1. Three polygraph records of a 9-day-old rat. Top: fast transition from AW to AS. Middle: two awakenings followed by QS periods. Bottom: AS. Note the increased MUA activity in AS as compared with QS. Insert: fast sweep oscilloscope display of superimposed units. Time marks: 1 s.

cerebellum, and in the brainstem from ca 1 mm ventral to the cerebellum. The electrode tip position was verified with the Prussian blue staining method.

During recording sessions the rats were connected by a light flexible cable to a turning contact, and reduction of movement artifacts in the MUA-line was achieved by passing the signal through a FET system mounted on the animal's head. Behavioral changes were scored during continuous direct observation, while the bioelectric signals were being stored on magnetic tape. MUA was analyzed off-line by passing it through an amplitude window discriminator, the threshold of which was set on twice noise level. In general not more than five units were estimated to be included in the quantitative analysis, which was carried out by a cumulative reset counter. Firing rates were then calculated for each of the four behavioral states.

Marked differences between QS and AS in mean firing rates were found at all ages studied (fig. 1). Most neurons showed their lowest dis-

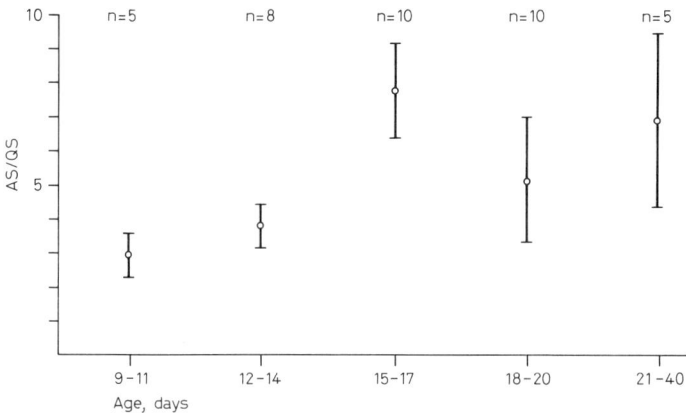

Fig. 2. The mean rate of MUA levels during AS over QS (AS/QS). Mean values (±SEM) are calculated for each developmental stage. Note the tendency to higher ratios in rats older than 14 days.

charge frequency during QS, whereas the highest rates occurred during AS and AW. This resulted in mean AS/QS values varying from 2.96 to 7.73 between the different age groups (fig. 2), whereas individual ratios ranged from 1.09 to 43.03. A slight tendency was observed for animals older than 14 days to have higher AS/QS values.

The firing level fluctuated greatly during all behavioral states with no particular pattern being evident except that for AS the highest frequencies were measured during motorically highly activated stretches (twitches and REMs) (fig. 1).

In conclusion: as early as 9 days of postnatal age, before the cortical EEG has become a useful indicator of physiological state, brainstem RF neurons in the rat show clear cut increases in firing rate correlated with the occurrence of AS, and particularly in its motorically most active positions. There is a slight tendency for this 'selectivity' to become greater after 14 days of age.

References

1 McGinty, D.J.; Siegel, J.M.: In Drucker-Colin, McGaugh, Neurobiology of Sleep and Memory, pp. 135–138 (Academic Press, New York 1977).
2 Hobson, J.A.: In Weizman, Advances in Sleep Research, pp. 217–250 (Spectrum, New York 1974).

H.L. Bour, MD, Netherlands Institute for Brain Research, Ijdijk 28,
NL-1095 KJ Amsterdam (The Netherlands)

A Central and Central-Temporal 3 c/s Rhythm Occurring during REM and Stage 2 of the Non-REM Sleep in Humans

A. C. Declerck, A. Wauquier, R. Sijben-Kiggen, M. Pans-Raedts

Epilepsy Centre Kempenhaeghe, Department of EEG and Clinical Neurophysiology, Heeze, The Netherlands

Sleep deprivation is becoming one of the important provocation methods for the diagnosis of epilepsy [2] and is routinely applied in our centre [e.g. 1]. During sleep recordings in patients following 1 night of sleep deprivation, carried out for diagnostic purposes, a specific and localized 3 c/s rhythm in the EEG was observed in part of the population (40% in a randomly selected group of patients, n = 104). This phenomenon was systemically studied in 40 patients, and the present paper gives a preliminary account of this finding.

Materials and Methods

Patients. EEG recordings were carried out in 40 patients, which came to our centre for diagnostic purposes. The median age was 24 years (range 10–47). All patients had a normal to slightly subnormal intelligence. They followed school or had a profession and none of them was institutionalized. 26 of them had some form of epilepsy for which they were taken medication.

Recordings. Polygraphic (16 channels) recordings were carried out, using the international 10–20 system of electrode placement, on the morning after a night of sleep deprivation. The electrode placement is specified in figure 1. The mean ($\pm$ SEM) duration of the recording was 189.5 (+ 4.6) min.

The EEGs were visually inspected and computer-based automatic sleep stage classification was carried out [3]. Hypnograms were made and the percentage partition of the time spent in the different stages of the non-REM sleep was calculated for each individual patient.

The EEG rhythm studied was 2.5–3.5 c/s of an amplitude from 25 to 60 µV, occurring in bursts of 1 s to more than 10 s.

All EEGs were visually inspected and the time (in seconds) of the occurrence of the rhythm was specified for each sleep stage and for each channel. Thereafter, the total time (in seconds) of the rhythm was calculated for each stage and of the time per sleep stage (table I).

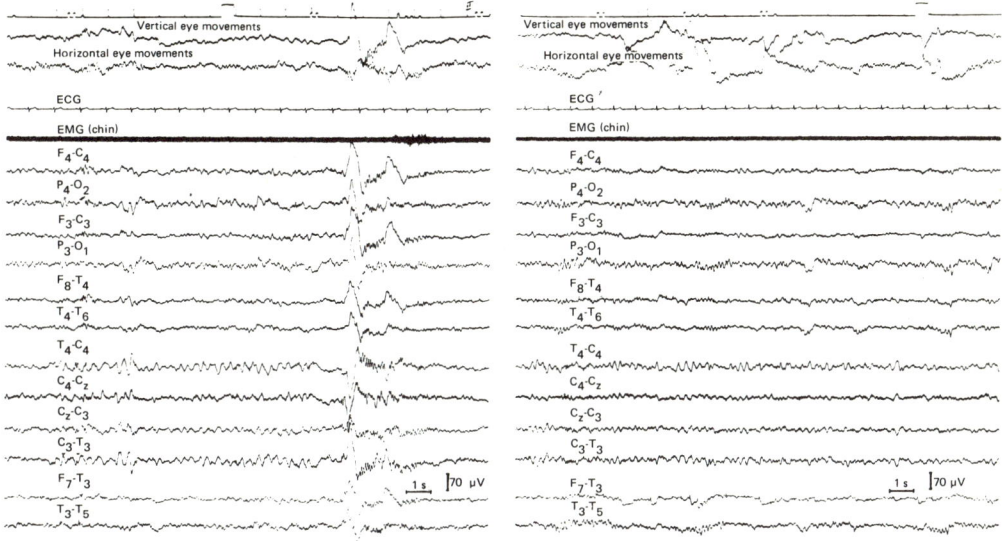

Fig. 1

Results

The 3 c/s rhythm was specifically localized in the central area, being found only in the derivations C_4-C_0, C_0-C_3, T_4-C_4 and C_3-T_3. The rhythm was usually of short duration, occurring in bursts averaging approximately 3 s, but with a range from 1 to more than 10 s.

The 3 c/s rhythm central, was seen during stage 2 of the non-REM sleep and during REM sleep (fig. 1). It was rarely seen in stage 1 or 3 and never in stage 4 of the non-REM sleep. During stage 2 and REM sleep it occurred in all patients and in 14 out of 40 in stage 1 and in 12 out of 40 in stage 3. Table I gives a survey of the mean duration and the percentage partition of this central rhythm for the group of 40 patients.

The amplitude of the rhythm was usually somewhat lower during REM than during stage 2 (25–35 µV). When the frequency was slightly lower, its amplitude was usually up to 60 µV.

Discussion

In the present paper, a preliminary description is given of a rhythm occurring in the central region, during stage 2 of non-REM sleep and during REM sleep, recorded after 1 night of sleep deprivation in humans. This rhythm had a frequency of 2.5–3.5 c/s and was usually 25–35 µV, some-

Table I. Partition of central and centro-temporal 3 c/s rhythm according to sleep stage, determined from polygraphic recordings following deprivation in 40 subjects

	Stage 1			Stage 2			Stage 3			REM			Total
	n	s	%	n	s	%	n	s	%	n	s	%	s/%
Versus total sleep													
Mean	14	13.9	0.14	40	157	1.35	12	4.35	0.03	40	266	2.32	440/3.90
SE		± 4.38	±0.04		± 14.3	±0.13		±1.71	±0.00		± 26.4	± 0.25	±31.9/+0.29
Per stage													
Mean	14		1.38	40		4.89	12		0.50	40		21.8	
SE			±0.44			±0.38			±0.21			± 1.79	

times up to 50–60 µV. It was specifically derived from the central-central and central-temporal derivations.

This rhythm appears different from the described 4 c/s rhythm vertex spindles by others, since it was seen during specific stages of the sleep, whereas the vertex spindles did not occur during drowsiness or sleep.

Though a large number of the patients (26 out of 40) in the present study were epileptics, it cannot be concluded that this rhythm neither the vertex spindles are specifically associated with epilepsy.

A question raised by the present study is the relation between this rhythm and the sleep deprivation. However, observations on normal sleep patterns without deprivation in some of the patients studied here suggest that this is not the case.

A particular characteristic of this rhythm is that it is seen as well during non-REM sleep as during REM sleep. One possible suggestion is that this rhythm is a sign of neurovegetative lability, but further analysis of the present results and additional studies are required to determine the significance of this rhythm.

References

1   Declerck, A.C.; Kuyer, A.: Clin. Neurol. Neurosurg. *81:* 69–140 (1979).
2   Schwarz, J.; Zomgemeister, W.H.: Neurology, Minneap. *218:* 179–186 (1978).
3   Wauquier, A.; Verheyen, J.L.; Broeck, W.A.E. van den; Janssen, P.A.J.: Electroenceph. clin. Neurophysiol. *46:* 33–48 (1979).

A.C. Declerck, MD, Epilepsy Centre Kempenhaeghe, Department of EEG and Clinical Neurophysiology, Heeze (The Netherlands)

Selective Paradoxical Sleep Deprivation in 6-OHDA-Lesioned Kittens[1]

J.P. Kahn, J. Adrien

INSERM U3, Paris, France

The locus ceruleus (LC) and noradrenergic systems have been implicated in paradoxical sleep (PS) [4, 7], but also in several other physiological and behavioral regulations [2]. So far, contradictory results are available. It actually appears that the LC, if it does not seem to play any essential role in PS, may be considered as having a modulatory function [2, 7], the finality of which – if any – has not yet been clearly determined. It has been suggested [2] that this modulatory role could consist in a 'stress-dampening function', allowing an adaptative behavioral adjustment to disturbing conditions. On the other hand, neonatal impairment of these systems, resulting in permanent catecholaminergic deficit, did not lead to any long-term effects [1]. Compensatory mechanisms such as denervation supersensitivity would account to some degree for functional restoration [5].

The aim of the present work was to test the importance of this presumed modulatory role of the LC, in the situation of PS deprivation, knowing that monoaminergic regulations are mature at age 3 weeks in the kitten [1] and that, in the adult cat, PS deprivation is followed by a 'rebound' [6, 8]. We investigated the ability of neonatally lesioned kittens to compensate for PS deprivation, as compared to age-paired controls.

Methods

One group of 11 kittens underwent between age 4 and 9 days a subtotal lesion of the noradrenergic systems. 8 µg 6-OHDA, solved in 8 µl ascorbic acid (1%), was stereotaxically injected at three different sites within each LC, in order to destroy the noradrenergic cell bodies. After completion of all experiments, the destruction was verified by means of the Glenner histological technique. 3 kittens which exhibited more than 75% destruction were

[1] This work was supported by DRET (79.1091) and by DGRST (79.7.1238).

selected, while the others where discarded from the data. After standard implantation for sleep monitoring and recovery from surgery, a lesioned kitten and a control littermate were placed in the experimental cage for a 2-day habituation period under standardized conditions. Both groups, lesioned and controls, were submitted to two 12-h selective PS deprivations, respectively at age 3 and 5 weeks. In the first experiment, at age 3 weeks, they were recorded in the litter whereas at age 5 weeks, after weaning, they were recorded alone. For both groups and experiments, the protocol consisted in 24 h baseline period, 12 h PS deprivation and 48–60 h recovery. The animals were recorded continuously throughout the entire protocol. PS deprivation took place between 11 a.m. and 11 p.m. and was performed manually by the experimenter who continuously watched the recording, and gently handled the kitten or tracted on the recording cable each time the animal entered PS.

Results

Injections of 6-OHDA in the LC resulted in definitive disappearence of the noradrenergic neurons, as revealed by histological data obtained 2 months after lesioning. At age 3 weeks, in the controls as well as in the early lesioned kittens, the 12-h PS deprivation was followed only by a slight PS rebound. However, at age 5 weeks, in both groups, an important rebound was observed during the first recovery of 12 h (table I). We may conclude that there was no difference between lesioned kittens and controls with respect to their ability to compensate for PS deficit. The twofold increase in rebound noticed from age 3 to 5 weeks might reflect the maturational pattern of PS compensatory regulations [3].

Table I. 12-h PS deprivation (age: 3–5 weeks)

	PS deficit min	Rebound: first 12 h min	Total recovery % of deficit
3 weeks			
Lesioned	274	+71	26
Controls	240	+48	20
5 weeks			
Lesioned	175	+89	51
Controls	201	+86	43

Selective PS deprivation in the kitten: The total PS deficit due to the deprivation was calculated from the baseline recording and expressed in minutes. The rebound represents the total number of additional minutes of PS during the 12-h period postdeprivation, as compared to the corresponding baseline 12 h. The decrement of PS deficit from 3 to 5 weeks reflects the normal evolution of sleep during ontogenesis. Note that in each condition, the total recovery is identical for the lesioned group (n = 3) and the controls (n = 4).

Discussion

From the present data, no significant difference in recovery was evidenced between lesioned and control animals. However, according to the hypothesis of *Amaral and Sinnamon* [2], the behavioral deficit due to LC lesion could be revealed under more drastic conditions; preliminary data obtained from 1 lesioned kitten and 1 age-paired control (2 months) in a PGO spike deprivation experiment are in agreement with their hypothesis: the control exhibited a strong PS rebound whereas the lesioned animal did not. If damage to the LC produces little apparent changes under 'normal' conditions, more extreme ones could represent a useful means to investigate modulatory influence of the LC in PS regulation.

References

1 Adrien, J.: Progr. Brain Res. *48:* 393–403 (1978).
2 Amaral, D.G.; Sinnamon, H.M.: Progr. Neurobiol. *9:* 147–196 (1977).
3 Berger, R.; Meier, G.: Psychophysiology *2:* 354–371 (1966).
4 Jacobs, B.L.; Jones, B.E.: in Butcher, Behavioral biology, pp. 271–293; (Academic Press, New York 1978).
5 Jonsson, G.; Hallman, H.: Neurosci. Lett. *9:* 27–32 (1978).
6 Jouvet, D.; Vimont, P.; Delorme, F.; Jouvet, M.: C.r. Soc. Biol., (Lyon) *158:* 756–759 (1964).
7 Ramm, P.: Behav. Neural Biol. *25:* 415–448 (1979).
8 Vimont-Vicary, P.; Jouvet-Mounier, D.; Delorme, F.: Electroenceph. clin. Neurophysiol. *20:* 439–449 (1966).

Dr. J.P. Kahn, INSERM U3, 91 Boulevard de l'Hôpital, F-75634 Paris Cedex 13 (France)

cAMP Concentration in the Hypothalamus and Cerebral Cortex of the Rat in Wakefulness and Sleep[1]

G. Zamboni, E. Perez, P. L. Parmeggiani

Istituto di Fisiologia umana, Università di Bologna, Bologna, Italy

The study concerns photoperiodic influences on the cAMP concentration in the preoptic region of the rat at different ambient temperatures.

420 male Sprague-Dawley rats, weighing 160–180 g, were used. After a week of adaptation (T_a 22 °C, LD 12 h–12 h schedule, 0700–1900 L, food and water ad libitum) in individual cages, the animals were placed in a thermoregulated room and studied under one of the following experimental conditions: *control* (T_a 22 °C, 3 days), *sustained wakefulness* (T_a −10 °C, 3 days), *sustained sleep* (T_a 22 °C, 1 day after 2 days' exposure to −10 °C).

Groups of 4–10 animals were killed by immersion in liquid nitrogen at fixed hours (L phase: 0900, 1500, 1900; D phase: 2100, 0300, 0630). A stereomicroscope in a dry-ice filled box was used to remove samples of the preoptic region and parietal cortex from head slices stored at −80 °C. After careful homogenization in cool TCA, cAMP and protein concentration were determined by means of the saturation binding [1] and photometric [2] techniques, respectively.

In the control condition the mean cAMP concentration in the preoptic region changed in relation to the LD cycle (fig. 1). cAMP concentration was lower during the L phase and higher during the D phase (table I). In contrast, no photoperiodic rhythm of mean cAMP concentration was found in the parietal cortex (fig. 1, table I). The preoptic cAMP photoperiodic rhythm disappeared in the condition of sustained wakefulness and of sustained sleep (fig. 2). In sustained wakefulness there was no increase in cAMP concentration during the D phase with respect to the L phase, whereas in sustained sleep the cAMP concentration of the L phase almost reached the D phase value (table I). The disappearance of preoptic cAMP

---

[1] Supported by grant No. 79.01946.04 from the National Research Council (CNR, Rome, Italy).

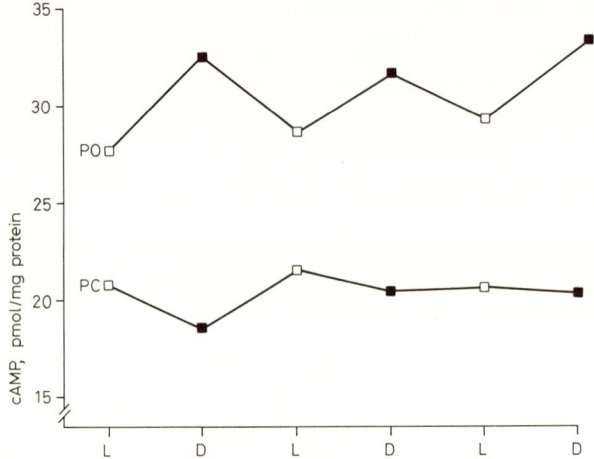

Fig. 1. Mean cAMP concentrations (pmol/mg protein) of the preoptic region (PO) and parietal cortex (PC) in the L and D periods in the control condition.

Table I. cAMP concentration (pmol/mg protein; mean ±SE) in the preoptic region and parietal cortex during the LD cycle in different experimental conditions

	L	D	t	df	p
Preoptic region					
C	28.76±0.69 (79)	32.57±0.81 (107)	3.409	184	<0.001
sW	26.49±0.78 (70)	27.71±0.82 (57)	1.074	125	NS
sS	31.10±0.80 (66)	32.23±1.47 (35)	0.744	99	NS
Parietal cortex					
C	21.03±0.51 (77)	20.11±0.54 (92)	1.219	167	NS
sW	19.36±0.48 (59)	20.68±0.49 (46)	1.917	103	NS
sS	19.28±0.54 (35)	20.30±0.46 (27)	1.391	60	NS

L = Light; D = dark; C = control; sW = sustained wakefulness; sS = sustained sleep. Number of animals in parentheses.

photoperiodicity is related to a progressive decrease and a rebound increase of the daily mean cAMP concentration of this region in sustained wakefulness and in sustained sleep, respectively. No changes in the daily mean cAMP concentration with respect to the control value (table I) were found in the parietal cortex under the same experimental conditions (fig. 2).

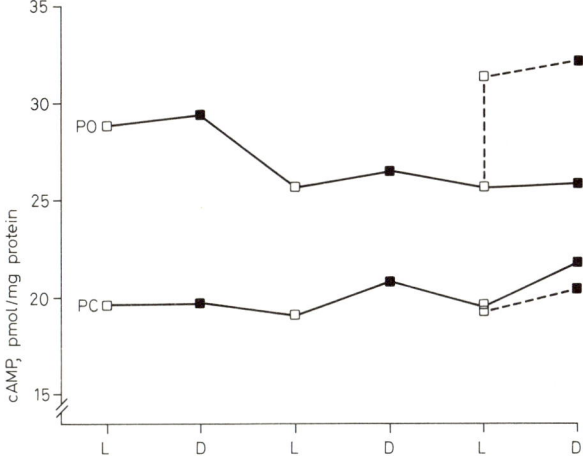

Fig. 2. Mean cAMP concentrations (pmol/mg protein) of the preoptic region (PO) and parietal cortex (PC) in the L and D periods in sW (——) and sS (---).

The results show that opposite functional conditions underlying either a decrease or an increase in preoptic cAMP concentration are also related to the suppression of cAMP concentration photoperiodicity.

References

1   Brown, B.L.; Albano, J.D.M.; Ekins, R.P.; Sgherzi, A.M.; Tampion, W.: Biochem. J. *121:* 561–562 (1971).
2   Lowry, O.H.; Rosebrough, N.J.; Farr, A.L.; Randall, R.J.: J. biol. Chem. *193:* 265–275 (1951).

G. Zamboni, MD, Istituto di Fisiologia umana, Piazza di Porta San Donato 2, I-40127 Bologna (Italy)

C. Physiology and Phenomenology of Sleep

Electrodermal Activity as a Function of Sleep Stages in the Cat

J. C. Roy, E. Freixa i Baqué, B. Delerm

Laboratoire de Psychophysiologie, Université de Lille I, Villeneuve d'Ascq Cedex, France

The spontaneous electrodermal activity (EDA) during sleep has been extensively studied in humans. According to most authors, the frequency of electrodermal responses increases from light sleep to reach a maximum during stage 4 [1]. During paradoxical sleep (PS) it decreases sharply or disappears. The present study was designed in order to discover whether the pattern found in humans is also observed in cats.

Material and Methods

Subjects were 5 adult cats weighing between 2.7 and 4 kg. Under chloralose anesthesia, EEG recording electrodes were chronically implanted on the surface of the cortex, bilaterally on the frontal and temporal areas. EMG was recorded by two silver electrodes sewn into the extensor muscles of the neck. The spontaneous skin potential responses (SSPR) were recorded differentially between two electrodes: the recording electrode (Beckman cup electrode) fixed on the pad of the hind paw and the reference electrode on a shaved part of the same paw. The EDA amplifier had a time constant of 3 s and a filter for frequencies above 15 Hz. Recordings were made from each animal during six sessions, each of which comprised several sleep cycles. During registration, the animals were placed in a sound attenuated room, where their behavior was monitored through closed-circuit television. Temperature in the cage was maintained constant around 20 °C.

Four stages of vigilance were differentiated according to the usual electrophysiological and behavioral criteria: waking (W), drowsiness (D), slow wave sleep (SWS), PS. The SSPR were taken into account only when their amplitude was equal or superior to 0.2 mV. For each stage of waking and sleep, the frequency (number of SSPR per minute) and mean amplitude (in mV) of responses were calculated. In order to minimize peripheral variations (due, for instance, to changes in skin-electrode contact), each stage is matched with that immediately preceding it. In this way 53 W-D pairs, 63 D-SWS pairs and 64 SWS-PS pairs were collected, both for frequency and amplitude. A t-test for matched data was applied to these pairs after a 'z'-transformation.

Results

It was found that EDA is high during waking, decreasing progressively during D and SWS. PS is usually characterized by a long electrodermal silence; the few SSPR then observed are generally concentrated into bursts.

The t-test applied to the pairs of values of SSPR frequency revealed significant differences between each contiguous stage: W-D: $t=8.9$, $p<0.001$; D-SWS: $t=4.45$, $p<0.001$; SWS-PS: $t=4.45$, $p<0.001$. The amplitude of SSPR does not fall off during D and SWS. The only significant difference observed is between SWS and PS ($t=2.07$, $p<0.05$), confirming that amplitude decreases during PS (fig. 1).

Discussion

The results show that the pattern observed during PS is identical for both humans and cats. The evolution of EDA during D and SWS is the opposite of that found in humans: the frequency of SSPR decreases from light sleep to stage with slow waves EEG in cats, whereas a progressive increase is always observed in humans.

SSPR can be negative, positive or diphasic in humans [3] whereas they are always negative in cats. A peripheral mechanism related to positive skin

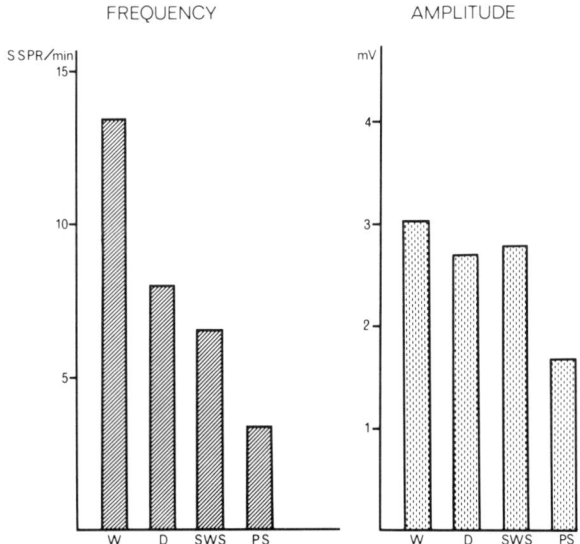

Fig. 1. Mean values of frequencies and amplitudes of SSPR during the stages of sleep and waking.

responses in humans has been suggested to interpret this difference [2]. Thus, it could be assumed that the global EDA increase in humans during stages 3 and 4 is essentially due to positive SSPR, which do not exist in cats, and that negative responses decrease in this stages as they do in cats. The discrepancy between humans and cats would be then related to a peripheral effector difference. However, as negative and positive SSPR have not been taken into account separately in human data, this hypothesis cannot be validated to date. Therefore, these discrepancies between humans and cats might be related to central differences as well.

References

1   MacDonald, D.G.; Shallenberger, H.D.; Koresko, R.L.; Kinzy, B.G.: Psychophysiology *13:* 128–134 (1976).
2   Venables, P.H.; Christie, M.J.: In Prokasy, Raskin, Electrodermal activity in psychological research, pp. 1–124 (Academic Press, New York 1973).
3   Yamazaki, K.; Tajimi, T.; Okuda, K.; Niimi, Y.: J. comp. physiol. Psychol. *89:* 364–370 (1975).

J.C. Roy, Laboratoire de Psychophysiologie, Université de Lille I,
F-59655 Villeneuve d'Ascq Cedex (France)

Nocturnal Sweat Secretion during the Dissociated Sleep Stages[1]

U. Weber, H. Haller, R. Boenke, K. Kendel

Neurological Department, Lahr, FRG

Dissociated sleep stages can be recognized with some diseases. A 'stage I REM with tonic EMG', which we call 'mixed phase', with delirium under alcohol and drug dependence and under Biperiden medication has been observed. With our patients we observed dissociated sleep stages with primarily 'mixed phases' in patients suffering from hypersomnia syndrome and narcolepsy. We were unable to detect any functional or lesional alterations causing this dissociated sleep stage. As a moderate contribution we describe vegetative phenomena accompanying this sleep stage.

On 6 patients, suffering from hypersomnia syndrome and narcolepsy, we conducted polygraphic night sleep recordings comprising EEG, EOG, EMG, respiration, body position and especially sweat secretion, all of which were registered continuously throughout each night.

We used our previously described registration system of sweat secretion [1]. The entire nocturnal sweat secretion with normosomnic patients being set at 100%, the entire night as well as the single night quarters have the following portion of sweat secretion: In stages W and 1 about 10%, in stage 2 about 15%, in stage 3 about 22% and in stage 4 close to 30%. During the REM stage its portion with 12% was a little bit higher than in stages W and 1, but definitely lower than during the other sleep stages. As we showed in a previous paper the definitely falling tendency of nocturnal sweat secretion in the course of the night can be visualized. Despite the falling tendency of sweat secretion towards the morning the relation of the rate of sweat secretion remains constant within the different sleep stages.

[1] Supported by a grant from the 'Deutsche Forschungsgemeinschaft' Bonn-Bad Godesberg.

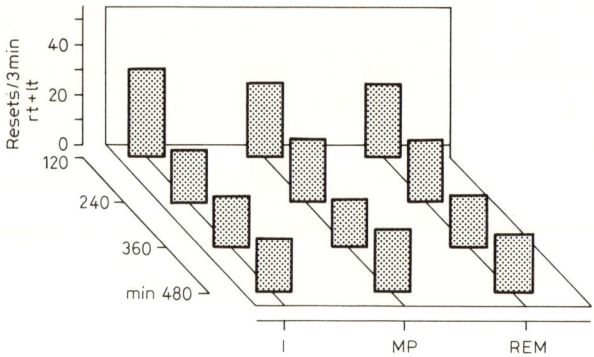

Fig. 1. Sweat secretion during dissociated sleep stages. The abscissa shows the sleep stages. The ordinate stands for the average number of air exchanges visualized as so-called resets of both right and left body halves within a 3-min tact. The Z-axis represents an 8-h night being subdivided into 4 quarters. I = Stage 1; MP = mixed phase.

Figure 1 demonstrates the sweat secretion during the 'mixed phases' compared with stage 1 and REM. It shows that the rate of sweat secretion during the 'mixed phases' is almost identical with that during stage REM. All patients showed a diminished rate of sweat secretion during 'mixed phases' in the first quarter of the night compared with stages W and 1, indeed even less than during the synchronized sleep stages. There was no significant difference between stages 1, 'mixed phase' and REM during night quarters 2 through 4. Whereas rare phasic increases of sweat secretion could be demonstrated during stage REM, sweat secretion remained constant during the 'mixed phases' without phasic increase. We furthermore examined whether we could see a regular alteration of sweat secretion 10 min prior to the beginning of 'mixed phases' and 10 min thereafter. We could show that up to the beginning and immediately after the end of the dissociated sleep stages any rate of sweat secretion can occur. Therefore, sweat secretion does not allow an early recognition of the beginning or the end of this sleep stage. The duration of the 'mixed phases' was irregular and lasted from 2 to 51 min.

The regularity of sweat secretion during the dissociated sleep stage is very likely an indicator that there is no influence whatsoever through anatomic structures rostrally from the hypothalamus on this sleep stage. The nocturnal rate of sweat secretion is bound to the circadian rhythm of the rectal temperature. The highest amounts of sweat secretion can be registered at the beginning of the night, when rectal temperature will be declining.

Whereas the average measures of sweat secretion of stage 1 in the first quarter of the night will be double those in the last quarter, the 'mixed phases' and stage REM show a slow seemingly continuous decline during the night. Since the rate of sweat secretion in stage 1 and the first quarter of the night is markedly higher, one can be led to the assumption that this sleep stage contrary to the 'mixed phases' and stage REM does participate with the thermoregulatory mechanism.

Reference

1   Weber, U.; Haller, H.; Boenke, R.; Kendel, K.: Z. EEG-EMG *10:* 226 (1979).

U. Weber, MD, Neurological Department, D-7630 Lahr (FRG)

Circadian Heat Transfer

Colin M. Shapiro, A. T. Moore

Department of Physiology, University of the Witwatersrand, South Africa

The energy requirements of normal man are still not fully understood. In particular few 24-h studies have been undertaken since the studies by *Benedict and Carpenter* [3] in 1910.

We had the opportunity of carrying out heat balance studies over three separate 24-h periods in a fit young man who was accustomed to spending long periods in the calorimeter [12]. This enabled us to measure body temperatures, partially partitioned heat loss, and to observe sleeping and waking periods. We have previously shown that REM sleep disrupts normal thermoregulatory processes [9].

Recent sleep studies have shown increasing support for the theory of sleep having a restorative function. This evidence includes the observations of increased release of anabolic hormones during sleep [4], particularly growth hormone [8], increased mitosis during sleep, faster bone growth during sleep [11] graded increase of slow wave sleep (stages 3 and 4) following graded increases in daytime exercise [10] and the relationship of mass to sleep both within human subjects [1] and across species [15].

Energy exchange as measured by oxygen consumption (VO_2) could be aligned during sleep depending upon the net balance of catabolism and anabolism, and may not be simply related to changes in body temperatures. Several studies have suggested that oxygen consumption is lower during sleep [5, 14]. It was therefore considered useful to measure heat balance throughout a 24-h period with a sensitive and accurate calorimeter.

Method

The subject was a 19-year-old male 154 cm in height and 64 kg weight who freely consented to the procedures. He was placed in a conditioning room at a dry bulb temperature similar to that of the calorimeter for 8 h prior to entering the calorimeter. The open circuit calorimeter allowed the following measurements to be made: (1) Rectal(T_{re}) and mean skin ($T_{\bar{s}}$) (Ramanathan 4 point) [7] temperatures. (2) Sensible (or dry) heat output, i.e. sum of conducted, convected and radiated heat gathered into the air stream. (3) Insensible heat output, i.e. the change in moisture content of the air stream. (4) Dry bulb (T_a) and dew point (T_d) temperatures of the incoming air. (5) Terminal oxygen consumption (VO_2) using a closed circuit system. All functions were recorded every 2 min but sensible and insensible heat outputs were continuously monitored and 2-min means recorded. These values were arranged into hourly values which were plotted against time (fig. 1).

The subject wore only a pair of gym shorts and had been conditioned by many other experiments in the calorimeter for a year prior to this series. The recordings were separated by at least 1 week. The subject lay down resting for the whole 24-h period in the calorimeter. Whilst in the calorimeter, water was given ad libitum and the subject's only meal was given in the evening. This consisted of 250 g of lean steak with or without 120 g of white bread. The subject had been allowed to eat and drink ad libitum until 4 h prior to entering the calorimeter. Whilst in the calorimeter the subject could both see and be seen and be heard by the authors, one of whom was always present. A general low level of illumination was maintained.

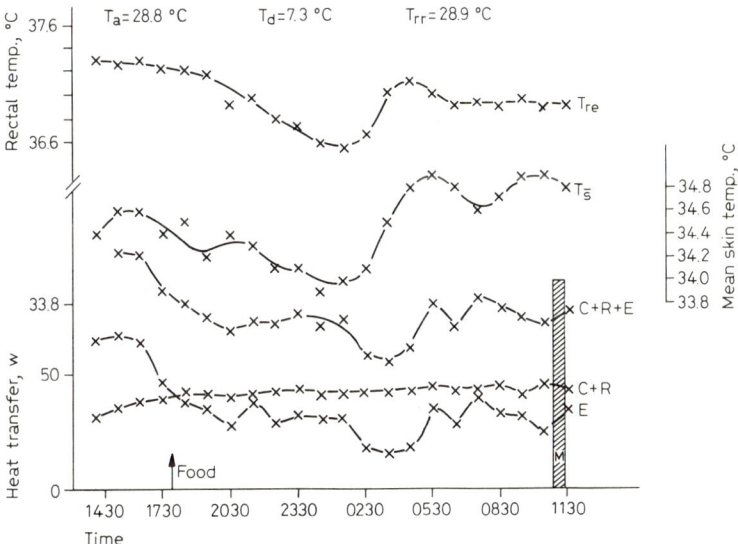

Fig. 1. Whole body heat transfer (watts) as evaporative (E), conductive plus convective plus radiative (C + R) heat transfer. Each point is the hourly average of successive 2-min integrations. Rectal (T_{re}) and mean skin ($T_{\bar{s}}$) temperatures are the average of spot readings each 2 min. Ambient dry bulb temperature (T_a) = 28.8 °C. Dew point temperature (T_d) = 7.3 °C, preconditioning room (T_{rr}) 28.9 °C. Mean total heat transfer = 75 W. Metabolic rate (M) = 90 W (from V_{O_2}) at the time indicated by the hatched column. T_a, T_d and T_{rr} were also the average values from successive 2-min spot readings.

Results

Three ambient temperatures (T_a) were used, one on each occasion. The first run was at 28.8 °C which was thermally neutral (as previously determined), the second was 26.8 °C, being slightly below the lower critical point, and the third was at 33.2 °C, somewhat above the upper critical point.

In figure 1 (thermal neutrality) the pattern of changes can be seen when no thermal stress was applied. T_{re} shows a clear circadian rhythm with a low point at 0230 hours. During the subsequent 3 h until 0530 hours the subject, who had been sleeping from about 2330, altered his thermal balance. The evaporative heat loss, E, fell sharply as T_{re} and $T_{\bar{s}}$ rose.

At subneutrality there were a number of less distinct waves. The low nadir in T_r was 36.6 °C (the same as that in fig. 1). During the early hours heat transfer parallelled T_{re}, in comparison to the counter-movement in figure 1. The metabolic rate was 10 W higher reflecting the mild cold stress.

At superneutrality T_{re} and $T_{\bar{s}}$ show a rhythm with a nadir at 0000 hours and a generally lower level of E and total output between 0000 and 0500. During this period T_{re} and $T_{\bar{s}}$ rise. A second dip occurred at between 0800 and 1000 hours during a morning sleep. The nadir of T_{re} is 0.6 °C higher than in the colder temperatures.

Discussion

The most interesting observations to emerge from these studies is the decrease in heat transfer during sleep. This is in agreement with the oxygen consumption data and adds further support to the theory that sleep is restorative and a period when energy stores increase. Earlier studies with a prototype calorimeter [2] and water cooling undergarment type calorimetry [13] are in agreement with above findings. A recent study [6] using a larger (and therefore slower reacting) calorimeter has also shown decreased heat transfer at night.

References

1 Adam, K.: Br. Med. J. *1:* 813–814 (1977).
2 Benedict, F.G.: Science *42:* 75–84 (1915).
3 Benedict, F.G.; Carpenter, T.M.: publ. No. 123 (Carnegie Institution, Washington 1910).
4 Boyar, R.M.; Rosenfeld, R.S.; Kapen, S.; Finkelstein, J.W.; Roffwarg, H.P.; Weitzman, E.D.; Hellman, L.: J. clin. Invest. *54:* 609–618 (1974).
5 Brebbia, D.R.; Altshuler, K.Z.: Science *150:* 1621–1623 (1965).

6 Dauncey, M.J.: Br. J. Nutr. *43:* 257–269 (1980).
7 Ramanathan, N.L.: J. appl. Physiol. *19:* 531–533 (1964).
8 Sassin, J.F.; Parker, D.C.; Mace, J.W.; Gotlin, R.W.; Johnson, L.C.; Rossman, L.G.: Science *165:* 513–515 (1969).
9 Shapiro, C.M.; Moore, A.T.; Mitchell, D.; Yodaiken, M.L.: Experientia *30:* 1279–1280 (1974).
10 Shapiro, C.M.; Griesel, R.D.; Bartel, R.R.; Jooste, P.L.: J. appl. Physiol. *39:* 187–190 (1975).
11 Valk, I.M.; Van den Bosch, J.S.G.: Growth *42:* 107–111 (1978).
12 Visser, J.; Hodgson, T.: S. Afr. Mech. Engng. *9:* 243–269 (1960).
13 Webb, P.: Int. J. Biometeor. *15:* 151–155 (1971).
14 Webb, P.; Hiestand, M.: J. appl. Physiol. *38:* 257–262 (1975).
15 Zepelin, H.; Rechtschaffen, A.: Brain Behav. Evol. *10:* 425–470 (1974).

C.M. Shapiro, MD, Department of Psychiatry, Royal Edinburgh Hospital, Morningside Park, Edinburgh EH10 4HF (Scotland)

Construction of REM-NREM Sleep Hypnograms from Body Movement Recordings[1]

J. Hasan, J. Alihanka

Department of Physiology, University of Turku, Turku, Finland

As all-night polygraphic sleep studies are expensive and time-consuming, attempts have been made to produce hypnograms with more simple alternative methods. Changes in body movement activity have been shown to correlate well with changes in the sleep-wakefulness cycle [4, 5], and when wrist actigraphic recordings have been combined with EOG and submental EMG recordings, even NREM and REM sleep were separated with relatively good accuracy [3]. In our laboratory a mattress, suitable for body movement recordings, has been under development during recent years [1]. The greatest advantage of this 'static charge-sensitive bed' (SCSB) method is that no electrodes need to be attached directly to the patient; the usual mattress is simply replaced by the SCSB. The method has been described in detail earlier [1]. The sensitivity of the system is good enough to record even the smallest movements of the fingers as well as the ballistocardiography and respiration. The present study is an attempt to construct a hypnogram using SCSB recorded body movements only. The goal was to distinguish between REM and NREM sleep.

Methods and Results

All-night SCSB recordings together with standard EEG-EOG recordings were performed on 10 healthy, 20- to 25-year-old male medical students on 2 consecutive nights. Only data from the second nights were included in the analysis. Sleep stages were scored in epochs of 20 s using standard criteria [6]. The number of body movements recorded with the SCSB was determined for each epoch.

The amount of body movements per minute in different sleep stages is shown in table I. It can be concluded that if REM sleep is to be separated

---

[1] Supported by a grant from the Juho Vainio Foundation, Helsinki.

Table I. Number of body movements per min in different sleep stages (n = 10 nights)

Movement duration, s	Sleep stage					
	W	S1	S2	S3	S4	SREM
< 5	0.21	0.23	0.10	0.05	0.01	0.38
5–10	0.10	0.05	0.05	0.02	0.01	0.10
10–15	0.08	0.05	0.04	0.02	0.01	0.06
>15	0.11	0.05	0.04	0.03	0.05	0.04
All movements	0.52	0.39	0.22	0.12	0.09	0.58

from the other stages, the best discrimination will probably be obtained by using only short movements of less than 5 s.

For describing body movement activity index values were calculated using the following procedure: the number of body movements lasting less than 5 s was determined for each epoch of 20 s. The time series thus obtained was smoothed using a window of 51 epochs (17 min), which was moved forward in steps of one epoch. The sum of the movements within the window was calculated and given as an index for the middle epoch. As expected, an almost total motor silence was found during stages S3 and S4. During S2 the index values were generally less than three, representing a general motor activity of less than three short movements per 17 min. During stage REM the index values were well above those of S2, the peaks being between 11 and 19. These values were correlated to the total amount of short movements per night so that subjects having a great amount of movements had higher peak index values than those having less short movements per night. Hypnograms were constructed using the index values: an index greater than or equal to three represented REM and an index less than three NREM sleep. Such a hypnogram is presented in figure 1, together with the hypnogram obtained from the original EEG-EOG recording of the same night. The mean agreement between the hypnograms of the 10 subjects was 75.3%; the range was between 62.6 and 90.0%.

Discussion

The agreement found in this preliminary study between the hypnograms constructed with body movement and conventional EEG-EOG recordings is good enough to justify further development of the system. As all-night, instead of 24-h, recordings were used, the amount of wakefulness (W) is small and no real estimation about the accuracy of W detection can be made. Our unpublished preliminary studies as well as those of others [2]

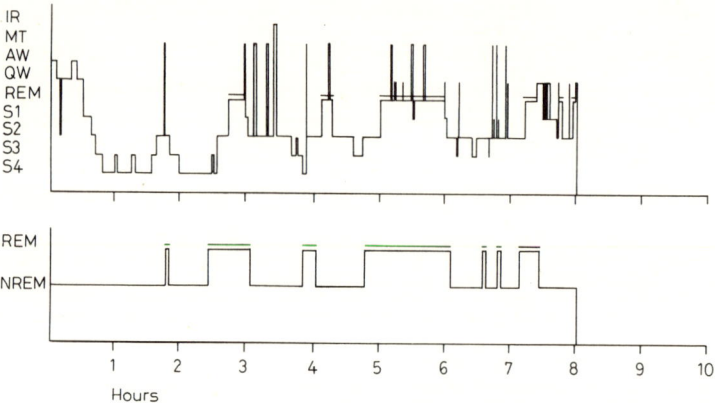

Fig. 1. A hypnogram constructed using body movements lasting less than 5 s (below) and a hypnogram from the same night constructed from a conventional EEG-EOG recording (above). IR = Interruption of the recording; MT = movement time (sleep stage not possible to determine because of movement artifact); AW = active wakefulness (eyes open); QW = quiet W (eyes closed); REM = stage REM; S1 = stage 1; S2, S3 and S4 accordingly; NREM = non-REM sleep.

seem to suggest that the proportion of body movements of different duration and amplitude is different in different sleep stages and W. Using the present version of the SCSB, the significance of amplitude measurement of the movements is difficult to interpret. A new version, where this aspect has been taken in consideration, is under development. By using all body movements instead of only those lasting less than 5 s and by taking also the amplitude of the movements into account, it seems probable that a recording and analyzing system based on the SCSB method will in the future serve as an inexpensive, simple and accurate method for preliminary home and laboratory sleep recordings and for example, pharmacological experiments. The body movements can easily be detected with an automatic analyzer and the calculation needed can be performed with a small, inexpensive microcomputer.

References

1. Alihanka, J.; Vaahtoranta, K.: Electroenceph. clin. Neurophysiol. *46:* 731–734 (1979).
2. Gardner, R.; Grossman, W.L.: in Weitzman, Advances in sleep research, vol. 2, pp. 67–107 (Spectrum Publications, New York 1975).

3   Kayed, K.; Hesla, P.E.; Rosjo, O.: Sleep 2: 253–260 (1979).
4   Kripke, D.F.; Mulleney, D.J.; Messin, S.; Wyboeaney, V.G.: Electroenceph. clin. Neurophysiol. 44: 674–676 (1978).
5   Kupfer, D.F.; Foster, F.G.: J. nerv. ment. Dis. 156: 341–348 (1973).
6   Rechtschaffen, A.; Kales, A.: A manual of standardized terminology, techniques and scoring systems for sleep stages of human subjects. (Brain Information Service/Brain Research Institute, University of California, Los Angeles 1968).

J. Hasan, MD, Department of Physiology, University of Turku, Kiinamyllynkatu 10, SF-20520 Turku 52 (Finland)

Is Human Sleep for Body Restitution?

J. A. Horne

Department of Human Sciences, Loughborough University, Loughborough, Leicestershire, England

Whilst at least some of human sleep (obligatory sleep) may be necessary for brain 'restitution', there is no firm evidence showing that sleep rather than rest is essential to body 'restitution'; sleep is probably only a convenient vehicle for whatever body restitutional processes may occur in sleep. Because of the usual night-time fast with man, and that the last meal of the day is usually several hours before sleep, supplies of amino acids are reduced during sleep [1], indicating that protein synthesis rates should be low during night-time sleep. The only sleep where these rates would be high is during a post-prandial nap. The energy cost of protein synthesis is difficult to assess. The cost of creating peptide bonds within the cell may account for about 17% of basal metabolism [2], but if all processes leading up this are included, then the figure is nearer 50% [3]. The low metabolism of night-time sleep would militate against high protein synthesis at this time. Recently [4] it has been estimated that protein synthesis rates correlate +0.9 with heat production (metabolism), again indicating low protein synthesis during night-time sleep.

Findings that night-time *sleep is a state of increased protein conservation* rather than growth, come from the only report to date [5], measuring protein metabolism during human sleep. Both protein synthesis and loss (urea formation) fell beyond 30% a few hours after the last food intake of the day and stayed at low levels through the night until morning re-feeding.

Cell adenylate energy charge (EC) may be lower during wakefulness than during sleep in rodents [6], permitting higher protein synthesis during sleep given adequate food substrates, but in man EC is high during wakefulness [7], even higher than that of sleeping rodents [6], presumably because of man's relative inactivity during wakefulness. Even if EC were higher in human sleep, this will not necessarily cause protein synthesis increase, given the amino acid constraint [8]. Also, the body is not oriented to maximising protein synthesis, only to covering requirements, leaving a reserve capacity.

The sleep-hGH release is probably a misleading factor in body restitution and sleep, as this hormone only *permits* increases in protein synthesis, and such a release may be more associated with the re-distribution of fats [9]. Few other mammals demonstrate such a release; indicating that body restitution and the sleep-hGH release are not causally related. Total sleep deprivation in man is remarkably uneventful for the body [10] and there are no indications, as yet, of impairments to mechanisms associated with body restitution (for further details see other abstract).

Although mitotic peaks occur during the sleep period, sleep alone is not essential, as such peaks are evident during sleep absence [11]. Further evidence that inactivity is the key comes from a study [12] demonstrating that waking exercise inhibits and delays such sleep-period peaks.

The EEG is a poor index of body restitution; it is the cerebrum which exhibits SWS, not the body. Thus, SWS increases following daytime exercise cannot point to elevations in body restitution without more direct restitutional measurement, which has not been performed. Also, the lack of clear SWS increases following exercise in unfit subjects has to be explained when using the more profound SWS increases of fit subjects as evidence favouring body restitution. Whilst fit subjects may incur more absolute wear and tear, proportionately with fitness this would be similar for the two types.

It is concluded that from the evidence to date, modern man's usual amounts of relaxed wakefulness and feeding suggest that for him sleep is not necessary for body restitution. In mammals unable to demonstrate significant periods of relaxed wakefulness, sleep may well be an enforced immobiliser, conserving energy and permitting increases in body restitution.

References

1 Wurtman, R.J.: in Munro, Mammalian protein metabolism, vol. 4, pp. 445–479 (Academic Press, New York 1970).
2 Waterlow, J.C.; Garlick, P.J.; Millward; D.J.: Protein turnover in mammalian tissues and in the whole body (Elsevier, Amsterdam 1978).
3 Webster, A.J.F.: Livestock Prod. Sci 7: 243–252 (1980).
4 Pullar, J.D.; Webster, A.J.F.: Br. J. Nutr. 37: 355–363 (1977).
5 Garlick, P.J.; Clugston, G.A.; Swick, R.W.; Waterlow, J.C.: Amer. J. clin. Nutr. 33: 1986 (1980).

6   Durie, D.J.B.; Adam, K.; Oswald, I.; Flynn, I.W.: IRCS med. Sci. *6:* 351 (1978).
7   Harris, R.C.; Hultman, E.; Nordesjo, L.-O.: Scand. J. clin. Lab. Invest. *33:* 109–120 (1974).
8   Atkinson, D.E.: Cellular energy metabolism and its regulation (Academic Press, New York 1977).
9   Horne, J.A.: Physiol. Psychol. *7:* 115–125 (1979).
10 Horne, J.A.: Biol. Psychol. *7:* 55–102 (1978).
11 Scheving, L.E.: Anat. Rec. *135:* 7–20 (1959).
12 Fisher, L.B.: Br. J. Derm. *80:* 75–80 (1968).

J.A. Horne, PhD, Department of Human Sciences, Loughborough University, Loughborough, Leicestershire (England)

D. Sleep, Perception and Memory

Influence of a Complex ('Enriched') Environment on the Sleep-Waking Patterns of Developing Rats

M. Mirmiran, H. Van den Dungen

Netherlands Institute for Brain Research, Amsterdam, The Netherlands

Rats reared in 'enriched' environments show in adulthood: (1) a heavier brain and cerebral cortex; (2) increased synthesis of brain protein; (3) greater cortical thickness due to an increase both in dendritic branching and in number of glial cells; and (4) an increased acetylcholinesterase activity throughout the brain. It has also been established that complex environments enhance subsequent performance of appetitive and aversively motivated behaviors. Along with these morphological, biochemical and behavioral changes, as a consequence of such treatment the total sleep time is also increased in later life [4, 5]. It is quite conceivable that sleep is affected also *during* the rearing period, and that such an effect contributes to the observed alterations in brain maturation. Since nothing is known on that score it was decided to investigate the quality and quantity of sleep during prolonged exposure to environmental complexity in juvenile rats.

Male Wistar rats delivered to our laboratory after being weaned at postnatal day 22, were implanted under chloral hydrate anesthesia with electrodes for recording the cortical EEG and the neck muscle EMG. These variables were used in order to distinguish three different states of vigilance: wakefulness (W), active and quiet sleep (AS, QS). At one month of age the animals were randomly assigned to either the experimental or the control group. Experimental animals (EC: n=5) were housed together in the complex large cage described by *Bennett et al.* [1], while the isolated controls (IC: n=5) were placed individually in standard laboratory cages. Another group ('standard social controls': n=6) was placed two by two in large empty cages. The animals were recorded in isolation for 8 h in a soundproof box on successive days during the light period, alternating between experimental and controls, starting from day 30 and continuing for 6 weeks.

Table I. Mean values (±SEM) over 5 animals, age 6–8 weeks, and per 8-h recording time

Conditions	%TST	QS dur. min	AS dur. min	QS lat. min	AS lat. min	AS ep. s	Cycle min	Cycles No.
Enriched	75 (±2)	312 (±5)	56.3 (±2.6)	20.4 (±4.8)	46 (±7)	104 (±5)	9.4 (±0.6)	30.8 (±1.9)
Standard	61[a] (±3)	260[a] (±7)	32.4[a] (±3.1)	42.1[a] (±6.1)	92[a] (±8)	89 (±13)	8.6 (±0.4)	22.0[a] (±2.0)
Isolated	63[a] (±1)	269[a] (±3)	43.1[a,b] (±1.0)	26.2[b] (±7.8)	114[a] (±16)	90 (±8)	8.2 (±0.6)	28.0 (±4.1)

[a] Significantly different at $p < 0.01$ from enriched condition.
[b] Significantly different at $p < 0.01$ from standard condition (Mann-Whitney U test).

Each animal was thus examined on three different occasions. Mean values were calculated for the variables presented in table I. It should be added that we measured the active sleep latency from the beginning of the recording session to the appearance of the first active sleep epoch, and the length of the sleep cycle measured from the beginning of quiet sleep to the end of the ensuing active sleep epoch if no awakenings lasting longer than 1 min intervened.

There were no significant differences in the amount of sleep in control and experimental animals during the 1st 2 weeks of the experiment (fig. 1), but both QS and AS were clearly augmented in the latter group during the subsequent 2-week period (table I). These differences persisted in the 3rd 2-week period (fig. 1). The latency to the 1st AS epoch was strikingly shorter in the experimental animals, and in addition there was a tendency for active sleep epochs to last longer than in either of the control groups (table I).

The sleep differences observed in the experimental group cannot be attributed to the lack of physical activities and social interactions in the control animals, since the standard controls failed to show any increase in either QS or AS. Furthermore, increased physical activity in rats having access to activity-wheels during rearing failed to show increased sleep time [6]. Stress also cannot account for the difference between experimental and isolated rats since sleep latency and the overall length and number of sleep cycles did not differ significantly in the two groups (EC versus IC).

Despite the fact that rearing in a complex environment influences the development of brain and behavior, the factors responsible for those effects are not yet established. *Ferchmin and Eterovic* [3] hypothesized that concurrent effects of excitation impinging upon the brain from environmental

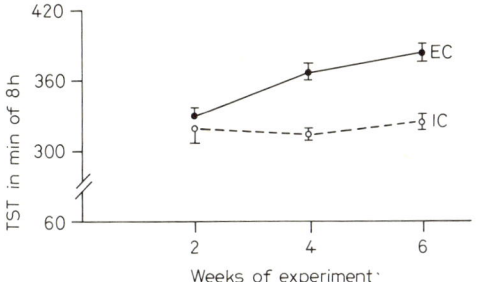

Fig. 1. Increase in the total sleep time (TST) as a function of rearing in a relatively complex environment. Each circle represents the mean ($\pm$ SEM) of 5 animals in each group.

(sensory) stimuli and from the animal itself (proprioceptive stimuli), coupled with motor activity and memory storage are responsible for the observed developmental differences. *Cummins et al.* [2], on the other hand, suggested that arousal responses evoked both by social interactions and novel object exploration cause a nonspecific stimulation of cortical elements which is transduced into biosynthetic activities affecting neuronal growth. However, the lack of such effects either in rats injected with amphetamine or in animals forced to perform complex acrobatic skills are in discrepancy with above-mentioned hypothesis.

The present results demonstrate that differences in sleep patterns can be produced by prolonged exposure to different sensory environments which are known to influence brain and behavior development. It is intriguing to speculate that sleep effects themselves contribute to brain development, a hypothesis which lends itself to experimental analysis [*Corner and Mirmiran,* this volume].

References

1   Bennett, E.L.; Diamond, M.C.; Krech, D.; Rosenzweig, M.R.: Science *146:* 610–619 (1964).
2   Cummins, R.A.; Livesey, P.J.; Evans, J.G.M.; Walsh, R.N.: Science *197:* 692–694 (1977).
3   Ferchmin, P.A.; Eterovic, V.A.: Physiol. Behav. *18:* 455–461 (1977).
4   Gutwein, B.M.; Fishbein, W.: Brain Res. Bull. *5:* 9–12 (1980).
5   Tagney, J.: Brain Res. *53:* 353–361 (1973).
6   Webb, W.B.; Friedman, J.: Physiol. Behav. *6:* 459–460 (1971).

M. Mirmiran, MD, Netherlands Institute for Brain Research, Ijdijk 28,
NL-1095 KJ Amsterdam (The Netherlands)

Dreams as Problem Solving
A Computer-Aided Dream Diary

George W. Baylor
Université de Montréal, Montréal, Canada

The goal of this research is to study dreams as a problem-solving process. This implies, first, to discern the problem the dream is attempting to solve; and, second, to characterize both the process and representational system the dream employs in its solution attempt. This paper reports on one facet of this research: namely the implementation of a tool that enables individuals to build up a dream bank by recording their dreams daily on an interactive computer system, encoding and indexing them for future retrieval. This helps, first, to get at the problems since it incorporates *Hall's* [3] dream series' method and *Faraday's* [2] maintenance of a glossary. Second, to properly encode the dream is, in the *Newell and Simon* [7] theory of human problem solving, the first step in characterizing the dream process and representational system. A properly encoded dream meets the conditions of their fundamental theoretical category: the problem space. 'A problem space consists of a set of symbolic structures (the states of the space) and a set of operators over the space...[where]...sequences of operators define paths that thread their way through sequences of states.' Thus, 'a problem in a problem space consists of a set of initial states, a set of goal states, and a set of path constraints. The problem is to find a path through the space that starts at any initial state, passes only along paths that satisfy the path constraints, and ends at any goal state' [6, p.5]. With the specification of the dreamer's problem space the path the dream travels can be charted.

The program is called CADD, a *c*omputer-*a*ided *d*ream *d*iary, written in the *s*cientific *i*nformation *r*etrieval language, SIR [8] and operates on a CDC CYBER-173 computer at the Université de Montréal. Figure 1 presents the general structure of the system. It consists of two parts: First, three dream

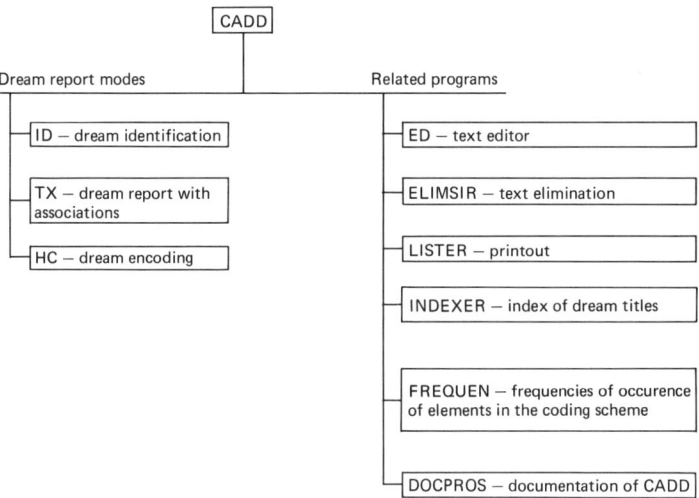

Fig. 1. Schematic structure of CADD (computer-aided dream diary).

text modes: ID for identifying the dreamer, date, place, and title of the dream; TX for writing out the dream report and associations in full; and HC for encoding the dream directly into *Hall and Van de Castle's* [4] content analysis format as modified by *Hall and Nordby* [5] and myself. The ten categories of the code are physical settings, objects, descriptive elements, theoretical scales, characters, interactions, activities, achievement outcomes, environmental press, and emotions. Eventually, other recording and encoding modes will be added to permit a fuller picture of the dream. Because of its high scorer reliability the Hall and Van de Castle system is the first storage and retrieval scheme to be explored and a first approximation to specifying the dreamer's representational system and certain processes in the problem space.

Second, the system contains a number of related programs for facilitating text editing and using the system: ED for modifying, ELIMSIR for eliminating, and LISTER for printing out any number of dreams. INDEXER creates an index of dream titles that enables the user to retrieve former dreams for comparative purposes and, perhaps most importantly, FREQUEN tabulates the frequencies of occurrence of the elements of the code over any range of dreams in the dreamer's bank. Finally, DOCPROS facilitates learning the system by giving the user access to a manual that explains CADD's structure and functioning. Additional statistical programs will be added as needed. The entire system is now documented in a user's manual [1].

References

1. Baylor, G.W.; Goulet, P.; Pinard, F.: Manuel d'utilisation du système CADD (Université de Montréal, Montréal 1980).
2. Faraday, A.: The dream game (Harper & Row, New York 1974).
3. Hall, C.S.: The meaning of dreams (McGraw-Hill, New York 1953).
4. Hall, C.S.; Van de Castle, R.L.: The content analysis of dreams (Appleton Century Crofts, New York 1966).
5. Hall, C.S.; Nordby, V.J.: The individual and his dreams (Mentor, New York 1972).
6. Newell, A.: Reasoning, problem solving and decision processes: the problem space as a fundamental category (Carnegie-Mellon University, Pittsburgh 1979).
7. Newell, A.; Simon, H.S.: Human problem solving (Prentice-Hall, Englewood Cliffs, 1972).
8. Robinson, B.N.; Anderson, G.D.; Cohen, E.; Gazdzik, W.F.: SIR scientific information retrieval user's manual (SIR Inc., Evanston 1979).

G.W. Baylor, PhD, Département de Psychologie, Université de Montréal,
Montréal, Que, H3C 3J7 (Canada)

Problem-Solving in Dreaming: an Empirical Test

Milton Kramer, Edward McQuarrie, Michael Bonnet

Dream/Sleep Laboratory, Veterans Administration Hospital, and
Department of Psychiatry, University of Cincinnati, Cincinnati, Ohio, USA

The problem-solving function of the dreams of a night has been described as a pattern of problem statement, mounting tension in the search for solutions, resolution and then decreased intensity or relaxation [1, 2]. The pattern suggested by the clinical illustrations is U-shaped or an inverted U.

If the dream serves a problem-solving function across the night, one would infer that the dreams of the night are ordered, i.e. non-random, and discriminable, i.e. each different from the other. Evidence does exist that dreams are not random events [3].

However, compelling evidence that *each* of the dreams across the night is discriminable is less clear [4]. Dreams early in the night are different in content from dreams later in the night [5] and dreams from the first half of the night can be discriminated from those of the second half [6] along a number of content dimensions.

We sought to examine the dreams of a large group of subjects, over many nights, using a great number of the most reliable of the *Hall and Van deCastle* [7] dream scales in the hope of finding the relationships which might support *most explicitly* a problem-solving function for dreams, namely each dream position is distinguishable.

Method

22 subjects, 11 male and 11 female, ages 20–25 slept in the laboratory for 20 consecutive nights. At the end of each of the first four REM periods of the night dream recall was obtained.

The dream content was scored for characters (10 subscales), activities (9 subscales) and modifiers (22 subscales) of the Hall/Van deCastle dream content scoring system [7]. Two additional scales were derived: one, a total content score representing the sum of all characters, activities and modifiers minus negatives, and second a word count score.

Analysis

The data was found to be non-normally distributed. Significant REM period content differences were determined using a Friedman analysis of variance by ranks. Post-hoc comparisons were made using the Wilcoxon signed ranks test. The word count data was analyzed first and then controlled in succeeding analyses.

Results

Recall. There were 1,760 possible awakenings (22 subjects times 20 nights, times 4 awakenings per night). In fact, there were 1,659 awakenings. Dreams were reported after 1,190 awakenings, and this resulted in a 72% recall rate.

Word Length Analysis. An ANOVA for word count was done and was significant ($p<0.004$). Post-hoc analysis revealed: (1) that there were fewer words in REM period I dreams than in REM period II ($p<0.01$), (2) fewer words in REM period III dreams than in REM period IV ($p<0.05$), and (3) that REM period II was not distinguishable from REM III.

Dream Content Variables. The ANOVA showed that only eight content variables, among 22 having sufficient content to be analyzed, varied significantly across the four REM periods (all $p<0.03$). Total, single and female characters, and total, physical and verbal activities and size minus and total scored content were the eight significant content variables.

For none of the content variables was there a significant change across each of the REM periods. Four contents increased in frequency from REM period I to II and five increased in frequency between II and III. Only total scored content changed across both of these. None of the variables increased in frequency from REM III to IV.

When one compares REM period I to the sum of II, III and IV, seven of the eight word corrected contents increased in frequency. And, when one compares the sum of REM I and II to the sum of REM III and IV, six of the eight increased.

Mean Scores. An inspection of the mean scores of the nine significant variables, word count plus the other significant variables, suggests that the primary pattern of change is a linear increase of content across the four

REM periods. For example, for total scored content, the ratio of appearance across the four REM periods is 7.41, 8.12, 8.95 and 9.04 across respective REM periods. A similar increase was found for all of the variables except for the three significant character variables, total characters, single characters and female characters. In these three variables slight (nonsignificant) reductions in content was found in the fourth REM period as compared to the third.

Discussion

The present data serve as evidence that there are consistant increases from REM 1 through REM IV (though those changes were not always significant). This pattern supported systematic changes across the night, but it did not fit the ideal pattern of increase followed by decrease proposed for a problem solving theory of dream function.

Several possible explanations exist for the failure to support the ideal form of the problem-solving hypothesis for those variables. It is possible (1) that the wrong variables were measured; (2) that the variables measured were measured incorrectly; (3) that the hypothesized function exists but that the down turn of the curve does not come until after REM IV or there is no down turn; (4) that problem-solving procedes at a different pace depending upon the night, problem or subject so that overall differences simply average out, and finally (5) that dreams in general do not serve a problem-solving function and that the characteristic changes found across the night reflect increased REM length or underlying circadian rhythms.

Three correlated variables showed the inverted U-shaped pattern that was predicted (although the final down-turn was nonsignificant). They were total characters, single characters and female characters. One could argue that this was a fortuitous change and occurred in these three variables because they were highly correlated (rho>0.70). On the other hand, the character changes might reflect a change of some importance because it is a change in the character dimension in our other work that has been the variable which discriminates in significant circumstances [8, 9].

In summary, the data from the present study do not provide the evidence for a specific pattern of dream content change which would support unequivocally the problem solving function of dreaming as it has been described clinically. However, the data does lend support to the notion that there is systematic dream content change across the night which could be the basis for a less specific theory of the nature of problem solving in dreams.

References

1 Cartwright, R.D.: Night life: explorations in dreaming (Prentice-Hall, Englewood Cliffs, 1977).
2 Kramer, M.; Whitman, R.M.; Baldridge, B.J.; Lansky, L.M.: J. nerv. ment. Dis. *139:* 426–439 (1964).
3 Kramer, M.; Hlasny, R.; Jacobs, G.; Roth, T.: Am. J. Psychiat. *133:* 778–781 (1976).
4 Cartwright, R.; Lloyd, S.; Butters, L.; Weiner, L.; McCarthy, L.; Hancock, J.: Psychophysiology *12:* 149–159 (1975).
5 Domhoff, B.; Kamiya, J.: Archs gen. Psychiat. *11:* 529–532 (1964).
6 Verdon, P.: Percept. Mot. Skills *20:* 1253–1268 (1965).
7 Hall, C.S.; Van deCastle, R.L.: The content analyses of dreams (Appleton Century Crofts, New York 1966).
8 Kramer, M.; Roth, T.: Compreh. Psychiat. *14:* 325–329 (1973).
9 Kramer, M.; Baldridge, B.J.; Whitman, R.M.; Ornstein, P.H.; Smith, P.C.: Dis. nerv. Syst. *30:* suppl., pp. 126–130 (1969).

M. Kramer, MD, Dream/Sleep Laboratory, Veterans Administration Hospital, University of Cincinnati, Cincinnati, OH 45220 (USA)

Recall and Retrieval Processes Related to Stage II Sleep[1]

C. Cipolli, P. Salzarulo, E. Calasso, S. Maccolini, R. Pani, G. Tuozzi

Istituto di Psicologia, Università di Bologna, Bologna, Italy, and
INSERM U3, Paris, France

Recall of mental sleep experience (MSE) can be obtained in verbal reports upon awakening during NREM sleep as well as REM sleep [2]. But it is well known that the observed frequency of NREM recall varies considerably from study to study, far more than is the case with awakenings from REM sleep. Various explanations have been proposed to account for this variation: the criteria used to define the MSE, the criteria used for the selection of subjects, interview modes, and the expectations of subjects and researchers [4]. In particular it appears that the schedule of awakenings used is of great importance in determining the frequency of recall. *Pivik and Foulkes* [5] showed different amounts of recall according to the stage of sleep (being greater for II than for III and for IV) and according to the cycle of sleep in which awakening takes place.

It may be asked to what extent the quantitative differences in frequency of MSE recall are due to memory factors. We have examined stage II, for which there exist data suggesting quantitative [5] and qualitative [3] differences between the first and the second sleep cycles. In our study we have made use of the fact that memory processes are reflected not only in the frequency of recall in verbal reports, but also in the organization of these reports as linguistic acts. We analyzed verbal reports for both recall frequency and for the linguistic indicators of retrieval processes.

Material and Methods

10 male University students, all native speakers of Italian, aged 19–27, were recorded for three non-consecutive nights. They were awakened 3 min after the onset (1) of stage II sleep in

---

[1] Supported by Grant No. 79.00530.08 of CNR (Italy).

the first cycle, (2) of the first REM, and (3) of stage II sleep in the second cycle, on each of the 3 nights.

Upon awakening, they were asked this question: 'What was going through your mind before awakening?' The contentful verbal reports obtained were analyzed according to linguistic criteria of generative-transformational grammar [1].

We compared only reports obtained during stage II of the two cycles, performing statistical tests on the individual mean values of linguistic indicators of recall and of retrieval processes. These were, for recall: the frequency of contentful reports and the mean length of these: and for retrieval processes: the frequency of waking-related 'kernel' sentences per report, mean sentence length, and the frequency of hesitional pauses [for a fuller description, see 6].

Results

Indicators of Recall. Subjects gave contentful reports in 76.6% (23/30) of the awakenings in stage II of the first cycle and in 60% of awakenings in stage II of the second cycle: this difference is not statistically significant: nor is the difference between the mean length of reports (19.69 vs 17.13 kernel sentences per report).

Indicators of Retrieval Processes. Mean sentence length and the frequency of waking-related kernel sentences (which constitute intrusion in the reports) were not significantly different in the two cycles. The first cycle reports, however, showed significantly greater difficulty in semantic planning of the contents than the second cycle reports. The pauses associated with waking-related kernel sentences, which are interruptions in the description of the contents, are significantly more frequent than those pauses associated with kernel sentences describing the MSE (table I). This is true both for complete and incomplete kernel sentences, but more markedly so for the former. Complete kernel sentences in fact principally occur at the beginning of sentences and the presence of pauses thus indicates major difficulties in the retrieval of the sentence topic. Pauses associated with waking-related kernel sentences are also significantly more frequent in the first cycle reports than in the second one.

Discussion

As far as recall was concerned, our data on the whole seem to be compatible with those of those studies in which the influence of the schedule of awakenings has been observed [4, 5]. However, since recall frequency varies from one study to another in relation to the method used, we can suppose that this is not sufficient to express effectiveness of memory processing during given stage of sleep.

Table I. Indicators of retrieval processes

Measure		Cycle I	Cycle II
Mean sentence length		7.32	7.36
% Waking-related kernel sentences		20.42	31.18
% Kernel sentences preceded/followed by pause	complete MSE–KS	30.14 ↕1	29.04
	complete WE–KS	70.07 ↑ ⎯⎯⎯3⎯⎯⎯	37.75 ↑
	incomplete MSE–KS	40.65 ↕2	44.37
	incomplete WE–KS	68.08	60.16

[1] Two-tailed Student test for independent data, t = 2.991, d.f. = 15, p<0.01.
[2] Two-tailed Student test for independent data, t = 2.368, d.f. = 18, p <0.05.
[3] Two-tailed Student test for independent data, t = 2.271, d.f. = 13, p <0.05.

The indicators of retrieval processes, in particular those of semantic planning difficulty, show that retrieval of contents is more difficult for the first cycle than for the second. We can account for this by assuming that the MSE contents are less well consolidated in memory during stage II of the first cycle than the second. If this is the case, it follows that at the moment of awakening in the first cycle there may be other contents in memory which do not get recalled and verbalized. We believe that we have here another possible explanation of variation on recall frequency across sleep cycles. If retrieval is more difficult where contents are less well consolidated, this may in certain cases result in total recall failure. This would mean that for the first cycle proportionately fewer of the original MSEs get recalled than for the second.

References

1 Chomsky, N.: Aspects of the theory of syntax (MIT Press, Cambridge 1965).
2 Foulkes, D.: J. abnorm. Psychol. 65: 14–25 (1962).
3 Foulkes, D.; Vogel, G.: J. abnorm. Psychol. 70: 231–243 (1965).
4 Herman, J.H.; Ellman, S.J.; Roffwarg, H.P.: in Arkin, Ellman. The mind in sleep: psychology and psychophysiology, pp. 59–92 (Erlbaum, Hillsdale 1978).
5 Pivik, T.; Foulkes, D.: J. Cons. Clin. Psychol. 32: 144–151 (1968).
6 Salzarulo, P.; Cipolli, C.: Percept. Mot. Skills 49: 767–777 (1979).

Dr. C. Cipolli, Istituto di Psicologia, University of Bologna, Viale Berti-Pichat 5, I-40127 Bologna (Italy)

REM Sleep and Reminiscence[1]

J. A. C. Empson, K. M. T. Hearne, A. J. Tilley

Department of Psychology, University of Hull, Hull, England

It has been shown that REM sleep deprivation reduces recall of prose material, both compared to yoked-control awakenings [1] and to stage 4 deprivation awakenings [2]. It is still unclear however whether these effects are due to a specific pro-active effect of REM sleep deprivation on recall performance, or whether normal consolidation processes have been disrupted. The present experiment was designed to assess the course of forgetting through the night, using a minimal amount of disturbance, merely by reducing the amount of REM sleep achieved in the first half of the night to less than 2 min, before the first attempt at recall after 3½ h sleep.

Method and Material

34 young adult females spent 3 nights in the sleep laboratory. After an adaptation night, the subject was allowed slightly restricted sleep (6 h) on the 2nd night, in order that she would be reasonably sleepy the experimental night.

The subjects learned two sets of prose material and word lists before going to bed at their preferred time. They were deprived of REM sleep for 3½ h, and woken, and tested for recall of half of the material learned. They were then allowed to sleep uninterrupted for another 3½ h, until woken for recall of the other half of the material.

The subject's state of cortical arousal was measured by frequency-domain-averaging FFT analysis of EEG samples before bed and after each test. At those times, the subject was also given a semantic memory recognition test. Words were displayed by a slide projector, and the subject's verbal responses were timed using a voice switch. The subject's self-assessment of alertness and temper were also recorded, using 10 cm analogue scales.

Word Lists. Two word lists, consisting of 15 unassociated pairs, equated in terms of imagery, meaningfulness and frequency [3], were learned by the method of paced anticipation, to a

---

[1] Supported by a grant from the Medical Research Council.

criterion of 10/15 correct verbal responses. List order was counterbalanced for learning and recall. Written recall was also obtained.

Prose Pieces. Two short stories were used, and were played to the subject on a tape-recorder. Order was counterbalanced across subjects for learning and recall. The number of 'cognitive items' and number of words in the written account were used as data, as in previous work [1, 2].

Results

10 subjects were excluded from the sample, and their results analysed separately, as they achieved more than 2 min REM sleep in the first half of the night, due to experimenter error. The sleep parameters are presented in table I. It had been predicted that reminiscence would occur, so that recall after 8 h (including 60 min or more of REM sleep) would be superior to recall after 3½ h with less than 2 min of REM sleep. In fact, in that group, there was no significant difference in recall scores, in either word lists or prose material, contrary to prediction (table II). In recall of the word lists,

Table I. Sleep parameters (min)

	Onset	1	2	3	4	REM	TST
(a) n = 24							
ca 2330–0300	23.3	15.5	66.6	12.2	55.5	0.65	150.5
ca 0330–0700	25.1	10.5	72.3	7.5	31.6	62.7	183.6
(b) n = 10							
ca 2330–0300	23.7	18.4	78.9	11.0	57.0	3.0	168.3
ca 0330–0700	30.6	8.6	69.1	6.3	27.1	70.0	180.8

Table II. Word lists and prose recall data (n = 24)

(a) Word lists		Verbal recall	Written recall
1st test		8.9	9.9
2nd test		8.6	9.2
	^{f}[df 1.22]	0.14	0.87

(b) Prose		Story items	Number of words
1st test		20.3	137.1
2nd test		19.8	138.9
	^{f}[df 1.22]	0.15	0.03

Table III. Word lists and prose recall data (n = 10)

(a) Word lists		Verbal recall	Written recall
1st test		10.3	10.6
2nd test		8.7	9.2
	^{f}[df 1.8]	5.33*	4.61
(b) Prose		Story items	Number of words
1st test		19.8	148.1
2nd test		24.8	162.7
	^{f}[df 1.8]	5.8**	1.22

*p<0.05 (2-tailed); **p<0.01 (1-tailed).

both groups showed some forgetting, with better initial recall by the over-2 min REM sleepers, although there was no significant interaction between groups and trials. Recall by over-2 min REM Ss for the two stories was significantly better after 8 h than after 3½ h (f = 5.8; df 1.8; p<0.01, 1-tailed; table III).

There is thus clear evidence of reminiscence occurring over the course of the night, but *only* in subjects who achieved a first REM sleep period, and even though this had no measurable effect immediately (as assessed by recall of prose material after the 1st sleep period).

A significant increase in alpha period and power was noted between bed-time and 1st test, but there was no difference in either measure between 1st and 2nd tests. There was similarly no difference in retrieval efficiency between 1st and 2nd tests, as measured by latencies in semantic retrieval.

These results seem to indicate that consolidation processes during sleep may rely on an orderly sequence of sleep stages, and that undisturbed REM sleep in the first half of the night is of crucial importance.

References

1   Empson, J.A.C.; Clarke, P.R.F.: Nature, Lond. *227:* 287–288 (1970).
2   Tilley, A.J.; Empson, J.A.C.: Biol. Psychol. *6:* 293–300 (1978).
3   Paivio, A.; Yuille, J.C.; Madigan, S.A.: J. exp. Psychol. Monogr. supp. *76* (1968).

Dr. J.A.C. Empson, Department of Psychology, University of Hull, Hull, HU6 7RX (England)

Picture Recall and Recognition following Total and Selective Sleep Deprivation

A.J. Tilley, J.A.C. Empson

MRC Applied Psychology Unit, Psychophysiology Section, Cambridge, and Department of Psychology, University of Hull, Hull, England

This experiment addressed itself to two aspects of the topic of sleep and memory.

First, most studies in this area have concentrated on verbal learning and behaviour, where it is by now a long and well-established fact that sleep improves retention [1, 4]. However, less attention has been payed to pictorially based material which, naturally, focusses on the primary and dominant modality of everyday learning experiences, namely vision. How, then, does sleep affect the retention of pictorially represented learning material?

Second, and more importantly, it is still unclear whether one form of sleep is more beneficial to memory than another [2, 3, 5, 6] or whether it is simply any kind of sleep that will do the trick. Equally important, how is the effect mediated? Does sleep simply reduce interference or slow down the rate of memory trace decay, or, as would seem more likely from recent evidence [5], does it play a more active role in memory processes by, for example, promoting or facilitating the course of memory consolidation?

Method

A total of 36 subjects (18 pairs) participated in the experiment. At 2300 hours on their second night in the sleep laboratory, a series of 30 pictures was presented at a rate of 1 every 5 s and immediate free recall attempted. They were then assigned, as pairs, to one of four conditions. One group of subjects underwent REM sleep deprivation (REMD); a second group underwent stage 4 sleep deprivation (S4D); a third group suffered total sleep deprivation (TSD); and a fourth group were allowed an undisturbed, full night's sleep (S).

REMD was achieved by waking the subject whenever the EEG, EOG and EMG showed signs of impending REM sleep. On each occasion, they were kept awake for at least 3 min. with the bedroom lights on, and asked to count backwards, in threes, aloud, at the same time being exhorted to keep their eyes wide open until allowed to resume their sleep. S4D was

Table I. Mean recall and recognition scores

	Sleep	TSD	REMD	S4D
Immediate recall (2300 hours)	17.00	16.25	16.50	16.40
Morning recall (0700 hours)	18.75	15.00	15.20	15.30
Morning recognition (0700 hours)	26.35	26.75	27.00	26.60

achieved in a similar manner, except that the number of awakenings, and hence the time spent awake, was matched, as far as possible, to the mean number of awakenings required for REMD. Once this limit had been reached, further S4D was achieved by gently rousing subjects to a lighter sleep stage, rather than fully awakening them.

At 0700 hours both recall and recognition were tested. The recognition task required subjects to select the appropriate pictures from a pool of 60.

Results

Table I shows the mean recall and recognition results. As can be seen, there is practically no difference between groups at immediate recall except for a small, non-significant advantage for the sleep group ($f = 1.4$, $p > 0.05$). At morning recall, however, there is a different pattern of results. Analysis of covariance, which adjusted for the minor differences between groups at immediate recall, revealed that the sleep group recalled significantly more pictures than the other three groups ($f = 3.69$, $p < 0.05$), indicating a relative improvement in recall after an undisturbed, full night's sleep compared to a relative recall decrement following total or selective sleep deprivation. There were no significant differences between groups with respect to their recognition scores ($f = 0.15$).

Discussion

At first sight, and taken in isolation from the recognition scores, the recall results may have suggested that total and selective sleep deprivation impedes memory storage or, conversely, that sleep facilitates memory storage, or perhaps both. The recognition results, on the other hand, may point to a different explanation. They suggest that the locus of the recall effect may lie with a retrieval failure following total or selective sleep deprivation, rather than a storage failure, since SD, REMD and S4D have no detectable adverse effects on memory when additional retrieval cues are provided in the form of a recognition task.

This apparent retrieval phenomenon may be mediated by subtle differences in the level of arousal between sleep deprived, or sleep disrupted, and non-sleep deprived subjects at the time of the morning tests. Alternatively, viewing the problem in attentional terms, it could be argued that total and selective sleep deprivation forces or encourages the adoption of a memory search strategy that focusses on the more dominant pictures in the series at the expense of less dominant pictures. That is, in seeking to reduce cognitive effort to a minimum, a strategy is adopted to optimise the recall of those items which are more readily accessible in memory for reasons, say, of familiarity, commonality or perhaps simply their serial position in the series. The less familiar, less common or less favourably positioned pictures in the series (for example, middle items) only become accessible when the search strategy is widened or modified with the aid of extra retrieval cues.

In conclusion, therefore, it would appear that perhaps sleep is less beneficial to memory storage than has hitherto been thought to be the case. It may be necessary to re-evaluate the sleep and memory effect from a different perspective. The retrieval component of memory function has been largely overlooked in this area of research.

References

1   Ekstrand, B.R.; Barrett, T.R.; West, J.N.; Maier, G.: in Drucker-Colin, McGaugh, Neurobiology of sleep and memory, pp. 419–438 (Academic Press, New York 1977).
2   Empson, J.A.C.; Clarke, P.R.F.: Nature, Lond. *227:* 287–288 (1970).
3   Fowler, M.J.; Sullivan, M.J.; Ekstrand, B.R.: Science *179:* 302–304 (1973).
4   Jenkins, J.G.; Dallenbach, E.M.: Am. J. Psychol. *35:* 605–612 (1924).
5   Tilley, A.J.; Empson, J.A.C.: Biol. Psychol. *6:* 293–300 (1978).
6   Yaroush, R.; Sullivan, M.J.; Ekstrand, B.R.: J. exp. Psychol. *88:* 361–366 (1971).

A.J. Tilley, PhD, MRC Applied Psychology Unit, Psychophysiology Section, Cambridge, CB2 2BW (England)

Reliability of a Structural Dream Analysis Method

R. Seitz, J. Bernhard, R. Burkhardt, M. Faeh, U. Moser, I. Strauch

Institute of Psychology, University of Zürich, Zürich, Switzerland

Problem

The SSLS (scoring system for latent structures) is a highly sophisticated method for analyzing content and structure of dream reports and related free associations. The system has been published by *Foulkes* [1]. *Foulkes* and co-workers reported very high intercoder reliability scores for the content analysis part of the system. Is it possible to reproduce such a high intercoder agreement by scoring dream reports written in the German language?

Method

3 coders, trained in SSLS, independently scored a sample of 20 laboratory dream reports (both REM and NREM dreams). Following the methodological and statistical analyses proposed by *Foulkes et al.* [2], several agreement measures were computed, covering soft and hard reliability criteria.

Results

Table I compares the intercoder reliability scores with the scores reported by *Foulkes et al.* [2]. *Significantly lower* reliability scores were achieved.

Discussion

Several factors may account for the relatively low reliability scores:

The Setting Factor. In contrast with *Foulkes et al.* our coders did not try to achieve a prescoring consensus on semantic ambiguity of the dream texts (e.g. to what refers 'she' or 'it'?). A verb dictionary for scoring German verbs

Table I. Comparison of reliability scores

Reliability measure	Reliability criterion	Reliability scores	
		Foulkes et al. [2] %	own study %
r_1	agreement on unitizing the dream reports	96.5	95.7
r_2	agreement on assigning interactive vs associative relationship	98.7	77.8*
r_3	agreement on verb scoring	96.8	66.1*
r_4	agreement on noun scoring	87.7	48.4*
r_5	agreement on total lexical scoring	87.2	46.9*

*Significant at the 1% level.

does not yet exist; by using the English SSLS dictionary some translation ambiguities could not be avoided.

The Textual Factor. Our coders scored both REM and NREM dream reports and special reliability scores for these dream types were not computed separately. German language brings about specific scoring difficulties because of its syntactic complexity, its tendency to compound words and to verbalize actions by nouns.

The Personal Factor. Our coders were less trained in SSLS: intercoder agreement increased significantly from the first to the second part of the dream sample. Our coders also had less implicit knowledge about SSLS than the co-workers of *Foulkes*.

Conclusion

From the fact that we were unable to confirm the extremely high reliability scores reported by *Foulkes et al.* [2], one cannot conclude that the reliability of SSLS itself is rather low. By controlling some of the factors discussed above, we probably would get higher reliability scores. Moreover, our reliability scores compare with results of other dream content analysis methods (e.g. *Hall and Van de Castle* [3]: scores between 54 and 100%). As far as we can see none of these other dream analysis methods compares with SSLS with regard to complexity and structural conception. We therefore regard our SSLS results as being rather encouraging for future research on dream formation and dream interpretation.

References

1. Foulkes, D.: A grammar of dreams (Basic Books, New York 1978).
2. Foulkes, D.; Butler, S.F.; Maykuth, P.L.: in Foulkes, A grammar of dreams, pp. 419–428 (Basic Books, New York 1978).
3. Hall, C.S.; Van de Castle, R.L.: The content analysis of dreams (Appleton Century Crofts, New York 1966).

R. Seitz, lic. phil., Institute of Psychology, University of Zürich, CH-8000 Zürich (Switzerland)

Freud and Chomsky: Our Linguistic Capacities during Dreaming

Frank Heynick

Department of Philosophy and Social Sciences, Eindhoven University of Technology, Eindhoven, The Netherlands

In the psychoanalytic dream model, a major feature distinguishing the 'secondary process' (predominant in waking thinking) from the 'primary process' (characteristic of dreaming thinking) is the former's respect for the laws of grammar and formal logic, in contrast to the latter's non-discursive, imagistic nature. The distinction is highlighted by a comparison of *Chomsky's* [3] model of sentence generation (a secondary process) with *Freud's* [5] model of dream synthesis (a primary process). Although the two systems bear definite similarities in their organization (see [4]), in the transformational generative (TG) model – as further developed by generative semanticists – rules for the mapping of underlying thought relations onto communication sentences are defined with strict mathematical formality, while in the psychoanalytic model the 'rules' for transforming underlying thoughts into the perceived dream are notoriously ill-defined and haphazardly applied. Some authors have recently suggested functional anatomical concomitance to the waking thinking/dreaming thinking distinction: higher versus lower brain activity and left versus right hemispheric dominance [11].

Yet, it has always been known that dreams need not be 'silent movies'; that (to continue the modern metaphor) they may contain 'sound tracks'. Indeed, specimens of dream dialogue appear in the *Traumdeutung*, and Freud was understandably concerned as to their place within his overall dream theory. Only recently, however, have formal psycholinguistic models – with their emphasis on the Chomskyan differentiation of actual 'linguistic performance' from the ideal, underlying 'linguistic competence' – afforded us a suitable tool for the comparative study of human language capacities in dreaming and waking. In the light of these newly developed models, we are also better able to evaluate the internal consistency of Freud's theory when it attempts to incorporate an apparently secondary (linguistic) process into the general framework of primary-process dreaming.

Syntactic Characteristics of Dream Speech

In a pilot experiment, we regularly awakened by telephone on random nights and at random hours 29 volunteer subjects in order to directly elicit verbal material recalled from their dreams (and to further elicit the general dream context). 106 awakenings yielded 32 REM-like reports, 18 of which contain immediately recalled dream dialogue. Upon linguistic analysis, all specimens proved to be fully grammatical, and most were complex, rather than simple or elliptic sentences, indicating – on the face of it – that linguistic competence is functioning with at least waking adequacy in dreaming, i.e. that complex transformations (or formally equivalent processes) are being well executed. The corpus is similar in its syntactic grammaticality and variety to the corpus assembled by *Kraepelin* [8] from his own dreams. Recent experiments by *Salzarulo and Cipolli* [10] have likewise yielded extensive corpora of dream utterances which display a degree of syntactic grammaticality at least comparable to that of waking speech.

Dream-Speech Recall and 'Secondary Revision'

Yet, no corpora of such linguistic performance, no matter how extensive, can in and of themselves prove that secondary process linguistic competence is indeed functioning properly in supposedly primary process dreaming. Alternative interpretations of the data are possible. The specimens of dream-speech which Freud presented (perhaps reluctantly) in the *Traumdeutung* are likewise fully grammatical and syntactically varied, but this he attributed in part to the process of 'secondary revision', a 'contribution from waking thought' involved in the recall of dream material (verbal or otherwise), which could serve to 'edit', as it were, grammatically defective 'sound tracks'. However, data from a previous experiment of ours somewhat weighs against the grammaticality-thanks-to-secondary-revision hypothesis. (For more details, see [2].)

In this take-home experiment, we 'directly monitored' on tape 83 instances of sleeptalking spoken aloud by 15 subjects while sleeping (and therefore *not* reported after awakening). No clear-cut instance of ungrammaticality was discernible. Furthermore, many distinct utterances were, like the specimens of recalled dream-speech, of considerable length and complexity.

Caution is called for in interpreting this data however. That 'no clear-cut case of ungrammaticality was discernible' is not the same as saying that 'all specimens were clearly grammatical'. The slurred articulation which characterizes so much of sleeptalking makes interpretation difficult, and *Arkin* [2], in assessing the corpora gathered from his own experiments, feels that the language faculty in sleeptalking functions only intermittently with waking adequacy. Furthermore, the status of sleeptalking vis-à-vis dreaming and 'true sleep' is still uncertain for sleep researchers (see [1]), as it was for Freud himself.

The 'Replay Hypothesis'

The functioning of an underlying linguistic competence in dreams, as evidenced by either of the above two types of performance, can be radically downgraded in yet another way. One may, as Freud did, invoke a 'replay hypothesis': that the 'sound tracks' of dreams are based on speeches heard or spoken by the dreamer previously in waking life and 'recorded' in memory to be 'played back' as whole utterances or as composites by the dreamwork proper before being submitted for any necessary secondary revision 'editing'. This hypothesis serves to deny to our dreaming faculties the 'linguistic creativity' so central to the Chomskyan models.

> In the above-mentioned sleeptalking experiment, each subject, upon hearing his sleeptalking utterance on tape for the first time, was asked if he could recall having said or heard the same or almost the same utterance previously in waking life. For 50 of the 83 utterances, the subjects responded affirmatively, while the remaining 33 specimens could not be 'placed' by their respective speakers.

These results seem not to be incompatible with Freud's model of dream-speech generation. (General caution should however be exercized as to the validity of the subjects' judgments.) The replay hypothesis can also derive potential support from the experiments of *Penfield and Roberts* [9] which show how short cortical electrode stimulation during brain surgery produces (or rather: seems to the subject to produce) the experience of a 'video replay' (complete with verbal dialogue) of a scenario experienced previously in real life. (See also in this connection *Hartmann's* [6] hypothesis likening the effects of PGO impulses during REM sleep to those of cortical electrode stimulation.)

Conclusion

This paper has endeavored to show in Chomskyan terms how the model Freud presents in the *Traumdeutung* for the generation of syntactically grammatical and varied utterances in dreams is consistent with his overall dream model, thanks to his radically downgrading – despite apparently contrary 'linguistic performance' evidence – the functioning of secondary process 'linguistic competence' in primary process dreaming.

Indication was given of how the linguistic aspect of the psychoanalytic dream theory is – with the help of modern psycholinguistic models of sentence generation – particularly amenable to testing, thanks to the uniquely quantifiable nature of verbal material. The meagerness of the data to date calls for more extensive and sophisticated experiments on dream-speech in the future. Needless to say, such experiments could be carried out to test the validity of not only the Freudian model of dream-speech generation (and by extension the overall theory), but of every other dream model which – for the sake of its internal consistency and elegance – can make definite predictions about the characteristics and generation of dream-speech (see *Heynick* [7]).

References

1 Arkin, A.M.: J. nerv. ment. Dis. *143:* 101–122 (1966).
2 Arkin, A.M.: Sleeptalking: psychology and psychophysiology; appendix (Erlbaum, Hillsdale 1980).
3 Chomsky, N.: Syntactic structures (Mouton, The Hague 1957).
4 Foulkes, D.: A grammar of dreams, chap. 2 (Harvester, Hassocks 1978).
5 Freud, S.: The interpretation of dreams (1900), chap. VI (F) (Pelican, Harmondsworth 1976).
6 Hartmann, E.: The functions of sleep, p. 135 (Yale, New Haven 1973).
7 Heynick, F.: Linguistic aspects of Freud's dream model. Int. J. Psycho-Analysis (in press).
8 Kraepelin, E.: Über Sprachstörungen im Traume (Engelmann, Leipzig 1906).
9 Penfield, W.; Roberts, L.: Speech and brain mechanisms (Princeton University Press, Princeton 1959).
10 Salzarulo, P.; Cipolli, C.: Biol. Psychiat. *2:* 47–57 (1974).
11 Abstracts of the 5th European Sleep Congress of the ESRS (Amsterdam 1980), pp. 5, 47, 100.

Drs F. Heynick, Department of Philosophy and Social Sciences,
Eindhoven University of Technology, NL-5600 MB Eindhoven (The Netherlands)

Differential Aspects of Self-Rated Sleep Latencies

I. Strauch, I. Soukos-Valavani

Institute of Psychology, University of Zürich, Zürich, Switzerland

Sleep latency is considered as the crucial variable to separate good from poor sleepers [2]. In therapy outcome research, subjects often are selected according to their sleep latencies, and reductions in sleep latency, self-rated or measured, are regarded as the only indicator of successful treatment [1]. How reliable are prolonged sleep latencies associated with subjective insomnia complaints?

Method

160 high school adolescents (mean age 15.4 years) completed a sleep inventory and a personality questionnaire. From this sample, 51 subjects (31.4%) reported an average sleep latency of 30 min or more. These subjects were divided into two groups (table I): group A: prolonged sleep latencies *without* difficulties falling asleep; group B: prolonged sleep latencies associated with difficulties falling asleep at least several times a month.

Results

Several items significantly differed between the groups. Group B more often stated to sleep worse when expecting an exciting day (75 vs 41%, $p<0.03$), is more often prevented from falling asleep by cognitive intrusions (83 vs 33%, $p<0.001$), and is presently more concerned with personal problems (83 vs 48%, $p<0.02$). Group B also stated incipient problems of maintaining sleep with more nightly awakenings (fig. 1) which were more often accompanied by anxiety ($p<0.005$). Group B more often gets less sleep than usual (fig. 2) and scores higher on the personality scale nervousness ($p<0.03$). There were no significant differences with regard to age, sex, bedtimes, sleep duration and post sleep behavior.

Table I. Difficulties falling asleep

		Never	Rarely	Several times a month	Several times a week	Daily
Group A	(n=27)	2	25	0	0	0
Group B	(n=24)	0	0	16	7	1

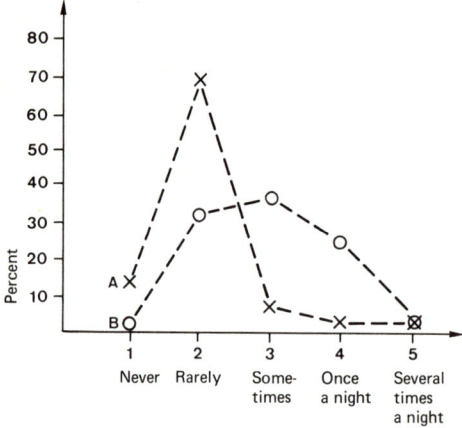

Fig. 1. Frequencies of nightly awakenings for groups A and B ($p<0.002$).

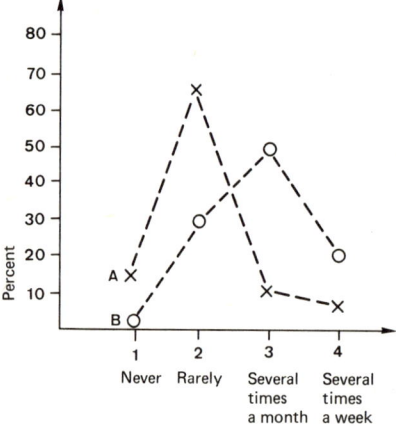

Fig. 2. 'How often do you get less sleep than usual?' ($p<0.002$).

Discussion

These results indicate that sleep latency alone may not represent the key variable to identify sleep onset insomnia. At least in adolescence, prolonged sleep latencies may not be associated with insomnia complaints and additional indications of sleep problems. More important than sleep latency per se seems to be the quality of cognitive activity and its subjective evaluation during delayed sleep onset. These criteria should be considered in therapy outcome research, for the selection of subjects and the evaluation of success as well.

References

1 Borkovec, T.D.: Adv. Behav. Res. Ther. *2:* 27–55 (1979).
2 Rechtschaffen, A.; Monroe, L.J.: in Kales, Sleep physiology and pathology, pp. 158–169 (Lippincott, Philadelphia 1969).

I. Strauch, PhD, Institute of Psychology, University of Zürich, CH-8000 Zürich (Switzerland)

E. Physiopathology of Sleep and Sleep Disorders

An Epidemiological Sleep Survey in a General Practice of a Middle-Sized Town

W. De Graaf, P. A. M. Poelstra, P. Visser[1]

Psychophysiology Laboratory, University of Amsterdam, Amsterdam, The Netherlands

What patients mean by sleep disorders is still a serious problem. To get a better insight into such problems, a survey was performed in a medium-sized practice of a family doctor in a middle-sized Dutch town with about 16,500 inhabitants. The total number of patients in the practice was about 2,300.

Method

A questionnaire covering information about personal, occupational, family and living circumstances, use of medicaments, influences of neighbourhood and sleeping habits was sent to all patients of 20 years and older. A 17-item sleep quality questionnaire [1] and the Profile of mood states (POMS) was inserted. In total 47 questions were to be answered. All questionnaires were numbered corresponding the clinical files of the family doctor. Provisions were made in order to guarantee anonymity to others than the family doctor.

Results

The response rate was 81%. About 11% of the questionnaires were answered incompletely, so that 70% of the questionnaires could be worked upon. This rather high response rate may be taken as a measure of the doctor-patient relationship. The respondents represent a reasonable sample of the Dutch population regarding sex and age. Some preliminary results of the most interesting questions are presented. Questions 29–45 are related with the sleep quality questionnaire, of which an earlier edition has been published [2].

The question if noise during sleep is often present in the neighbourhood was answered affirmatively by 25%, n = 1,332. Hindrance by noise was indi-

[1] The code book was elaborated by *J. Tulen* and *J. M. Pluymen*.

cated by 677 subjects, 12% of which indicated it as often, and 42% as sometimes present. The question about doing something to sleep better was answered by almost all subjects, n = 1,321, of which 68% never and 24% sometimes did something. Inquired after what they do was answered by 661 subjects: self-care 14%, warm milk 27%, medicines 14%, earplugs 1%. Alcoholics were indicated by 1,338 subjects, of which 15% always, 35% sometimes, 50% never.

Sleep after sexual intercourse, n = 1,258, is better 25%, worse 2%, the same 51%, no judgement 19%. Inquired after causes of worse sleep 18% indicated psychological factors, 41% coffee, 14% alcoholics, 1% pain, n = 386.

Answer to when not sleeping pleasantly: 0–1 year 24%, 1–2 years 11%, 2–5 years 12%, more than 5 years 11%, n = 291. The sum scores of the sleep quality questionnaire were divided by the number of answered items and multiplied by ten. The total group of respondents was divided into five about equal-sized groups to get comparable cell frequencies in the cross-tabulation of sleep quality with other questions (table I).

From the cross-tabulation of the five sleep quality groups with the other questions, it appeared that clear-cut relations existed for age, mood and occupational group. Of the subjects of 60 years and older 50% belong to the groups I and II ('bad sleepers'), whereas the 20- to 40-year-old subjects belong for 50% to the groups IV and V ('good sleepers'). The subjects with a small mean mood disturbance according to the POMS belong for more than 50% to the groups IV and V, whereas subjects with a high mean mood dis-

Table I

Sleep quality	Absolute number	Frequency %				
0	50	3.6				
1	64	4.6				
2	53	3.8	19% =	262	I	very bad
3	37	2.7				
4	58	4.2				
5	79	5.7				
6	114	8.2	20% =	278	II	bad
7	85	6.2				
8	236	17.1	17% =	236	III	average
9	348	25.2	25% =	348	IV	good
10	258	18.7	19% =	258	V	very good
Total	1,382			1,382		

turbance belong for more than 50% to the groups I and II ('bad sleepers'). The housewives, n=346, are found for 53% in the groups I and II ('bad sleepers') in contradistinction to clerks of which 56% are found in the 'good sleepers' groups IV and V.

Conclusions

The well-known relation between age and sleep quality is confirmed in this study. This finding lends support to the validity of the sleep quality questionnaire we used in this study. Mood and sleep quality relate in an expected manner; however, no conclusions can be made about the causal link. Housewives are overrepresented in the 'bad sleepers' group. This finding may give an indication of the stressfulness of this occupation.

References

1 De Diana, I.F.P.: Sleep Res. *1976:* 101.
2 Visser, P.; Hofman, W.F.; Kumar, A.; Cluydts, R.; De Diana, I.P.F.; Marchant, P.; Bakker, H.J.; Van Diest, R.; Poelstra, P.A.M.: in Priest, Pletscher, Ward, Sleep research, pp. 135–145 (MTP Press, Lancaster 1978).

W. De Graaf, MD, Psychophysiology Laboratory, University of Amsterdam, Amsterdam (The Netherlands)

Sleep Habits and Sleep Disorders in 2,537 Young Finnish Males

M. Partinen, P. T. S. Putkonen

Departments of Neurology and Physiology, University of Helsinki, Helsinki, Finland

Surveys assessing the amount and quality of sleep have been performed as parts of larger epidemiological studies (e.g. [2]). More extensive data on different vigilance parameters have been derived from patients complaining about excessive daytime sleepiness (EDS) e.g. [1]. The present study was designed to collect normative sleep-wakefulness data from the Finnish population and to evaluate the possibilities of questionnaires in indicating persons with specific sleep pathology.

Methods

A detailed questionnaire with 110 items concentrating on sleep, vigilance and sleep disturbances was answered by 2,552 recruits shortly after entering service. Only 15 questionnaires had to be discarded, leaving 2,537 for the data processing. The sample was representative of the socioeconomic and regional distribution of the Finnish population. The mean ($\pm$SD) age was 20.2 ± 2.8 years (range 17–29). This report deals with questions concerning the time before entering military service.

Results and Comments

The mean reported sleeping time in our subjects was 8 h 15 min $\pm$ 1 h 14 min per night. This is more than the 7 h 41 min mean (including naps) in 24-h sleep charts of 20- to 29-year-old British subjects [5], or the mean 'sleep period times' of 7 h 37 min and 7 h 4 min of 16- to 19- and 20- to 29-year-old American males in a sleep laboratory [7]. In a survey of 2,369 17-year-old students from Florida 8% claimed less than 6.5 h and 13% more than 8.5 h of night sleep [6]. The 6.5-h limit set for 'natural short sleepers' would in our material give a corresponding percentage of 6.2. The 8.5-h limit for exceptionally long sleep, however, could not be applied in Finland as 37.5% would pass this criterion and 15.8% claimed more than 10 h of sleep per

Table I. Sleep disturbances

Percentage of answers	+++	++	+	−	?
Snoring	2.9	6.6	30.8	38.8	20.9
Sleeptalking	0.8	5.2	33.3	45.2	15.5
Oppressive nightmares	0.4	2.7	39.1	54.8	3.0
Hypnagogic hallucinations	0.3	1.9	14.3	77.2	6.3
Bruxism	0.6	1.8	8.2	74.6	14.8
Sleep onset paralysis	0.3	1.1	7.4	84.6	6.6
Sleepwalking	0.0	0.6	6.9	88.8	3.7
Enuresis	0.1	0.2	1.2	97.2	1.3

+++ = 'always'; ++ = 'often'; + = 'sometimes'; − = 'never'; ? = 'can't answer'.

night. While the cited studies are not strictly comparable, the evidence points to longer sleep in Finnish youths than in recent Anglo-American data. On the other hand, our figures fit better older British observations, which indicated a circa 8-h 'normal' sleep demand [4]. An earlier assessment of sleep in the American population [3] arrived at a mean of 8 h 23 min. These figures suggest a possible decrease in sleep in recent decades. Hypothetically this might result from sleep-depriving tendencies of an increasingly urbanized and accelerated way of life, impact of TV, etc. The longer sleep in our subjects may reflect a less urban population rather than biological or geographical differences. We found, however, no significant differences between students, non-manual, and manual workers and farmers.

Chronic or frequent insomnia was a complaint in 5.4% of our subjects and 3.8% regarded their sleep as disturbed by frequent awakenings. In comparison, 9.5–8.3% of 18- to 19- and 20- to 29-year-old American males reported 'trouble with sleeping often or all the time' [2]. Frequencies of various sleep-related disturbances are given in table I.

Our question: 'Do you consider yourself daily more sleepy than your friends or workmates' was answered in the affirmative by 9.5%. This was taken as a criterion for EDS together with a 'yes' to 'Are you often sleepy during the day'. In the EDS group the frequency of all sleep-related disturbances, except snoring, increased dramatically (table II).

In an attempt to sort out the subjects most liable to suffer from the narcolepsy-cataplexy or hypersomnia-sleep-apnea syndromes two subgroups were derived from the EDS+SA on the basis of answers indicating cataplectic symptoms (CPL) or persistent snoring (SNR). The numbers are too small for conclusions, but some interesting trends may be noted (table II).

Table II. Reported disturbances in different groups

Sleep disturbances	Total (n = 2,537)	EDS (n = 240)	EDS+SA +CPL (n = 17)	EDS+SA +SNR (n = 15)
Snoring	9.5/40.3	12.4/39.2	17.6/64.7	100 /100
Sleeptalking	6.0/39.3	17.7/52.1	23.6/76.5	26.7/73.3
Nightmares	3.1/42.2	13.8/57.4	17.6/58.9	0.0/66.7
Bruxism	2.4/10.6	4.1/17.5	17.6/29.4	13.3/33.3
Hypnagogic hallucinations	2.2/15.5	4.1/38.8	11.8/53.3	0.0/26.7
Sleep onset paralysis	1.4/ 8.8	6.1/23.5	11.8/23.5	20.0/40.0
Sleepwalking	0.6/ 7.5	3.1/13.5	11.8/41.2	6.7/20.0
Enuresis nocturna	0.3/ 1.5	4.2/ 6.3	0.0/ 0.0	13.3/13.3

Percentages of answers: often or always/sometimes (+often or always).
SA = 'sleep attacks'; CPL = cataplexy; SNR = snoring.

There is a fairly parallel rise in both subgroups in nocturnal agitation indicated by increased sleeptalking and bruxism. Sleepwalking increases more with cataplexy than with snoring. Hypnagogic hallucinations clearly dominate in cataplectics. The high percentage of sleep onset paralysis in snorers is due to two subjects overlapping with the cataplectic group.

Taking into account all the answers of the questionnaire, 4 subjects of the EDS+SA+CPL group appear highly suspect of narcolepsy. In 1 of them the diagnosis has been confirmed, whereas others wait follow-up studies.

References

1 Guilleminault, C.; Carskadon, M.: in Koella, Levin, Sleep 1976, pp. 95–100 (Karger, Basel 1977).
2 Karacan, I.; Thornby, J.I.; Anch, M.; Holzer, C.E.; Warheit, G.J.; Schwab, J.J.; Williams, R.L.: Soc. Sci. Med., vol. 10, pp. 239–244 (Pergamon Press, Oxford 1976).
3 Laird, D.A.: Med. Rec. *139:* 169–170 (1934).
4 Lewis, H.E.; Masterton, J.P.: Lancet *i:* 1262–1266 (1957).
5 Tune, G.S.: Br. med. J. *ii:* 269–271 (1968).
6 Webb, W.B.; Agnew, H.W.: Science *168:* 146–147 (1970).
7 Williams, R.L.; Karacan, I.; Hursch, C.J.: EEG of human sleep. Clinical applications (John Wiley & Sons, New York 1974).

Dr. M. Partinen, Department of Neurology, University of Helsinki,
SF-00290 Helsinki 29 (Finland)

All-Night Polygraphies for Healthy Aged People (2nd Report)

Y. Hayashi, R. Yoshida, H. Watanabe, S. Endo

Department of Neurology, Second Hospital, Tokyo Women's Medical College, Tokyo, and Department of Psychophysiology, Psychiatric Research Institute of Tokyo, Tokyo, Japan

This report was designed to compare the sleep characteristics of aged people in their 70s with those people in their 80s in order to study the effects of aging.

Material and Methods

Polygraphic sleep patterns for 26 healthy aged people (6 males and 20 females) in the Yokufukai nursing home, aged between 70 and 89 years, were observed for 3 consecutive nights. 13 male university students were used as a control group. The aged group was divided into two groups according to their age. Then the sleep characteristics of the three groups (two aged and one control) were compared.

Results

The Length of REM Periods. The length of each REM period in the young adults appeared to increase progressively towards the end of the sleeping period, while that of those people in their 70s was constant, and that of those in their 80s was long in the first cycle then became shorter in the second and the third cycles (table I).

The Temporal Distributions of REM Sleep. The composite histogram of REM sleep of the group in their 70s is shown in figure 1. REM sleep of the people in their 70s and 80s showed increases during the earlier part of the night as compared to the young adults, and then remained almost constant throughout the night.

Table I. Comparison of the sleep characteristics between the young adults, the people in their 70s and the people in their 80s

	Young adults	People in their 70s	People in their 80s
Number	13	15	11
Average age	20.9± 0.8	74.6± 2.9	82.6± 2.5
Range of age	19–22	70–79	80–89
Percent SW	2.6± 2.6	18.6± 9.3	24.8± 7.9*
Percent SREM	23.2± 3.9	15.1± 6.1	14.7± 5.0
Percent S1	5.5± 2.3	14.4± 5.0	16.9± 7.1
Percent S2	50.1± 5.2	50.5± 9.3	41.7± 8.4**
Percent S3	8.3± 4.3	1.3± 1.8	1.9± 2.8
Percent S4	10.4± 5.1	0.1± 0.3	0.0± 0.2
Average length of REM period			
1st	14.8±11.0	17.3± 9.4	21.5±11.1
2nd	27.4±14.8	20.1±13.4	19.1±10.3
3rd	28.6±14.6	16.9±10.5	15.8± 9.6
4th	33.0±17.0	20.6±14.8	20.5±12.0

*$p<0.05$; **$p<0.01$: by using t-test (differences between the people in their 70s and the people in their 80s).

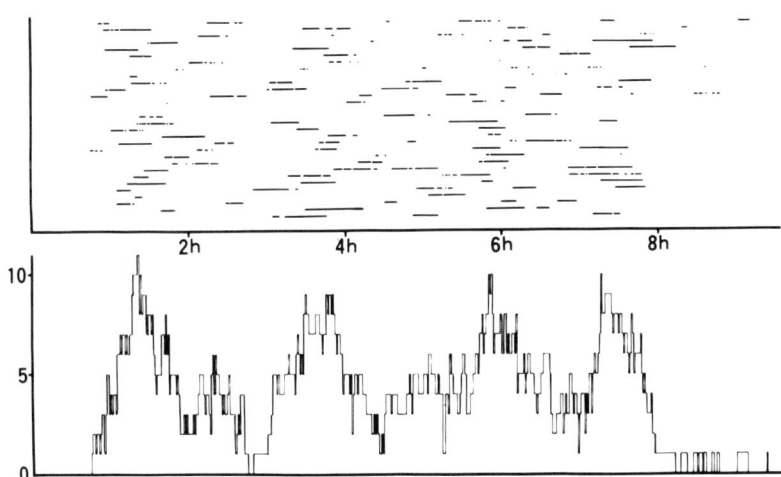

Fig. 1. The composite histogram of REM sleep of the group in their 70s. The REM periods of the people in their 70s are shown in a form of a histogram.

The Hourly Percentages of S3 and S4. The S3 and S4 figures of the young adults showed very high percentages in the earlier part of the sleep, then decreased along a hyperbolic curve. However, those of the people in their 70s and 80s were very low throughout the night.

Discussion

The study of the sleep characteristics in aged people is thought to be useful in further understanding the sleeping-waking mechanisms. By age 70, S3 and S4 are already greatly diminished and therefore show no further changes with age.

It has been reported that the first REM period in elderly people were longer compared with those of young adults [1, 2]. It is noticeable that among the three groups, the first REM period is the longest in the group in their 80s. It is suggested that the shift of REM sleep towards the early portion of the night for aged people is due to the disturbed mechanism of slow wave sleep during the earlier part of the night. The histogram of REM sleep of the group in their 70s showed four apparent peaks, but those peaks became obscure after the age of 80, suggesting that the cyclic changes of REM sleep decreased for people 80 years and older.

References

1 Feinberg, I.: J. psychiat. Res. *10:* 283–306 (1974).
2 Kahn, E.; Fischer, C.: J. nerv. ment. Dis. *148:* 477–494 (1969).

Y. Hayashi, MD, Department of Neurology, Second Hospital,
Tokyo Women's Medical College, 2–1–10 Nishi-Ogu, Arakawaku, Tokyo (Japan)

Excessive Daytime Somnolence in Patients with Chronic Respiratory Failure Showing Prolonged Episodes of Severe Oxygen Desaturation during REM Sleep

M. Billiard[a], F. B. Michel[b], A. Bertrand[c], J. Milane[c], P. Passouant[a]

[a] Service de Physiopathologie des Maladies Nerveuses, Centre Gui-de-Chauliac,
[b] Service des Maladies respiratoires, Clinique de l'Aiguelongue,
[c] Clinique des Maladies Infectieuses A., Centre Gui de Chauliac, Montpellier, France

Prolonged episodes of severe O_2 desaturation ocurring during sleep in some patients with chronic obstructive pulmonary disease were first reported by *Flick and Block* [2]. Subsequently it was shown that such episodes exclusively occurred during REM sleep and did not result from apneas or hypopneas [3]. Finally *Douglas et al.* [1] suggested that these episodes could only be encountered in the 'blue and bloated' patients and not in the 'pink and puffing' patients. Why this phenomenon occurs is not yet completely understood. Nevertheless, it is generally assumed that it could depend on a combination of alteration in the ventilation-perfusion ratio and of alveolar hypoventilation.

We now report on patients with chronic respiratory failure showing such prolonged episodes of severe oxygen desaturation during REM sleep and complaining of excessive daytime somnolence. The subjects were 5 males and 1 female, mean age 57.4 ± 4.3, who had been referred to our department with a putative diagnosis of hypersomnia-sleep apnea syndrome. Each subject had a complete work-up including detailed clinical history, physical examination, pulmonary function tests, all-night polygraphic recordings with standard procedures and control of respiratory patterns by means of nasal and oral thermistors, thoracic and abdominal strain gauges and a Hewlett-Packard 47201 A ear oximeter. In addition, 3 subjects (Nos. 2, 3, 6) had 5 daytime sleep latency tests.

According to Lorentz's formula, 4 subjects were overweight (56–162%) and 2 were of normal weight (13 and 16%). All subjects had mild or conspicuous signs of heart failure. Lung disease was either of the restrictive

Table I. Baseline SaO₂ values, almost absence of sleep apnea and high percentage of REM sleep with $SaO_2 \leq 70\%$

Case No.	Baseline SaO_2 %	Sleep % in apnea		Sleep % with $SaO_2 \leq 90\%$		Sleep % with $SaO_2 \leq 70\%$	
		NREM	REM	NREM	REM	NREM	REM
1	83	0	0	100	100	0.6	100
2	90	0	0	100	100	0	12.1
3	92	0	6.7	0	34.4	0	19.3
4	90	0	0	100	100	0	66.8
5	86	0	0	100	100	0	45.7
6	92	0	0	29.6	48.4	0	9

(subject Nos. 1–3) or of the obstructive (subject Nos. 4–6) type. Determination of blood gas tensions revealed hypoxemia ($\bar{x} = 57.6$ mm Hg ± 2.8) and hypercapnia ($\bar{x} = 53.0$ mm Hg ± 1.7).

All night polygraphic recordings showed disturbed sleep patterns including decreased total sleep time (342.8 ± 59.3), increased intervening wakefulness (188.8 ± 36.5), increased number of awakenings ≥ 2 min. (11.8 ± 2.0). Moreover, subjects No. 1, 3, and 6 had sleep onset REM periods at the beginning of the night and, among the 3 subjects who had daytime sleep latency tests, 2 of them (Nos. 2 and 6) had daytime sleep onset REM periods. Respiratory monitoring showed few or no sleep time in apnea contrasting with a high rate of severe O_2 desaturation ($<70\%$) during REM sleep (table I).

Thus, some patients with chronic respiratory failure, with or without obesity, and with prolonged episodes of severe oxygen desaturation occurring in REM sleep, may complain of daytime excessive somnolence. We are now attempting to compare the degree and duration of nocturnal oxygen desaturation in hypersomnolent and non-hypersomnolent patients with chronic respiratory failure.

References

1 Douglas, N.J.; Calverley, P.M.A.; Leggett, R.J.E.; Brash, H.M.; Flenley, D.C.; Brezinova, V.: Lancet *i:* 1–4 (1979).
2 Flick, M.R.; Block, A.J.: Ann. intern. Med. *86:* 725–730 (1977).
3 Wynne, J.W.; Block, A.J.; Hemenway, J.; Hunt, L.A.; Shaw, D.; Flick, M.R.: Chest *73:* suppl., pp. 3015–3035 (1978).

M. Billiard, MD, Service de Physiopathologie des Maladies Nerveuses,
F-34059 Montpellier (France)

Age and Intervening Wakefulness in Chronic Primary Insomnia

L. Garma, G. Bouard, O. Benoit

Laboratoire d'Etude du Sommeil, U3 INSERM, Hôpital de la Salpêtrière, Paris, France

The present study deals with the respective role and importance of the number of awakenings and of their length in the increase of the intervening wakefulness amount. The data were collected from a group of adult insomniacs aged 17–70 years. The group was composed of 38 subjects suffering from chronic primary insomnia, showing no other pathology including sleep apnea, who came to the sleep laboratory's consultation on sleep pathology. Patients were examined in neuro-psychiatric consultation prior to being recorded. All hypnotic medications were withdrawn at least 3 weeks before the recordings.

Patients were polygraphically recorded 2–4 consecutive nights according to standard methods. Only those whose total sleep time exceeded 375 min were retained. The analysis was made on the second night taking as wakefulness time, after the beginning of the first stage 2, any period of 20 s showing, during at least 10 s, either alpha at the wakefulness frequency or a desynchronized record. Four different age groups were defined, G1: 11 patients aged 17–27 years (mean 22.1 ± 3.4); G2: 10 patients aged 30–39 years (mean 35.6 ± 2.9); G3: 10 patients aged 42–46 years (mean 44.5 ± 1.3), and G4: 7 patients aged 53–70 years (mean 61.1 ± 6.7). In order to study the effects of age on the intervening wakefulness the 4 groups of insomniacs were compared to one another.

The *total amount of intervening wakefulness* increased progressively with age (G1–G4: 16.9 ± 12.3; 30.5 ± 34.7; 67.2 ± 69.9; 105.2 ± 53.3 min; $p \leq 0.01$ G4 vs G1 and vs G2).

The *total number of awakenings*, whatever their length, increased regularly with age, but differences between the groups were not as strong as they were for the wakefulness amount (G1: 9.4 ± 7.5; G4: 17.9 ± 9; NS). Conse-

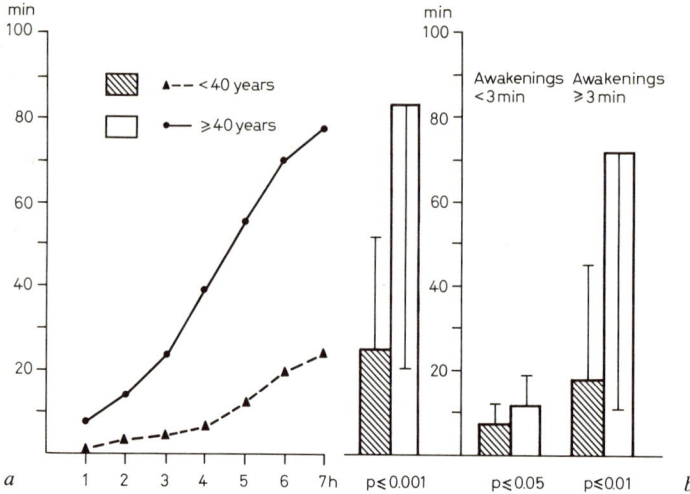

Fig. 1. a Hourly total mean cumulated amount of intervening wakefulness during the first 7 h of sleep. *b* Total wakefulness amount and partial amounts due to awakenings of <3 min and of ≥3 min during total sleep time.

quently, if age influences the intra-sleep wakefulness, it probably influences the length more than the number of awakenings.

The analysis of the awakenings lasting <3 min and ≥3 min made these modifications more clear. The *number of awakenings ≥3 min* increased with age: 0.8 ± 0.6 in G1 and 4.8 ± 3 in G4 ($p \leq 0.01$); its relative proportion in the 4 successive groups was 9, 12, 17 and 27%.

With increasing age, the *amount of wakefulness due to awakenings ≥3 min* constituted the greatest part of the total amount of wakefulness: 65, 74, 84 and 91% in groups 1–4. The difference was significant ($p \leq 0.01$) between the two youngest groups (G1: 10.9 ± 10.4 min; G2: 22.5 ± 34.6 min) and the oldest (G4: 96.2 ± 51.7 min). When the insomniacs were grouped in two age categories of less than and more than 40 years, the same differences were observed (fig. 1).

According to *Gaillard's* [2] work, if slow wave sleep (SWS) has a driving effect on sleep, one could thus suppose that its decrease or its suppression would cause primary insomnia. In comparing patients of less than and more than 40 years of age, we found that SWS decreased from 110.2 ± 35.2 to 79.2 ± 36.7 min for stages 3+4 and from 48.4 ± 27.6 to 29.8 ± 18.9 for stage 4 ($p \leq 0.05$). SWS decreased but it did not disappear.

Our hypothesis is that among our old patients insomnia would be linked to an excess of wakefulness, itself linked to an increase of wakefulness continuity. The concept of wakefulness stability used by *Brezinova et al.* [1] can be applied to our study. Among the oldest insomniacs the amount of intervening wakefulness due to long awakenings increases more than the total number of awakenings. The number of awakenings and their length are to be considered as two distinct physiological mechanisms: the awakening and the length of time necessary to go back to sleep. With increasing age, the distribution of awakenings according to their length changes: the wake time due to long awakenings becomes more important. Getting older, insomniacs wake up a little more often but, above all, they go back to sleep less quickly. In other words the probability of awakening expressed by the total number of awakenings does not change notably in old insomniacs. On the other hand, the probability of going back to sleep expressed by the awakenings length is strongly influenced by age.

The present data underline the importance of intervening wakefulness in the physio-pathology of insomnia.

References

1 Brezinova, V.; Oswald, I.; Loudon, J.: Br. J. Psychiat. *126:* 439–445 (1975).
2 Gaillard, J.M.: Sleep *1:* 133–147 (1978).

Dr. L. Garma, Laboratoire d'Etude du Sommeil, U3 INSERM, Hôpital de la Salpêtrière, 47 Boulevard de l'Hôpital, F-75651 Paris Cédex 13 (France)

Clinical and Polygraphic Studies of Sleep Drunkenness

S. Nevšímalová, B. Roth, V. Ságová, D. Paroubková, A. Horáková

Department of Neurology, Charles University Medical Faculty, Prague, Czechoslovakia

Sleep drunkenness is a phenomenon most frequently seen in patients with idiopathic hypersomnia. Getting woken up takes a long time in these patients, up to several hours, during which the patients are incapable of logical thinking and likely to perform automatically only a few well-established stereotypes. Movement is largely uncoordinated with a prominent component of ataxia. This independent clinical entity was described by *Roth* [2] and *Roth et al.* [3]. In healthy individuals this type of condition is quite exceptional – after physical exertion and sleep deprivation. Polygraphic studies of sleep drunkenness in healthy subjects woken up from deep sleep were undertaken by *Broughton and Gastaut* [1]. We would like to present the results of our polygraphic, clinical and psychological studies of this condition in patients with idiopathic hypersomnia and controls.

Material and Methods

The study was based on afternoon sleep recordings from 8 patients suffering from idiopathic hypersomnia with sleep drunkenness. 24 h prior to the test, the patients stopped receiving all stimulant drugs. Each patient was examined repeatedly under the following three conditions: after normal 8-h nocturnal sleep, after merely 4 h of nocturnal sleep and after 12 h of nocturnal sleep. A total of 24 recordings were evaluated, and the results were compared with an equal 24 recordings from control subjects free from sleep disorders: (a) after 1 night-long deprivation, (b) after 4 h of nocturnal sleep, (c) after normal 8-h nocturnal sleep. The polygraphic examination consisted of the recordings of EEG, eye movements, EMG activity from m. mentalis, respiration, and ECG. Following an undisturbed recording of approximately 45 min duration, during which most patients and many controls fell asleep, the probands were woken up and made subject to a series of psychological tests lasting 10 min with polygraphic recording proceeding simultaneously. This was followed by a series of neurological tests with detailed cerebellar function examination. The psychological and neurological examinations were repeated after an interval of 20–30 min.

Results and Discussion

(a) Findings Prior to Waking. The amount of sleep activity in recordings of patients with idiopathic hypersomnia does not depend on total duration of preceding nocturnal sleep, whereas in control subjects it is in correlation with the degree of sleep deprivation.

(b) Findings on Waking. Polygraphic correlates of sleep drunkenness were found in the occurrence of periods of microsleep lasting 2–5 s or in more or less uninterrupted stage 1 sleep activity detectable both after waking and between the psychological tests (fig. 1). Manifestations of sleep activity were present after awakening in 80% of the recordings of hypersomniacs, but only in 8.3% of cases and only after night-long sleep deprivation in the control group. There was, however, a marked connection between EEG manifestations of sleep drunkenness and being woken up from the deeper stages of NREM sleep. In our patients, these signs were present at all times so long as the waking up took place in stage 2, 3 or 4. There was a marked contrast between the findings in patients with idiopathic hypersomnia and the continuous alpha activity in most control subjects even after being woken up from the deepest stages of sleep [3 and 4].

The EEG manifestations of sleep drunkenness were in full correlation with the clinical neurological findings. These were present in the patients in 80% of all the tests. The most frequent clinical manifestations were neocerebellar (89.4%) and paleocerebellar (84.2%) symptomatology. Among other neurological signs were proprioceptive hypo- or even areflexia (68.4%), vestibular symptoms (36.8%) and disseminated pyramidal irritation symptoma-

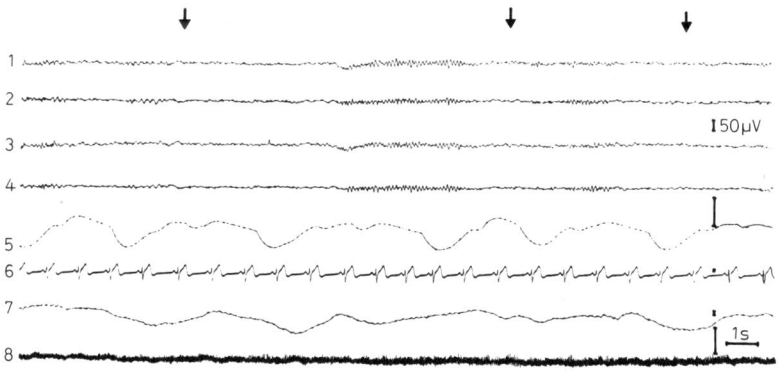

Fig. 1. Signs of microsleep during state of sleep drunkenness.

tology (10.5%). In the control group, clinical signs of sleep drunkenness were observed only in two instances.

Modified psychological tests were designed to monitor and evaluate memory, attention, visuomotor co-ordination, fine and gross motoricity. It appeared that in the patients group all the results immediately after waking up were inadequate, showing mild improvement after 20 min. In the control group the results were much better and showed no significant changes at the second checking after 20 min. A statistically significant difference was found between values recorded in the patients and healthy controls examined under analogical circumstances, namely at a 1% level of significance.

To conclude, states of sleep drunkenness are due to a retarded and protracted transition from sleep to wakefulness accompanied by some signs of sleep dissociation. The electrophysiological correlate of these states are manifestations of microsleep. In our opinion sleep drunkenness in patients with idiopathic hypersomnia is caused by chronic relative sleep deprivation. These patients need substantially more sleep than their life situations permit them. Signs of sleep drunkenness, according to our experience, disappear if the patients can sleep as much as they need – 12, 14 or even more hours a day.

References

1   Broughton, R.; Gastaut, H.: in Levin, Koella, Sleep 1974, pp. 82–91 (Karger, Basel 1975).
2   Roth, B.: in Guilleminault, Dement, Passouant, Narcolepsy, pp. 333–351 (Spectrum, New York 1976).
3   Roth, B.; Nevšímalová, S.; Rechtschaffen, A.: Archs gen. Psychiat. *26:* 456–462 (1972).

S. Nevšímalová, MD, Department of Neurology, Charles University Medical Faculty, CS-12000 Prague 2 (Czechoslovakia)

Nocturnal Asthma: Sleep and Dream Analysis

J. Montplaisir, J. L. Malo, J. Walsh, J. Monday

Hôpital du Sacré-Cœur, Centre d'Etudes du Sommeil, Montréal, Qué., Canada

Asthma may show marked circadian variations, often symptoms worsen at night and asthmatic death is more common toward the end of the night and early morning [1]. Several incriminating factors for nocturnal attacks have been proposed: allergy, posture, hypoxia, carbon monoxide retention, catecholamines and cortisol levels.

Few sleep studies have been done on nocturnal asthma. *Kales et al.* [2, 3] found no correlation between sleep stages and asthma occurring predominantly in the second half of the sleep period. But this study is open to question because patients were on continuous medication and their asthmatic symptomatology was based on subjective reports of dyspnea unconfirmed by objective measurements.

Methods

The present study was based on 4 nights of polygraphic recording of 12 patients with symptoms of nocturnal attacks for several consecutive prior nights and 12 normal controls matched for age and sex. Besides the polysomnography for sleep stages [4], respiration and ear oxymetry were monitored. The forced expiratory volume/s (FEV_1) was measured just before retiring, during attacks and upon final awakening.

The third night was concerned with whatever dream recall there might be from awakenings in REM sleep (subjects were awakened after each REM period of 5 min or more duration). After a lapse of approximately 2 weeks, a 4th-night recording was done in asthmatics while patients were asymptomatic and treated only with bronchodilators lacking significant systemic effects.

Results

Nocturnal Attacks. 26 nocturnal attacks were documented. None were recorded during the first hour sleep and 85% occurred in the last ⅔ of the night. Nocturnal attacks were more severe than diurnal attacks with a fall of the FEV_1 to 52% of the presleep value. No dream recall was reported upon awakening at the time of an attack.

Table I. Sleep analysis in asthmatics

	Sleep latency (to stage 2) min	Total sleep time min	Wake time after sleep onset min	Stages 3 and 4 min	REM latency min
Night 1					
Controls ($\bar{x}$ SE)[1]	48.1 ± 20.4	348.9 ± 27.2	116.4 ± 11.4	49.8 ± 13.1	165.7 ± 25.2
Asthmatics	32.0 ± 8.2	333.7 ± 17.5	74.3 ± 12.3	49.1 ± 12.4	181.7 ± 30.5
Anova[2]	n.s.	n.s.	$p<0.05$	n.s.	n.s.
Night 2					
Controls	17.9 ± 3.8	411.9 ± 13.8	31.2 ± 6.7	62.6 ± 10.4	78.7 ± 7.7
Asthmatics	20.9 ± 4.1	375.7 ± 16.4	87.6 ± 14.6	59.1 ± 11.6	120.9 ± 17.3
Anova	n.s.	n.s.	$p<0.001$	n.s.	$p<0.01$
Night 4					
Asthmatics	10.8 ± 2.4	467.2 ± 32.3	53.9 ± 14.3	60.9 ± 17.6	86.7 ± 16.4
Anova (night 4 vs night 2 of asthmatics	$p<0.05$	$p<0.05$	n.s.	n.s.	$p<0.01$

[1] Results represent the mean and the standard error.
[2] One way analysis of variance (Anova) was used to compare controls and asthmatics for each night and to compare night 4 with night 2 of asthmatics.

Apneas and Hypopneas. O_2 saturation fell on occasion by 5 to 15% of the presleep value. Apneas and hypopneas were present in both groups but were more numerous and severe in asthmatics.

Sleep Analysis. Sleep was disrupted in asthmatics at times of frequent nocturnal attacks (table I). Asthmatics have an increase in number of awakenings, wake time after sleep onset, and REM latency. REM latency was barely present in the first ⅓ of the night and long REM periods were seen in the second or third ⅓ of the night. However, all parameters return to normal values when asthma improves with treatment.

Dream Analysis. Upon spontaneous awakenings in the morning of nights 1 and 2, 32% of controls and only 8% of asthmatics could recall a dream. Night 3, asthmatics were awakened 43 times in REM sleep and could recall a dream in 61% of the awakenings while controls were awakened 46 times in REM sleep and could report a dream in 57%. In addition, after 9 awakenings (20%), asthmatics reported they were dreaming just prior to being awakened but could not recall any detail of the dream content, what might be called 'white dreams'.

Discussion

This study confirms the previous clinical observation that nocturnal attacks are more severe than diurnal ones. There was no correlation between asthma attacks and sleep stages. However, such a correlation is difficult to make since asthma attacks take several minutes to develop. The respiration changes may start during a stage of sleep and the patient may wake later in another stage, reporting the attack.

Changes in sleep parameters in asthmatics were significant but not specific. They resemble what was previously described as the 'first night effect'. Values in asthmatics and controls were similar on night 1 but while controls improve their nocturnal sleep on night 2, asthmatics continue to show the same disruptions with frequent awakenings and long REM latencies. All sleep parameters returned to normal when asthma improves with treatment (night 4).

At times of nocturnal attacks asthmatics have long REM period towards the end of the night. The REM density (number of eye movements per minute of REM sleep) was also increased in these patients. However, they recall fewer dreams upon spontaneous awakenings and report several 'white dreams' (dreams without content) upon awakenings in REM sleep. This is in agreement with psychological assessment of these patients and a current hypothesis in psychosomatic medicine suggesting that asthmatics and other psychosomatic patients have more difficulty to report and elaborate on their fantasy life.

References

1   Cochrane, G.M.; Clark, T.J.A.: Thorax *30:* 300–305 (1975).
2   Kales, A.; Beall, G.N.; Bajor, G.F.; Jacobson, A.; Kales, J.D.: J. Allergy clin. Immunol. *41:* 164–173 (1968).
3   Kales, A.; Kales, J.D.; Sly, R.M.; Scharf, M.B.; Tan, T.L.; Preston, T.A.: J. Allergy clin. Immunol. *46:* 300–308 (1970).
4   Rechtschaffen, A.; Kales, A.: A manual of standardized terminology, techniques and scoring system for sleep stages and human subjects. Publ. Hlth Service Publication 204 (US Government Printing Office, Washington).

J. Montplaisir, MD, Hôpital du Sacré-Cœur, Centre d'Etudes du Sommeil, Montréal, Qué. H4J 1C5 (Canada)

Sleep Apneas and Mental Deterioration in Elderly Subjects

M. Billiard, J. Touchon, P. Passouant

Service de Physiopathologie des Maladies Nerveuses, Centre Gui de Chauliac, Cliniques St. Eloi, Montpellier, France

Up to now, sleep-related breathing abnormalities have been extensively investigated in adults, children and infants and two main related syndromes have emerged: the hypersomnia sleep apnea syndromes [2, 3] and the sudden infant death syndrome [4]. In contrast, sleep-related breathing abnormalities have not been thoroughly looked at in elderly subjects, except in two recent studies by *Carskadon and Dement* [1] and by *Krieger et al.* [5] showing a rather high percentage of elderly subjects with apneas and hypopneas. Now as excessive daytime sleepiness, intellectual deterioration and personality changes are common features of the sleep apnea syndromes, we questioned whether progressive mental deterioration could be related to sleep apneas among the aged.

Subjects and Methods

The subjects were 12 females and 12 males over the age 65, recruited from a hospital for the aged, who were assigned to 2 groups of normal and mentally deteriorated subjects, by means of tests of general mental capacity including tests of orientation, memory, and reasoning. Group 1 included 11 normal subjects, 8 females and 3 males, mean age 79.4 ± 2.2. Group 2 included 13 mentally deteriorated subjects, 4 females and 9 males, mean age 75.9 ± 2.6. According to their origin (hospital for the aged) most of the subjects, normal and deteriorated, had positive past medical history such as heart failure, Parkinson's disease, hemiplegia, etc.

Due to the difficulties encountered with some severely mentally deteriorated subjects sleep was recorded on a single night. Standard polygraphic technics were used together with control of respiration by means of nasal and oral thermistors, thoracic and abdominal strain gauges, and a Hewlett-Packard ear oximeter. Apneas (stop of airflow for >10 s) and hypopneas (amplitude of respiration channels lowered by 50% for >10 s) were scored together. Mean apnea index (No of apneas per sleep hour), total sleep time spent in apnea or hypopnea and total sleep time with $<90\%$ O_2 saturation were calculated in each subject.

Results

Apneas and/or hypopneas were recorded in both groups but with a greater incidence in mentally deteriorated subjects ($\bar{x} = 109.4 \pm 38.5$) than in normal subjects ($\bar{x} = 17.5 \pm 8.7$, $p < 0.001$).

In the non deteriorated subjects the mean apnea index ranged from 0 to 14.3 with 8 subjects (72%) showing less than 5 apneas per sleep hour (a generally admitted upper physiologic limit) (fig. 1), the total sleep time spent in apnea or hypopnea ranged from 0 min to 41 min 21 s and O_2 saturation dropped to a level lower than 90% in only 1 subject for a very brief total time of 0 min 28 s.

In the mentally deteriorated group, the mean apnea index ranged from 0.2 to 86.2, with only 4 subjects (30%) having less than 5 apneas per sleep hour and 4 subjects having over 40. Among the latter was a 75-year-old woman rating extremely low on the tests of general mental capacity and showing 378 apneas during 4 h 23 min spent asleep (apnea index 86.2) (fig. 1). The total sleep time spent in apnea or hypopnea ranged from 0 min 22 s to 175 min 38 s with 3 subjects over 2 h. Now only 5 subjects had their SaO_2 dropping several times under 90%, for a total time ranging from 1 min 09 s to 13 min 29 s.

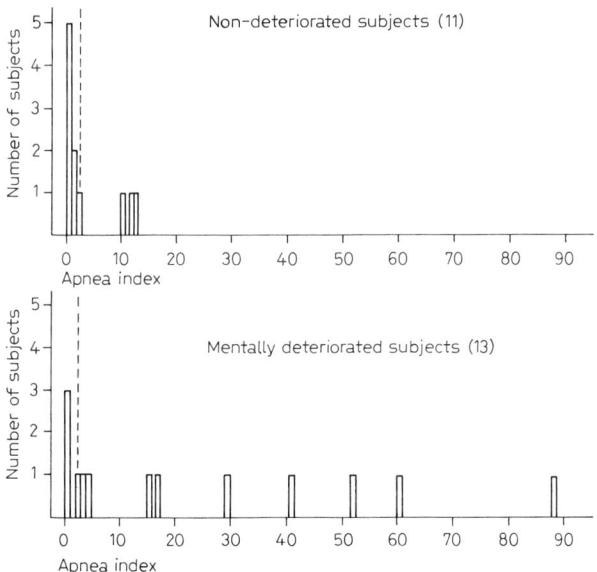

Fig. 1. Mean apnea index in normal and mentally deteriorated subjects. The dotted line indicates the upper physiologic limit in adults.

In addition, significant negative correlations (Spearman rank correlation) were found between general mental capacity and the mean apnea index ($r = -0.54$, $p < 0.001$) as well as between general mental capacity and the total sleep time in apnea ($r = -0.52$, $p < 0.01$).

Discussion

Therefore, in spite of the well-known difficulties in classifying subjects as normal and mentally deteriorated, there is some evidence suggesting a relationship between mental deterioration and sleep apneas in the aged over 65. However, whether apneas may cause mental deterioration or apneas and mental deterioration may depend upon a common origin cannot be ascertained from our data. We suggest that further studies should test the daytime alertness of these subjects in order to evaluate whether a correlation can be found between the number of apneas, the degree of daytime sleepiness and the mental deterioration.

References

1 Carskadon, M.A.; Dement, W.C.: Sleep Research 9 (in press, 1980).
2 Guilleminault, C.; Dement, W.C. (eds): Sleep apnea syndromes (Liss, New York 1978).
3 Lugaresi, E.; Coccagna, G.; Mantovani, M.: in Weitzmann, Advances in sleep research, vol. 4 (Spectrum, New York 1978).
4 Steinschneider, A.: Pediatrics, Springfield *50:* 646–654 (1972).
5 Krieger, J.; Mangin, P.; Kurtz, D.: Rev. EEG Neurophysiol. *10:* 177–185 (1980).

M. Billiard, MD, Service de Physiopathologie des Maladies Nerveuses,
Centre Gui de Chauliac, Cliniques St. Eloi, F-34059 Montpellier (France)

A Comparative Study of the Frequency of Sleep Apnea during the First Part of the Night and the Whole Night in Infants 2–12 Months of Age

B. Nogues[a], N. Monod[b], D. Samson-Dollfus[a]

[a] Neurological Testing, Rouen University Hospital, Rouen
[b] Center for Biological Research into Fetal and Neonatal Development, Paris, France

The calculation of the frequency of sleep apnea per unit of time (apnea score) in the different phases of sleep – quiet and REM sleep – is one of the criteria studied in the siblings of infants dying from sudden infant death syndrome (SIDS) and in near-miss infants. Some authors calculate the apnea score from the recording of a whole night's sleep, while others use the first two or three sleep cycles only. The study undertaken here will attempt to determine the value of each of these methods.

Material and Method

30 whole-night polygraphic studies of infants ranging from 2 to 11 months, recorded at Port Royal (Paris) and at the Rouen University Hospital, were analyzed as follows: the frequency of apneas lasting longer than 2 s was calculated per 100 min of total, quiet, indeterminate and REM sleep, first on the basis of a whole-night recording, and secondly on the basis of the first part of the night only. This was defined as sleep time preceding the spontaneous awakening closest to midnight.

We then calculated the correlation coefficient between the two groups of results. The degree of significance was estimated, taking into account the value of the correlation coefficient and the number of subjects studied in each group. The threshold of significance was set at 0.05.

The subjects were classified: (a) according to age (2–4 months, 5–7 months, 8–11 months), without taking clinical differences into account; (b) according to clinical factors – control group, siblings, 'near-miss', without taking age into account.

Table I. Levels of significance p and correlation coefficients r for frequency of apneas, between results obtained from whole night recordings and recording of the first part of the night only, in the different groups of infants

	Calm sleep	Indeterminate sleep	REM	Total sleep
Near miss (n=8)	r=0.76	r=0.96	r=0.79	r=0.91
	p<0.05	p<0.001	p<0.02	p<0.01
Siblings (n=18)	r=0.73	r=0.83	r=0.55	r=0.87
	p<0.001	p<0.001	p<0.02	p<0.001
Control subjects (n=4)	r=0.82	r=0.98	r=0.95	r=0.95
	n.s.	p<0.05	p<0.05	p<0.05
2≤ <5 months (n=15)	r=0.66	r=0.79	r=0.93	r=0.91
	p<0.01	p<0.001	p<0.001	p<0.001
5≤ <8 months (n=9)	r=0.74	r=0.91	r=0.45	r=0.77
	p<0.05	p<0.001	n.s.	p<0.02
8≤ <12 months (n=6)	r=0.88	r=0.95	r=0.68	r=0.91
	p<0.02	p<0.01	n.s.	p<0.01

Results

(1) The value of the apnea score was greater when calculated on a whole night than when calculated on the first part of the night only. (2) However, there were significant correlations between the two groups of results (table I): (a) for all subjects, taken as a whole; (b) for infants of less than 5 months of age; (c) for infants of more than 5 months in quiet, REM and total sleep, and (d) for high-risk infants, all ages. (3) There was no significant correlation (table I): (a) for infants of more than 5 months, in active sleep (without taking clinical factors into account), or (b) for the control group, during quiet sleep (but this group included only 4 infants).

Discussion

Age seems to be an important factor in determining the usefulness of recording sleep over a whole night, since the threshold of significance is not reached in the case of active sleep in infants over 5 months, while for infants of less than 5 months the correlations are all significant (p<0.05). It is possible that this indicates that after 5 months of age there is a circadian organization of sleep apneas. It would be necessary to verify this through daytime recordings of sufficiently long duration. The sleep apneas are much more frequent during REM sleep in the second part of the night, and this study shows that after 5 months of age there is no correlation between the results obtained from the first part of the night and from the night as a whole.

It would therefore be possible to limit oneself to short sleep recordings (covering only the first two or three sleep cycles) to evaluate the frequency of apneas in infants under 5 months, but beyond this it appears to be necessary to make whole-night recordings, at least when the frequency of apneas appears to be low during the first part of the night.

B. Nogues, MD, Explorations neurologiques, Centre Hospitalier Universitaire de Rouen, F-76031 Rouen (France)

Exercise and Sleep Behaviour: a Questionnaire Approach

J. M. Porter, J. A. Horne

Department of Human Sciences, Loughborough University, Loughborough, Leicestershire, England

Research into exercise and sleep has been, from a sleep-EEG standpoint, oriented towards stages 3 and 4 sleep and, although it is commonly believed that exercise is 'good' for sleep, there have been no systematic studies of these claims. Such sleep studies give little or no details of the effects of raised or reduced daytime physical activity levels upon subjective feelings of tiredness before and after a subsequent night's sleep, or whether subjects felt that they slept better. The aim of this study was to do this, using self-assessment questionnaires.

Method

Questionnaires. Pre and post-sleep questionnaires were carefully designed and also incorporated the Stanford Sleepiness Scale. The pre-sleep questionnaire asked for: daytime activity level (5 point scale), time of day of exercise (4 point scale), time when subjects felt ready for bed and bed-time, other daytime factors which might affect sleep, and time and duration of any naps. The post-sleep questionnaire asked for: sleep onset time, total sleep time, quality of sleep (5 point scale), whether more sleep was desirable and whether there was enforced awakening (e.g. alarm clock). In addition, a screening questionnaire was completed by each subject in order to assess subject fitness (2 point scale: fit, unfit) and to detect any health or sleep problem.

Subjects. Suitable volunteers were interviewed and recruited from a local sports centre. They were not used to daily exercise and had varied daily activity levels. They were asked to complete pre and post-sleep questionnaires for 7 consecutive days. No expectancy about possible sleep changes were conveyed to the subjects. 51 out of 80 volunteers returned a full quota of questionnaires.

Results

Out of a total of 332 acceptable (free of illness, etc.) data days collected, 46 were classified as having activity levels lower or much lower than normal, 139 higher or much higher, and 147 as normal. This distribution was similar in both the fit (n = 27) and the unfit (n = 24) groups. Statistical analysis did not identify significant relationships between daily activity levels and the majority of the tiredness and sleep variables. Activity level was not significantly related to time ready for bed or bed-time, unless exercise was taken late in the evening when bed-time was delayed by an average of 40 min. Activity level was significantly related to subsequent sleep quality, although both higher and lower than normal activity levels led to a deterioration.

Overall, regardless of activity level, those subjects claiming to be fit reported shorter sleep onset and total sleep times averaging 12 and 20 min less, respectively, than those of the unfit subjects. They also had significantly less post-sleep tiredness and better sleep quality.

Conclusion

These findings give little support to the belief that increased exercise 'improves' sleep that night. However, the findings indicate that fit people may need less sleep and/or sleep more efficiently than unfit people. This would suggest that regular physical exercice (becoming fit) may influence sleep. This limited study has several shortcomings, including possible biased subject sampling, only subjective estimates of fitness, and as there were no poor sleepers in the sample, these findings may not be applicable to them. In an earlier pilot study it was found that poor sleepers expected exercise to improve sleep, unlike the more open-mindedness in this respect, of normal sleepers.

J.M. Porter, PhD, Department of Human Sciences, Loughborough University, Loughborough, Leicestershire (England)

Bed-Time Carbohydrate Ingestion and Sleep

J. M. Porter, J. A. Horne

Department of Human Sciences, Loughborough University, Loughborough, Leicestershire, England

Whilst there have been several studies on the sleep motility effects of a bedtime snack, generally finding improvement, there has only been one EEG study [1]; a malted milk drink reduced restlessness in young subjects and reduced interim wakefulness in older subjects. Intravenous glucose during sleep [2, 3] indicates changes with SWS and REM sleep, suggesting that ingested carbohydrate (CHO) may well affect these types of sleep. This study assessed sleep-EEG effects of ingesting different CHO loads at bedtime, and whether CHO presented in a snack-like form 'improves' sleep.

Method

Three food supplements were designed giving an equal degree of subjective fullness without discomfort. Two contained small but equal quantities of protein and fat but differed in CHO by a factor of three, and the third consisted of methyl cellulose. Blood glucose levels during sleep were determined in a pilot study (table I). In the main study, 6 normal weight, healthy young male adults underwent all conditions in a different sequence. Over a 3-week period, each Monday served as adaptation. Each bed-time food supplement was given Tuesday to Thursday inclusive, 45 min prior to lights out. Full sleep-EEG recordings were performed on these nights. Daytime diets were controlled and replicated day for day in each run. Subjects completed a post-sleep inventory. Sleep data were compiled according to the row

Table I. Blood glucose levels for the 3 food supplements; means for 2 subjects

Food supplement	Blood glucose level, mg/100 ml			
	2400 hours	0200 hours	0400 hours	0800 hours
High CHO	240	159	107	90
Low CHO	99	109	93	86
Zero CHO	93	90	89	85

Table II. Group means over 3 nights for each of the food supplements

Whole night min	Carbohydrate level			SE of differences between diet means	Significance (all d.f.$_s$ = 2.40)
	zero	low	high		
Stages					
3+4	89.6[1]	104.3	92.1	4.46	diet F=6.7, p<0.005
4	46.5	53.6[2]	40.8	3.90	diet F=5.5, p<0.025
3	43.1	50.7	51.3	3.73	n.s.
REM	101.8	103.9	113.1	4.92	n.s.
2	221.6	211.1	220.4	5.50	n.s.
0+1	37.0	30.7	24.4[3]	4.50	diet F=4.0, p<0.05
0	3.6	6.6	0.8	0.91	n.s.
Sleep onset	16.3	12.9	14.9	2.42	n.s.
REM latency	81.2	91.0	69.8	10.78	n.s.
REM periodicity	94.9	96.2	101.5	16.11	n.s.
Number of REM periods	4.2	4.3	4.8[3]	0.93	diet F=4.4, p<0.05

[1] Significant difference between zero and low carbohydrate supplements.
[2] Significant difference between low and high carbohydrate supplements.
[3] Significant difference between zero and high carbohydrate supplements.

headings of table II, using 450-min TST. Two-way Anovas were performed with Studentised range tests as appropriate.

Results

There were no significant night effects or conditions x nights interactions for whole-night data, but significant findings between conditions. Whilst neither sleep onset nor the amount of interim wakefulness showed any significant changes, the amount of stages 0+1 combined was significantly affected, particularly a reduction under hCHO when compared with zCHO. Although there was no significant whole-night effect with REM sleep, there was a trend for REM to increase with increasing CHO; this reached significance for the first half of night data, particularly for hCHO. There was a significant increase in the number of REM periods under hCHO when compared with zCHO. Although stages 3+4 were lowest with zCHO, stage 4 sleep was lowest under hCHO. Thus, the main findings were with hCHO, where the decrease in stages 0+1, indicating some sleep 'improvement', seem to be reciprocated by a REM sleep increase. SWS became less deep under this condition, with a shift from stage 4 to stage 3. There were no significant changes for the post-sleep inventory.

Discussion

Blood glucose level under lCHO was nearer that of zCHO than that of hCHO. Because most of the significant differences between pairs of conditions involved hCHO it appears that a small increase in blood glucose as reflected by lCHO has little sleep effect. These hCHO sleep findings are similar to those of other studies [3, 4] even though diet was completely changed to a hCHO/low fat diet for 4 days in one study [4], without bedtime feeding. Although the present findings may simply be due to 'the more the food, the better is sleep', it is possible that hCHO had a more central effect relating to serotonin synthesis [5].

References

1 Březinová, V.; Oswald, I.: Br. med. J. *ii:* 431–433 (1972).
2 Lacey, J.H.; Stanley, P.; Hartmann, M.; Koval, J.; Crisp, A.H.: Electroenceph. clin. Neurophysiol. *44:* 275–280 (1978).
3 Parker, D.C.; Rossman, L.G.: J. clin. Endocrinol. *32:* 65–69 (1971).
4 Phillips, F.; Crisp, A.H.; McGuiness, B.; Kalucy, E.C.; Chen, C.N.; Koval, J.; Kalucy, R.S.; Lacey, J.H.: Lancet *ii:* 723–725 (1975).
5 Fernstrom, J.D.; Wurtman, R.J.: Science *174:* 1023–1025 (1971).

J.M. Porter, PhD, Department of Human Sciences, Loughborough University, Loughborough, Leicestershire (England)

Interpretation of Night Terrors in Adults Based upon Observations in the Dark

Dietrich Schneider-Helmert

Research Department, Psychiatric Clinic, Königsfelden, Switzerland

Psychopathological considerations, genetic studies and even experiments with a provocation of night terror attacks date back into the last century [1, 3, 6]. Night terrors were considered as a sleep pathology occurring on a possibly genetic disposition, the attacks being especially induced by physical and emotional stress, alcohol or sleeping in a new environment. Half a century ago, psychoanalysis [7] contributed the view that repressed sexual conflicts were the background, the attacks being an equivalent of anxiety. Polygraphic sleep recordings, dating from the last 15 years, enabled a more exact differentiation of night terrors (pavor nocturnus) and similar events like sleep-talking, sleep-walking, anxiety REM dreams, sleep paralysis. It was clearly shown that the attack starts out of delta sleep with an abrupt change to waking EEG, with initial screaming and vegetative activation. With regard to these psychophysiological observations, *Broughton* [2], referring implicitly to the James-Lange theory of emotions, postulated that night terrors were a disorder of arousal: An initial, pathological vegetative activation would lead to peripheral 'symptoms', and their perception would elicit the typical emotions. Cannon had taken issue with the James-Lange theory, suggesting that a strong emotion would cause simultaneously CNS arousal and vegetative activation (= emergency reaction). It was in this sense that *Fisher et al.* [4, 5] conceived the night terror attacks as a fight-flight reaction to preceding NREM-mentations. This view is in accordance with the old phenomenological descriptions but does not offer a new pathogenetic perspective.

Problems

This brief summary shows that a few crucial questions still remain unanswered: How is the divergency explained between a waking state (according to electrophysiological criteria) on the one side and dream-like experience and behaviour with retrograde amnesia on the other side? What is the role of the vegetative activation? What is the final cause for the attack? Is there any sense in the attack for a patient who is subject to these events each night? In my opinion, these questions can only be answered properly if an integrative model of night terrors is developed by considering the various aspects of this complex phenomenon including behaviour, a dimension more or less neglected so far. However, the recently developed technique of video-recording with infrared light renders now possible to observe a patient in complete darkness. Using a combination of this technique and polygraphy in the case of a 31-year-old patient, simultaneous video and polygraphic recordings of 17 attacks and 5 sleep utterances were obtained. This intensive case study, which was described elsewhere [8], yielded interesting findings in the behavioural dimension [9], mainly: The *non-verbal* behaviour is extremely expressive, giving a clear message already within the very first moment (<1 s) of the attack, and follows then a stereotyped sequence of fright–defense–counteraggression–relaxation. This material serves as a basis for answering the above-listed questions by proposing an integrative interpretation of night terrors in adults.

Interpretation

As a neurotic component of adaptation to social and moral (super-ego) demands, the night terror patient is continuously *repressing* affects and needs and *controlling* consciously his *hostility* during the waking state. By an unknown, possibly genetic factor the unresolved conflict is activated especially during NREM sleep instead of, or additional to, its manifestation as a REM dream. A dream-like threatening *experience during delta sleep* (this follows from recalls [5] as well as from the full-blown behaviour [9] in the first moment of the attack) provokes a startle reaction and intense emotions that induce simultaneously a cortical arousal (abrupt change from delta sleep to waking EEG) and vegetative activation towards ergotropy (rapid fall of basal skin resistance level, immediate vasoconstriction, gradual acceleration of heart rate, but not prior to the outbreak [8]). These activations, a typical *emergency reaction* in the sense of Cannon, are the *necessary psychophysiological conditions for the psychopathological events* in the further course of the attack, especially for the intense verbalization and motor actions: With the initial startle reaction the patient enters a dissociative state with lack of reality testing and of conscious control over actions and with hallucinatory experiences (menacing, scoffing, insult, as was concluded from the patient's verbal and non-verbal responses [8]) that cause a *dramatic acting-out* of repressed impulses [9]. The main contents of this acting-out are defense, counteraggressivity and self-assertion, thus compensating the suppression of vital needs in the preceding day. Finally, rapid reduction of excitement leads to psychophysiological *relaxation* (calming behaviour, decrease of vegetative activation and cortical arousal, rapid return to sleep behaviourally and electrophysiologically [8]). A hallucinatory acting-out in a dissociative state like this allows release of emotional tenseness but does nothing at all to solve the basic conflict which therefore must be completely *repressed again*. Two factors, which are typical for dissociative states in general, alleviate this process: disturbed consciousness during and retrograde amnesia after the attack. So, neither reason nor opportunity are given to the patient to confront himself with the attack and its contents.

With this interpretation, *teleologically an objective* can be attributed to night terror attacks in neurotic patients. The attack enables the patient to alleviate his emotional tenseness, especially hostility, which was accumulated during the day by repression, without confronting himself in reality with the underlying conflict. This mechanism thus helps to readjust and stabilize a neurotic balance.

References

1 Börner, J.: Das Alpdrücken, seine Begründung und Verhütung; Diss., Würzburg (1855).
2 Broughton, R.J.: Science *159:* 1070–1078 (1968).
3 Ebstein, E.: Z. ges. Neurol. Psychiat. *62:* 385–401 (1920).
4 Fisher, C., et al.: J. nerv. ment. Dis. *157:* 75–98 (1973).
5 Fisher, C., et al.: J. nerv. ment. Dis. *158:* 174–188 (1974).
6 Gudden, H.: Arch. Psychiat. Nerv. Krankh. *40:* 989–1015 (1905).
7 Jones, E.: On the nightmare (Hogarth Press, London 1931).
8 Schneider-Helmert, D.: Schweizer. Arch. Neurol. Neurochir. Psychiat. *126:* 155–177 (1980).
9 Schneider-Helmert, D.: Sleep Res. *9* (in press).

D. Schneider-Helmert, MD, Research Department, Psychiatric Clinic,
CH-5200 Königsfelden (Switzerland)

Clonidine Alleviates Cataplectic Symptoms in Narcolepsy[1]

P. T. S. Putkonen, L. Bergström

Department of Physiology and Neurology, University of Helsinki, Helsinki, Finland

Cataplectic attacks in narcoleptics are usually resistant to central stimulants prescribed against fatigue and sleep attacks. Tricyclic antidepressants and MAO inhibitors are effective against cataplexy and other symptoms of dissociated sleep but their usefulness is limited by frequent side-effects [4]. A recent clinical trial [5] with a selective MAO-B inhibitor, which specifically increases cerebral dopamine, gave negative results. Impressed by the potency of clonidine (CL) to suppress paradoxical sleep in the cat [2], we tried it in a narcoleptic patient especially annoyed by her cataplectic attacks. Following a first favourable impression, a single-blind cross-over study using alternating CL and placebo (PL) was started. A preliminary report was presented at the European Neuroscience Meeting in Rome [3].

Patients and Methods

Of 10 patients with a long history of narcolepsy with cataplectic symptoms, and willing to participate in the trial, 5 (3 women and 2 men) were chosen on the basis of their symptom charts showing sufficient self-reported cataplectic episodes. Their ages ranged from 32 to 55 years. They were free from major additional clinical disease. 4 reported the full narcoleptic tetrad of *Yoss and Daly* [6] whereas one did not have hypnagogic hallucinations, but had fits of sleep paralysis. Two complained of disturbed and irregular night sleep. 2 patients were taking a central stimulant (phentermin). Imipramine had been tried on 1 patient but he felt that its therapeutic effect was questionable and that it induced impotence.

During baseline and the trial, previous medication was withheld. The patients filled daily sleep and symptom charts marking sleep up to the nearest 0.5 h by lines on a 24-h scale and shorter sleep attacks or naps as vertical strokes (counted as 10 min in the analysis). Cataplectic attacks and other symptoms, such as errors at work or fits of automatic speech or behaviour

[1] Clonidine (Catapresan®) and placebo tablets supplied by Boehringer, Ingelheim.

were marked with different symbols. After a 2-week baseline the patients received either CL (150 µg) or PL tablets, which they were instructed to take daily at 1000 and 1400 hours for 2 weeks. The order of PL and CL periods (known to us but not to the patient) was varied from case to case. In 1 patient (L.P.) the dose was halved after she had failed to take our standard dose because of increased fatigue. She completed only 14 days on 2×75 µg of CL (compared with 14 days on halved PL tablets).

Results

The results, based on comparison of 1–2 alternating 14-day periods on PL and on CL are summarized in table I. The mean frequency of cataplectic episodes under PL varied from 0.8 to 15.6 per day and their nature from complete paralysis (lasting up to 14 min in T.H.) to buckling of knees or dropping objects. Loss of facial or vocal control was common, often in repetitive clusters, especially in T.A. The standard dose of CL reduced the daily frequency of cataplexy by 77–49% and the half dose by 40%. The residual attacks were milder, usually limited to the face, and no fits of complete paralysis were reported under CL. Patient T.H., who under PL had episodes of confused speech and automatic behaviour, reacting to hypnagogic reveries, lost these symptoms under CL.

Differences in reported sleep duration under CL and PL were negligible in 2 patients who slept more than 10 h per night. In others sleep under CL was increased by 12–35%. This was experienced as favourable by K.L., in whom most of the 35% increase was night sleep, and who slept only about

Table I. Mean ($\pm$SD) daily frequency of cataplexy and duration of sleep under placebo and clonidine

Patient	Sex	Dose	Days	Cataplexy			Sleep, min		
				PL	CL	Δ%	PL	CL	Δ%
T.H.	F	150×2	28/28	0.8 ±1.2	0.2** ±0.6	−77	625 ± 97	661 ±112	+ 6
U.L.	M	150×2	28/28	6.3 ±4.0	1.9*** ±2.5	−70	642 ± 50	614 ± 79	− 4
K.L.	F	150×2	14/14	0.9 ±0.8	0.3* ±0.8	−67	297 ± 61	401*** ± 51	+35
T.A.	M	150×2	14/14	15.6 ±6.4	8.0*** ±2.4	−49	419 ±131	506* ± 92	+21
L.P.	F	75×2	14/14	3.4 ±0.9	2.1*** ±0.8	−40	513 ± 41	574** ± 67	+12

*p<0.05; **p<0.01; ***p<0.001 (t-test).

5 h per night under PL. Naps were moderately increased in several patients, but only one chart (L.P.) showed a clear increase in sleep attacks and errors at work during CL. As the cataplectic attacks of this patient were mild she chose to resume her former stimulant medication at the end of the trial, whereas the others preferred to continue CL treatment. They reported even better therapeutic effects during continued open medication than during the trial. Patient U.L., who complained impotence during earlier imipramine treatment was satisfied with his sexual performance under CL.

Conclusions

The results show that the potent REM sleep-suppressing effect of CL, documented also in humans [1], may be put to use in treating cataplexy and probably also other symptoms of dissociated REM sleep. Its reputed sedative side-effect (likely to have been exaggerated in the acute on-off design of our trial) may, however, limit its usefulness in sensitive individuals.

References

1 Autret, A.; Beillevaire, T.; Cathala, H.-P.; Schmitt, H.: Eur. J. clin. Pharmacol. *12:* 319–322 (1977).
2 Putkonen, P.T.S.: Sleep research, pp. 19–34 (MTP Press, Lancaster 1979).
3 Putkonen, P.T.S.; Bergström, L.: Neurosci. Lett.: suppl. 3, p. 270 (1979).
4 Roth, B.: Sleep disorders. Diagnosis and treatment, pp. 29–59 (Wiley, New York 1978).
5 Schachter, M.; Price, P.A.; Parkes, J.D.: Lancet *i:* 831–832 (1979).
6 Yoss, R.E.; Daly, D.D.: Mayo Clin. Proc. *32:* 320–328 (1957).

Dr. P.T.S. Putkonen, Department of Physiology, University of Helsinki, SF-00170 Helsinki (Finland)

Effects of DSIP Applications in Healthy and Insomniac Adults

D. Schneider-Helmert, F. Gnirss, G. A. Schoenenberger

Research Department, Psychiatric Clinic, Königsfelden, Switzerland

Delta EEG sleep-inducing peptide (DSIP), a nonapeptide, was first isolated, characterized, sequenced and synthesized in 1976/77 [3, 4]. The sleep-promoting and related physiological effects of the compound after intraventricular, i.v. and i.p. application in different mammalian species, a bell-shaped dose-response curve, immediate differing from delayed effects, the endogenous distribution in the brain as well as the passage through the blood-brain barrier have then been reported [2, 5]. A critical comment on the state of the experimental findings and problems has recently appeared [2]. The first *human studies* with synthetic DSIP were designed in three consecutive steps of investigation: (1) Short- and long-term activities and eventual side-effects in healthy adults. (2) The dose-response with respect to varied i.v. injection rates. (3) Treatment of insomnia.

Daytime Administration to Healthy Volunteers

4 male and 2 female healthy volunteers underwent a double-blind, placebo-controlled cross-over study. After 1 adaptation night, 2 experimental days + nights in the sleep laboratory followed. The subjects received in the morning either 25 nmol/kg DSIP or placebo infused i.v. over 20 min under continuous, broad polygraphic and TV recordings and with extensive testing of psychophysiological functions. In addition, the night sleep following the daytime experiments was investigated [6].

Results. DSIP was tolerated without severe side-effects. Immediately after the infusion the subjects felt a pressure to sleep and some mild vegetative symptoms. Total sleep time (TST) increased within 130 min after infusion from (median) 29.75 min in placebo conditions to 47.25 min with DSIP. No impairment of alertness during waking was found according to subjective evaluations, performance measures in a stress test and power-spectral analysis of EEG. Furthermore, no changes in mood, concentration or sleepiness were found. A correspondence analysis on power spectra of 4 h EEG recordings after treatments revealed no DSIP effect on cortical electrophysiology. In the nights after DSIP, however, significant increases of TST, SWS and REM sleep as well as reductions of sleep latency and stage 1 occurred. It was concluded that DSIP sustains sleep, provided that physiological sleep conditions are given, but does not show the profile of pharmacological sedation [1, 6].

Varied Injection Rates

The injection times, i.e. instantaneous concentrations, of DSIP at constant dose (25 nmol/kg) were varied in a repetition of the above-described experiment with 3 subjects [1].

Results. As shown in figure 1, a 1-min injection time decreased the immediate TST to below placebo level whilst 2.5-min brought about an increase of 30.5 and 7.5-min even of 51 min as compared to a 20-min injection time. In analogy to the inverted 'U' dose-response curve in animals [5], the injection rate plays an important role in humans with an optimal effectiveness of DSIP in the range of approximately 2–10 min. It is likely that rapid injection results in an overflow at the site that gets lost because of rapid degradation of DSIP by enzymes in the blood and the brain [2].

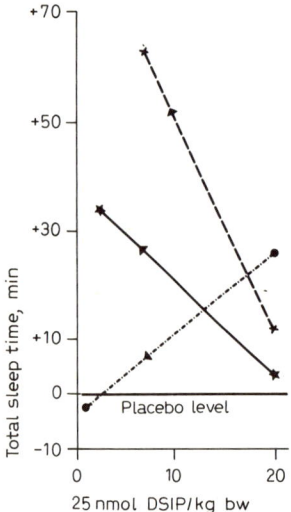

Fig. 1. Changes of sleep time with different infusion rates of DSIP.

Table I. Significant improvements of night sleep in insomniacs by DSIP

	Mean change	p<
Total sleep efficiency index	+ 0.05	0.1
Number of sleep cycles	+ 0.33	0.1
Stage 1% of sleep	− 4.4	0.025
Sleep efficiency excl. S1	+ 0.08	0.01
Min total sleep excl. S1	+32.3	0.01
Number of awakenings	− 6	0.1
Number of arousals	−13.8	0.1
Arousals/h sleep	− 3.6	0.05
REM sleep % of sleep	+ 2.3	0.1
Arousals in REM sleep	− 5.8	0.1

Night Sleep Study with Insomniacs

In a third, placebo controlled, double-blind, 5 consecutive nights experimental set, 25 nmol/kg DSIP was injected i.v. in 4 min prior to bedtime to 4 female and 2 male insomniacs aged 36-54 years (mean 45.2). The first 2 nights (with placebo injections) served for adaptation. The sequence of injections in the 3 treatment nights was either placebo–DSIP–placebo, or DSIP–placebo–DSIP, randomly attributed within each pair of subjects. With this design the results would not be affected by systematic trends that might occur within the treatment sequence.

Results. The significant results (t-test) are listed in table I. They indicate that DSIP induced longer and more efficient sleep, more REM sleep, and better sleep stability. There was no indication of a 'hangover' effect.

General Conclusions

DSIP is well tolerated at a dosage of 25 nmol/kg i.v. No complication occurred with totally 25 injections hitherto given. Infusion rate seems to be a crucial factor for the activity of DSIP. DSIP enhances sleep in conditions which are physiological under different viewpoints such as environment and time of the day, but DSIP seems not to be able to force sedation or sleep against high psychophysiological arousal. Within the aforementioned limits DSIP improves sleep of insomniacs primarily by a qualitative improvement or restoration, whereas the quantitative effect (increase of TST) is less pronounced. These results, which also include a delayed effect within the circadian wake-sleep rhythm, sustain the hypothesis that DSIP is rather a (naturally occurring) 'programming' than a modulator or transmitter peptide [5].

References

1 Gnirss, F., et al.: Sleep Res. *9* (in press).
2 Kastin, A., et al.: TINS *25:* 163–165 (1980).
3 Monnier, M.; Schoenenberger, G.A.: in Koella, Levin, Sleep 1976, pp. 257–263 (Karger, Basel 1977).
4 Schoenenberger, G.A.; Monnier, M.: Proc. natn. Acad. Sci. USA *74:* 1282–1286 (1977).
5 Schoenenberger, G.A.; Monnier, M.: IUPAC Medical Chemistry Proc., pp. 101–116, 6th Int. Symp. Med. Chem., Brighton 1978 (Cotswold Press, Oxford 1979).
6 Schneider-Helmert, D., et al.: Clin. Pharmacol. (in press).

D. Schneider-Helmert, MD, Research Department, Psychiatric Clinic, CH-5200 Königsfelden (Switzerland)

Successful Treatment of Insomnia by Interval Therapy with *L*-Tryptophan

D. Schneider-Helmert, F. Gnirss, J. Schenker

Research Department, Psychiatric Clinic, Königsfelden, Switzerland

The efficacy of *L*-tryptophan (Try) for the treatment of insomnia is not yet fully established despite numerous investigations. While Try generally showed a tendency to reduce sleep latency, not all studies reached significant results. This might be due to a specific mode of action of Try on the disturbed sleep regulation as we suggested upon the results of a study with the combination of Try and a β-adrenergic blocker (Oxprenolol) where we found a sequential effect in the sense that sleep was especially improved in the drug-free interval after the application [1]. A long-term interval therapy in a case of severe psychosomatic insomnia further supported this view [2]. Assuming that the crucial substance was Try, we designed a predictive double-blind laboratory study for testing the following hypothesis: Sleep of various types of severe insomniacs will be better in a 4-night drug-free interval after 3×2 g Try than in baseline conditions.

Methods and Patients

The experimental protocol comprised 12 consecutive nights: 5 nights in the laboratory with placebo treatment, the first being for adaptation, nights 2–5 for pretreatment baseline measures; nights 6+7 at home with Try, nights 8–12 in the laboratory, night 8 being the third treatment night and allowed for readaptation to the laboratory, nights 9–12, again with placebo, for the post-treatment interval measures. The daytime activity was measured with continuous ratings on the Stanford sleepiness scale.

A special statistical procedure enabled one to derive the limits of significance for the parameters of each patient which were selected on the baseline data and were individually predicted to improve in the Try-free interval prior to recording this period. The overall level of significance for this predictive experiment was 0.05.

The patients were 8 severe chronic insomniacs, otherwise healthy, 5 males and 3 females in the age of 32–47, mean 38.4 years. The patients with previous hypnotic drug therapy had a wash-out period of at least 5 weeks.

Results

Each of the 8 patients completed the study with significant improvements of the individually predicted sleep parameters. Thus, *the hypothesis that 3 times 2 g Try application improves sleep of the 4 post-treatment nights is accepted* at the predefined level of 0.05.

The intra-individual comparison of significantly improved parameters with the pattern of sleep in baseline conditions shows that the effect of this treatment refers to the most disturbed aspects of sleep in each individual. In an attempt to get data that would allow some generalization of the effects of this interval treatment, an a posteriori analysis of variance was made excluding the 1 subject with extreme variances of sleep data. The following results were found significant with $p<0.00001$: Increases of: total sleep time, time of sleep stages $2+3+4+$ REM. Decrease of: total waking time, stage 1 percentage of sleep, arousal density during sleep. The ratio of REM sleep and NREM sleep remained unchanged. The Stanford sleepiness scale showed no difference in the subjective evaluations of daytime alertness.

Discussion

In this predictive study, Try proved efficacious for the treatment of different types of severe, chronic primary insomnia, if given at a dosage of 3×2 g, with the positive effects in the 4 nights of the drug-free interval following Try application. Continued interval therapy of these and other insomniacs confirmed the conclusions of this study: A gradual and after some months definitive recovery was observed in most of the patients.

In contrast to the evident advantage for the practice, the results of this study on interval therapy pose problems as to the mode of drug action. Upon our results we can imagine that Try evokes two types of feedback mechanisms on the sleep regulatory system: immediate negative feedbacks long-term (up to several days) positive feedbacks. In a single-dose ap- ion, the negative feedbacks would neutralize more or less the direct :tion and the positive feedbacks. Repetitive application would build erful positive feedbacks that could display their full potency in the after Try application, i.e. when no immediate negative feedback 's the sleep improvement. This speculation and other possible ex- need to be further considered.

References

1 Gnirss, F.; Schneider-Helmert, D.; Schenker, J.: Pharmakopsychiatr. Neuro-Psychopharmakol. *11:* 180–185 (1978).
2 Gnirss, F.; Schneider-Helmert, D.; Schenker, J.: in Popoviciu, Sleep 1978, pp. 675–678 (Karger, Basel 1980).

D. Schneider-Helmert, MD, Research Department, Psychiatric Clinic, CH-5200 Königsfelden (Switzerland)

Effect of We-941 (a New Thienodiazepine) on Sleep Patterns

M. Velasco, F. Velasco, R. Romo, M.A. Perez

National Medical Centre, IMSS, Mexico City, Mexico

In a previous study, *Lohman* [3] reported subjective and EEG signs of sleep in 3 volunteers after a single oral dose of 0.5 mg of We-941 and thereafter *Grunberger et al.* [1], *Saletu et al.* [5] and *Kubicki* [2] confirmed its strong sedative effect on normal volunteers under a series of psychometric and pharmaco-EEG-clinical evaluation tests applied during daytime. The primary aim of the present investigation was to evaluate systematically the effect of We-941 on wakefulness-sleep parameters for all-night polygraph recordings in normal and insomniac subjects.

Material and Methods

This work was performed on a group of 10 normal male volunteers and another group of 20 ambulatory insomniac subjects associated to endogenous and exogenous depression. They were studied during 6 consecutive nights.

The first 3 nights were used for adapting the subjects to the laboratory and the last 3 nights for sleep studies as follows:

Normals. Night 4: no drug was administered (baseline); night 5: We-941 or placebo; night 6: We-941 or placebo.

Insomniacs. Night 4: initial placebo; night 5: We-941 (20 subjects) or flurazepam (10 subjects); night 6: placebo. We-941 vs placebo and We-941 vs flurazepam were administered under double- and single-blind procedures, respectively. The technique for recording and scoring wakefulness-sleep parameters was identical to that of *Rechtschaffen and Kales* [4].

Results

We-941 abolished the spontaneous arousal episodes shown by some normal subjects under baseline and placebo conditions. However, no significant differences in wakefulness-sleep parameters were found in normal subjects under We-941 against baseline or placebo. We-941 decreased duration of wakefulness and number of arousal episodes and increased duration of sleep, decreased latency of slow wave sleep I and II and increased duration of slow wave sleep III. These results were significantly different to initial but not to final placebos in the same insomniac subjects.

Flurazepam decreased duration of wakefulness and number of arousal episodes and increased duration of sleep and duration of rapid eye movement sleep. These results were significantly different to initial but not to final placebos in the same insomniac subjects. No significant differences were found between We-941 and flurazepam on slow wave sleep III and rapid eye movement sleep in different insomniac subjects.

References

1 Grunberger, J.; Saletu, B.; Lingamayer, L.; Kales, A.; Bemer, P.: Curr. ther. Res. *24:* 427–440 (1978).
2 Kubicki, S.: EEG-EMG *1979:* 95–100.
3 Lohman, H.: We-941 central depressant investigational brochure (Böhringer Sohn, Ingelheim 1976).
4 Rechtschaffen, A.; Kales, A.: A manual for standardized terminology, techniques and scoring systems for sleep states of human subjects. Publ. Hlth Service (US Government Printing Office, Washington 1968).
5 Saletu, B.; Grunberger, J.; Volovka, J.; Bemer, P.: Arzneimittel-Forsch. *29:* 700–704 (1979).

M. Velasco, MD, National Medical Centre, IMSS, PO Box 73–032,
Mexico City (Mexico)

Long-Term Effects of *L*-Tryptophan and Oxprenolol in Hyposomnia

R. Steinberg, N. Nedopil, E. Rüther

Psychiatrische Klinik der Universität München (Director: Prof. Dr. *H. Hippius*), München, FRG

A hypnotic effect of the serotonin precursor *L*-tryptophan (Try) seems to be well established [1] and has been offered as evidence for central serotonergic sleep mechanisms [2]. The combination of Try with oxprenolol (Oxp), which is a central β-receptor antagonist, increases slow wave sleep [3]. This may be due to a facilitation of serotonergic mechanisms in addition to a reduction of the activity of the catecholaminergic system. Data will be presented here showing the long-term effects of combined administration of Try and Oxp in cases of severe insomnia.

20 patients (15 females, 5 males) attending our sleep laboratory were selected for an open study (1–2 g Try and 80 mg Oxp per day). In order to exclude a mental or physical origin of insomnia, an interview and a medical examination were performed. The examinations included EEG, ECG, skull and chest X-rays, and routine blood and urine analysis. From the beginning of treatment, patients were asked to take only the prescribed Try and Oxp medication for sleep. Apart from contraceptives, no other medication was allowed; alcohol was restricted to 20 g daily. They had to fill in a questionnaire each morning in which sleep duration and sleep quality turned out to be the items which were answered most regularly. The 20 patients continued the treatment for at least 1 month, after which half felt much better and went back to their family doctor. However, 11 patients stayed for 4 months, and from these 6 were followed up for 7 months. The two subgroups did not differ essentially from the entire group with respect to their symptoms of insomnia.

The age of the patients was $46.6 \pm SD\ 9.2$ years, the history of insomnia was 12.2 ± 9.7 years. They had taken up to three different hypnotics daily over years, mainly of the tranquillizer type. Almost all patients reported a hypnotic effect of the drugs, which, however, decreased with time. They frequently suffered from 'hangover effects' and very often felt strongly psychologically dependent. 75% had difficulties in falling asleep and the same percentage complained about disturbances in maintaining sleep. At the beginning of the treatment the patients reported a sleep duration of 2.9 ± 1.8 h without medication. The subgroups did not differ significantly. In a control group of 20 subjects matched for sex and age (43.6 ± 11 years), sleep duration was 7.2 ± 0.9 h without any medication, which did not differ from the desired sleep duration (7.9 ± 1 h). Only 10–15% of the controls reported difficulties in falling asleep and maintaining sleep. Surprisingly the desired sleep duration of the patients (7.6 ± 1 h) was in the same range as that of the controls. With respect to other characteristic disturbances of insomnia, such as long sleep latency, frequent awakenings, and daytime sleepiness, the patients showed severe symptoms. Unfortunately, over the course of the study they failed to answer consistently questions concerning these items, so that apart from a tendency to improve – known from follow-up interviews – no data can be offered. Probably all these particular questions seem to be of minor importance in the subjective estimation of sleep compared to sleep duration and quality.

Figure 1 illustrates the reported estimation of sleep duration over the time of treatment. The first values, underlined with 'rating' give the sleep duration without any medication before treatment. Subsequent values are the mean of the 5 consecutive nights, assessed after various time periods up to 7 months. Statistically significant increases in sleep duration are evident

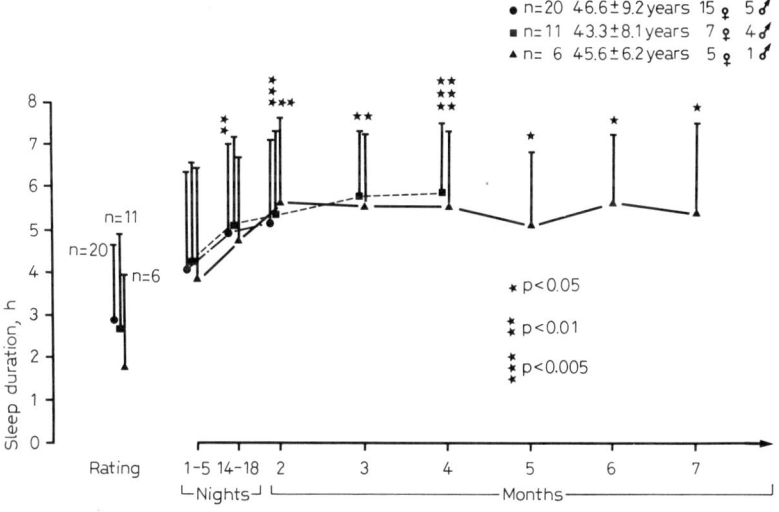

Fig. 1. Sleep duration during treatment.

at all times after 5 days of treatment. It is interesting to note that values reported after 1 day's treatment were equal to the pretreatment values (not illustrated).

The patients were asked in the questionnaire for a 'judgement of sleep quality'. Of the entire group of 20 patients only 16 answered this question regularly, of the first subgroup only 6. A qualitative estimation on a five-point scale (1 = very good to 5 = very bad) was made. Although the accuracy of such an assessment is questionable, a tendency towards better sleep quality could be seen. Before treatment began, patients reported a sleep quality value of 3.3 ± 1.2 (medium to bad); after a month's treatment this was significantly improved to 2.7 ± 1.1 (good to medium). The correlation coefficient between sleep duration and estimation of sleep quality was $r = 0.66$. The subgroup of 6 patients, which was followed up for 4 months, judged sleep quality at the beginning of the treatment at 3.4 ± 0.9 (medium to bad); 4 months later at 2.5 ± 0.9 (good to medium).

No side effects of the combination of Try and Oxp have been reported. All patients were outpatients and continued their daily work. Although because of the nature of the open study we are not able to exclude placebo effects, the medication with Try and Oxp seems to have a significant effect in the treatment of severe insomnia.

References

1 Hartmann, E.; Spinnweber, L.C.: J. nerv. ment. Dis. *167:* 497–499 (1979).
2 Koella, W.P.: Adv. Biochem. Psychopharmacol. *11:* 181–186 (1974).
3 Rüther, E.; Davis, L.; Piergies, A.: Sleep, 1976, pp. 388–391 (Karger, Basel 1977).

Dr. R. Steinberg, Psychiatrische Klinik der Universität München,
Nußbaumstraße 7, D-8000 München 2 (FRG)

F. Psychiatry, Neurology, Sleep Deprivation

Evaluation of Automatic Sleep Spindle Detection in Epileptic Patients

A. C. Declerck, A. W. Lustenhouwer

Epilepsy Centre Kempenhaeghe, Department of EEG and Clinical Neurophysiology, Heeze, The Netherlands

Polygraphic recording is one of the appropriate methods to study sleep-wakefulness patterns and the physiological and pathophysiological changes associated with it. One of the most characteristic and constant EEG phenomena occurring during sleep is sleep spindling [3]. It is usually described as a sigma-form rhythm with a frequency of 11.5–14.5 Hz, an amplitude of 20 µV and a duration of 0.5–1.5 s. Sleep spindles are already present at the age of 3 months [2]. Deviations from this pattern, qualitatively as well as quantitatively, would be indicative of altered brain function [1].

In our laboratory, approximately 300 long-term sleep-wakefulness polygraphic (16 channels) recordings (3–10 h) are carried out per year. An automatic hardware system was built to detect spindles and, in the present study, a clinical evaluation of this detector is described and is compared with a visual analysis.

Materials and Methods

Sleep recordings were taken from 100 patients who had epilepsy or who were suspected to have epilepsy. Recordings were done on a 16-channel Elema Schönander mingograph. 12 channels served for the EEG whereas EMG and EOG were recorded on the 4 remaining channels. The electrode placement was done according to the international 10–20 system. Spindles were also visually analyzed.

The hardware detection system based on that described by *Smith et al.* [4] was built. The detection criteria applied to the frontal-central area were: a frequency of 11–14 Hz and an amplitude of 20 µV or more. The frequency was detected via a zerocrossing detection after analogue filtering of the signal. The amplitude of the filtered signal is checked against the criteria. If a number of waves between two positive zerocrossings conform the criteria, the density is measured via a moving-window register (0.64 s). A spindle is detected when the density is above a prefixed value.

Results

Use of a hardware detection system allowed to obtain a quantitative analysis of sleep spindles. In 60% of the patients there was a relatively good agreement between the visual analysis and the hardware analysis. In 40% of the patients there were large differences.

The detection was not appropriate where spindles did not fulfill the preset criteria. Importantly, the amount of spindles detected was largely different depending on the derivation taken. The inter-electrode distance appears to be an important parameter for the quantification of the spindle detection. Another reason for the lack of agreement was that in certain EEGs there was a strong background activity of frequencies within the range of the spindling, though lacking the specific characteristics of spindles.

Discussion

A hardware detection of spindles is an appropriate quantification method but obviously depends on the quality of the sleep recordings itself. Specifically preset criteria might not be sufficient since they can deviate from the norms according to the individual person analyzed. This could be improved by adapting the criteria according to the individual.

It is important to achieve an optimal detection system especially in patients in which the diagnosis of epilepsy is not yet completed. This is particularly true in view of the fact that a deviating spindle activity is related to the degree and nature of disturbances in brain functioning and to the use of certain medication.

The failure to detect spindles in a large part of the present population is probably due to the fact that they were epileptics with additional brain dysfunctions. It seems to us that a hardware system does not give optimal detection and in view of its clinical significance, an automatic system based on software analysis seems more appropriate.

References

1 Gloor, P.: Epilepsia *20:* 571–588 (1979).
2 Pètre-Quadens, O.: Ned. Tijdschr. Geneesk. *121:* 323–326 (1977).
3 Silverstein, L.D.; Levy, C.M.: Electroenceph. clin. Neurophysiol. *40:* 666–670 (1976).
4 Smith, J.R.; Funke, W.F.; Yeo, W.C.; Ambühl, R.A.: Electroenceph. clin. Neurophysiol. *38:* 435–437 (1975).

A.C. Declerck, MD, Epilepsy-Centre Kempenhaeghe,
Department of EEG and Clinical Neurophysiology, Heeze (The Netherlands)

Night Sleep after Recovery from Brainstem Traumatic Injury

J. Landau-Ferey, O. Benoit, B. George

Laboratoire d'Etude du Sommeil (Unité 3 INSERM), Paris, France

Sleep organization in patients with acute or chronic post-traumatic coma is now well known. But few studies have been concerned with sleep evolution after recovery from comas with brainstem dysfunction at the acute stage. The purpose of this work was to establish whether sleep modifications persisted in such subjects when their clinical state improved.

Material and Methods

16 subjects, with ages ranging from 15 to 25 years, were studied either 1 month (T1 group, 10 cases) or 6 months (T6 group, 6 cases) after regaining consciousness. All of them had had severe head injury followed by coma with primary or secondary lesion of brainstem. Clinically, in the two groups, T1 and T6, both primitive level of brainstem dysfunction and duration of coma were studied.

EEG sleep recordings were visually scored according to standard criteria [9]. The different sleep stages were easily identified in all patients who were each recorded during 2 consecutive nights. The present study concerns only the intrasleep organization of REM sleep and intervening wakefulness since slow-wave sleep amounts were in the range of normal values.

Results

Clinical. Among the patients recorded after 1 month (T1) two subgroups were separated according to both duration of coma and primitive level of dysfunction as shown in a previous study [8]. The first subgroup (T1A) was composed of 4 patients whose duration of comatose state was the longest (15–40 days) and level of dysfunction the lowest (1 mesencephalic, 2 mesodiencephalic, 1 only diencephalic). The second subgroup (T1B) had 6 patients with a higher level of primitive dysfunction (5 diencephalic, 1 mesodiencephalic) and a shorter duration of coma (4–12 days).

Group T6 included patients with initial clinical histories similar to those of the two preceding subgroups. However, based on the sleep recordings, 6 months after the initial trauma the previous subgroups were no longer identifiable.

EEG Sleep Recordings. (1) Wakefulness (table I). The duration of intrasleep wakefulness as well as the number of awakenings increased in all three groups. T1A had the highest level of wakefulness, with 31.35% of total sleep time (TST) and 24 awakenings by night (mean by subject) with 27.4% of them being less than 1 min. When compared to T1A, T1B presented a smaller amount of wakefulness, 10.85% of TST, with a higher number of awakenings of short duration (61% being inferior to 1 min). Therefore, in group T6 the total amount of wakefulness was significantly higher than in normal subjects and similar to values found for subgroup T1B.

(2) REM sleep (table II). The percentage of REM sleep was lower than normal in both subgroups of T1 (9.57% of TST in T1A, 12.65% in T1B) but partial recovery was noticed in T6: 16%. This decrease of REM sleep amounts corresponded mainly to short durations of REM sleep episodes (T1A, 9 min 8, T1B, 11 min 9; but T6, 17 min), while latencies of first REM

Table I. Wakefulness and awakenings. Characteristics of the three groups of injured patients compared to a normal group of middle sleepers

	T1A	T1B	T6	Control
TST	481.75 ± 4.99	482.40 ± 55.26	432.70 ± 59.95	451 ± 41.16
% W	31.35 ± 3.56	10.10 ± 1.58	11.80 ± 4.69	2.20 ± 2.52
Number of awakenings	24.25 ± 16.7	19.77 ± 5.01	19 ± 5.39	10.1 ± 4.93
Number <1 min	6.5 ± 9.03	11.44 ± 4.18	13.08 ± 7.58	9.1 ± 4.25
Number 1–15 min	15.25 ± 10.14	7.78 ± 3.23	5.58 ± 3.55	1 ± 1.83
Number 15–30 min	1.25 ± 0.95	0.22 ± 0.44	0	0
Number >30 min	1.25 ± 1.25	0.33 ± 0.5	0.25 ± 0.45	0

TST = Total sleep time; W = wakefulness.

Table II. REM data. Characteristics of REM sleep in the same groups as shown in table I

	T1A	T1B	T6	Control
% REM	9.57 ± 3.13	12.65 ± 3.34	16.15 ± 4.37	20.52 ± 5.50
1st REM period's latency	91.6 ± 54.75	66.22 ± 24.68	102.75 ± 47.19	92.1 ± 42.24
Mean duration of periods	9.8 ± 0.6	11.9 ± 3.96	16.82 ± 2.82	20.67 ± 13.78
Rhythm	103.37 ± 34.95	82.29 ± 8.4	86.16 ± 11.2	91.31 ± 10.61
1st period's duration	7.33 ± 4.92	9.30 ± 2.59	8.38 ± 5.36	15.34 ± 11.90

period and durations of REM sleep cycles were in the normal range. Finally, in T1B as well as in T6, REM sleep was frequently interrupted by short awakenings which was not the case in T1A.

Discussion

An important observation was a net increase of the amount of wakefulness in these young injured patients when compared to normal subjects of the same age. Such a disturbance, but less important, is well known in insomniac subjects: about 11% of intrasleep wakefulness was reported by *Frankel et al.* [5] and only 4% by *Garma et al.* [2]; a mean value of 11 awakenings was given by *Gaillard* [6] in patients of the same age. But latencies of insomniac sleep onset are very long, which was not the case in this study.

On the other hand, in normal old people, results very similar to these have been published by *Březinová* [1] and *Feinberg et al.* [3] who found about 16% of intrasleep wakefulness in elderly people. *Williams et al.* [10] and *Foret and Webb* [4] observed a progressive increase with age in the number of awakenings.

Concerning REM sleep, its decrease was rather specific with values much smaller than those given by *Frankel et al.* [5] for young insomniac subjects. Even in elderly people REM is not clearly decreased, 18–22% having been cited by the authors mentioned above. On the other hand, frequent interruptions of REM sleep by short awakenings were observed in older subjects by *Kahn and Fisher* [7] which is similar to that found in the present study.

Thus, it might be concluded that, in the first month after regaining consciousness, the higher the level of dysfunction, the lower the disturbance of intrasleep organization. On the other hand, the lower the initial dysfunction, the greater the sleep disturbance.

The principal observations were a REM sleep decrease and a wakefulness increase, with a non-REM sleep which remains normal. Thus, a catecholamine dysfunction might be hypothesized in order to explain these data after lower brainstem trauma. Even after 6 months, sleep parameters of the patients seemed to be closer to those of elderly group values than either normal or insomniac people of the same age.

References

1 Březinová, V.: Electroenceph. clin. Neurophysiol. *39:* 273–278 (1975).
2 Garma, L.; Bouard, G.; Benoit, O.: Rev. EEG Neurophysiol. (in press).
3 Feinberg, I.; Koresco, R. L.; Heller, N.; Steinberg, H.: Excerpta Med. Int. Congr. Ser. No. 150, pp. 2928–2930 (1966).
4 Foret, J.; Webb, W. B.: Rev. EEG Neurophysiol. *10:* (in press, 1980).
5 Frankel, B. L.; Coursey, R. S.; Buchbinder, R.; Snyder, F.: Archs gen. Psychiat. *33:* 615–623 (1976).
6 Gaillard, J. M.: Eur. Neurol. *14:* 473–484 (1976).
7 Kahn, E.; Fischer, C.: J. nerv. ment. Dis. *148:* 477–494 (1969).
8 Landau-Ferey, J.; George, B.; Benoit, O.: Rev. EEG Neurophysiol. (in press).
9 Rechtschaffen, A.; Kales, A. (eds): A manual of standardized terminology, techniques and scoring system for sleep stages of human subjects. Publ. Hlth Service (US Government Printing Office, Washington 1968).
10 Williams, R. L.; Karacan, I.; Hursch, C. J.: EEG of human sleep. Clinical applications (Wiley & Sons, New York 1974).

J. Landau-Ferey, MD, Laboratoire de Physiologie, CHU Pitié-Salpêtrière, 91, Boulevard de l'Hôpital, F-75634 Paris Cedex 13 (France)

Sleep Study in Patients with Spinocerebellar Degeneration and Related Diseases

T. Shimizu, Y. Sugita, S. Iijima, Y. Teshima, Y. Hishikawa

Department of Neuropsychiatry, Osaka University Medical School, Osaka, Japan

It has been demonstrated by animal experiments [1] that neural mechanism controlling sleep is present in the brain stem: the locus coeruleus is regulating the occurrence of REM sleep and raphe nuclei are related to the occurrence of NREM sleep. Little is known, however, about the sleep of patients after destruction of these neural structures [3]. We performed polygraphic examination of nocturnal sleep in patients with degenerative disease involving the brain stem and/or cerebellum.

Subjects and Method

10 patients, 8 males and 2 females, aged 48–73 years, were examined. They were clinically divided into the following 4 groups. Group 1 consisted of 2 patients with parenchymatous cerebellar degeneration (PCD). Group 2 consisted of 4 patients with OPCA, and group 3, of 3 patients with Shy-Drager syndrome (SDS). Only 1 patient with progressive supranuclear palsy (PSP) belonged to group 4.

Results and Discussion

In 2 patients belonging to group 1, disturbance of nocturnal sleep was very mild or absent. However, in all patients belonging to the other groups, nocturnal sleep was highly disturbed and their sleep disturbance had the following characteristics. Nocturnal sleep was very unstable with frequent interruption by awakening, and the percentage of deep NREM sleep was significantly decreased or absent. The percentage of REM sleep was also markedly decreased and REM-NREM sleep cycle was disturbed in most patients. A peculiar sleep state characterized by concomitant occurrence of low voltage mixed frequency EEG, relatively elevated tonic discharge in the mental muscles, and frequent rapid eye movements was observed in all patients be-

Table I. Total sleep time and percentage of sleep stages in 10 patients with brainstem degeneration

	Case	Total sleep time, min	Percentage of different sleep stages						Delirious behavior during stage 1-REM
			1	2	3	4	REM	1-REM	
Group 1 (PCD)	1	420	17.5	63.4	0.4	0	18.7	0	
	2	515	17.5	58.3	8.2	0.6	15.4	0	
Group 2 (OPCA)	3–1	206	15.8	66.7	7.8	0.7	7.4	1.6	
	3–2	147	12.7	30.5	20.6	0	7.5	28.0	+
	4	507	35.8	48.0	0.2	0	15.6	0.4	
	5	406	44.0	41.0	0.6	0	6.4	8.0	+
	6	266	30.7	48.7	1.2	0	8.0	11.4	+
Group 3 (SDS)	7	354	54.2	25.7	0.1	0	0	20.0	
	8	324	20.6	72.2	0	0	2.3	4.9	
	9	499	55.4	27.5	0	0	0.7	21.4	+
Group 4 (PSP)	10	395	28.2	35.3	0.3	0	19.2	17.0	+

PCD = parenchymatous cerebellar degeneration; OPCA = olivo-ponto-cerebellar atrophy; SDS = Shy-Drager syndrome; PSP = progressive supranuclear palsy; + = occurrence of delirious behavior during stage 1-REM.

longing to groups 2, 3 and 4. This peculiar sleep state had been found also in alcoholic patients with delirium tremens and it was named stage 1-REM with tonic EMG (stage 1-REM) [7]. Total sleep time and percentage of different sleep stages in the 10 patients are shown in table I.

In patients with a high percentage of stage 1-REM, delirious and queer movements of the hands and arms as if fumbling for hallucinated objects were sometimes observed during stage 1-REM (table I).

REM density during stage 1-REM with delirious behavior was much higher (26.4/min) than that during stage 1-REM without such behavior (12.0/min) or during REM sleep (10.9/min).

1 patient with OPCA (case 3) was twice examined with an interval of 8 months. The development of clinical symptoms during this period was accompanied by worsening of sleep disturbance with an increased percentage of stage 1-REM and appearance of delirious behavior during stage 1-REM (table I).

Sastre and Jouvet [4] found in cats that, after localized destruction in the caudal parts of the locus coeruleus in the pons, muscle atonia did not occur during REM sleep, and that their cats repeatedly showed 'oneiric behavior' during REM sleep without muscle atonia. REM sleep without muscle atonia and oneiric behavior in their cats well resemble stage 1-REM and concomitant delirious behavior in our patients. It has been demonstrated by histological examination of the brain that the locus coeruleus is destroyed in patients with OPCA [2], SDS [5], and PSP [6].

Based on these findings, it is strongly suggested that the occurrence of stage 1-REM and concomitant delirious behavior in our patients are due to degeneration of the locus coeruleus or its related structures in the brain. High REM density during stage 1-REM with delirious behavior may indicate excitation of the neural mechanism underlying the generation of dream. Reduction of deep NREM sleep and marked disturbance of the REM-NREM sleep cycle in our patients may be due to degeneration of the neural structures for REM and NREM sleep in the brainstem.

References

1 Jouvet, M.: Science *163:* 32–47 (1969).
2 Lüthy, F.; Mumenthaler, M.: Arch. Psychiat. Z. ges. Neurol. *197:* 327–334 (1958).
3 Mouret, J.: Electroenceph. clin. Neurophysiol. *38:* 653–657 (1975).
4 Sastre, J.; Jouvet, M.: Physiol. Behav. *22:* 979–989 (1979).
5 Shy, G.M.; Drager, A.G.: Archs Neurol., Chicago *2:* 511–527 (1960).
6 Steel, J.C.; Richardson, J.C.; Olszewski, J.: Archs Neurol., Chicago *10:* 333–359 (1964).
7 Tachibana, M.; Tanaka, K.; Hishikawa, Y.; Kaneko, Z.: Advance in sleep research, vol. 2, pp. 177–205 (Spectrum, New York 1976).

T. Shimizu, MD, Department of Neuropsychiatry, Osaka University Medical School, Osaka 553 (Japan)

Effects of Different Paradoxical Sleep Deprivation Treatments on Locomotor Activity in Rats

A. M. L. Coenen, Z. J. M. Van Hulzen, E. L. J. M. Van Luijtelaar

Department of Comparative and Physiological Psychology,
University of Nijmegen, Nijmegen, The Netherlands

In order to obtain insight into the behavioural role of paradoxical sleep (PS), the method of PS deprivation is often employed. Due to methodological problems, however, deprivation studies have provided results which are difficult to interpret. For progress in this research area, it is important that attempts be made to solve these problems.

The first instrumental PS deprivation technique is the 'arousal' technique, originally used in human studies [3]. As soon as PS is identified by means of electrophysiological parameters, subjects are aroused from sleep. From animal studies, it turned out that this technique can effectively be applied for only a few hours. After that time, the number of awakenings increases rapidly making the technique unsuitable for long-term deprivation studies [7].

A popular technique allowing the simultaneous deprivation of a large number of animals is the 'watertank' technique [6]. Animals are placed on small platforms surrounded by water. On these platforms they can obtain sleep, but PS is restricted because of the loss of muscular tone associated with this type of sleep. Usually, in order to control for non-specific effects of the technique, animals are placed on large platforms permitting PS to occur. Although this control may be appropriate for most of the confoundings [10], it is generally felt that the watertank technique is controversial in regard to its behavioural consequences. For example, some doubt exists as to whether the enhancement of activity in rats after long-term application of this technique [1, 5, 8] is produced by PS deprivation per se [4].

Two ways may be chosen to eliminate these controversies. One way is to develop alternative PS deprivation techniques and search for common effects; another way is to improve the watertank technique by controlling or eliminating the confoundings of the technique.

The Pendulum Technique

A technique was developed [9] based on the observation that the occurence of PS is generally preceded by slow wave sleep (SWS). An apparatus slowly moves the animals' cages backwards and forwards like a pendulum, producing postural imbalance in the animals when the cages are at the extremes. This allows the animals to sleep for only brief periods, which prevents them from entering into PS. Rats quickly learn to avoid the point of imbalance by walking downwards to the other side of their cages.

The availability of two different PS deprivation techniques allows a closer investigation of the question of whether PS deprivation does lead to an increase of activity. A study of the effects of 72 h of PS deprivation on locomotor activity was carried out in rats [2]. The number of crossings in a shuttlebox was determined for a period of 15 min immediately following the PS deprivation treatment. The deprivation treatment was ended at the beginning of the dark period. During the first 3 h of the recovery sleep in the dark, a large increase in the amount of PS and no change in the amount of SWS were found in evaluating the pendulum technique [9]. Two PS deprivation groups were studied: a 'pendulum experimental' group and a 'watertank experimental' (small platform) group. In addition, three control groups were run: a 'pendulum control' group in which rats were placed in a moving pendulum adjusted in a way that permits PS in the animals, a 'watertank control' (large platform) group, and a 'home cage' control group.

In agreement with the literature findings, the watertank experimental group showed significantly more crossings than the watertank control group. On the other hand, no differences were found between the pendulum experimental and the pendulum control group. The most obvious interpretation of these results is that a non-specific effect of the small platform, probably its restriction of movement [4], was responsible for the increased locomotor activity of the watertank experimental group.

The Modified Watertank Technique

An experiment was conducted to investigate more systematically the effects of PS deprivation and movement restriction on general activity [*Van Hulzen and Coenen,* in preparation]. Rats were assigned to four conditions. In the 'classical platform' condition rats were placed on a small platform inducing both PS deprivation and movement restriction. In the 'multiple platform' condition, rats were maintained in a watertank with five small platforms positioned closely together. This condition is comparable to the classical platform in its effects on sleep, but allows animals to move about. Rats of the 'cuff platform' condition were placed on a single platform provided with a cuff, allowing animals to sleep while restricting their activity. In the 'home cage' condition, rats experienced neither PS deprivation nor movement restric-

Table I. Means and standard deviations of the number of cage crossings, rearings, and of the weight changes for the four conditions (n = 16)

Condition	PS deprivation	Movement restriction	Crossings	Rearings	Weight change, g
Classical platform	+	+	39.6 ± 4.5	21.8 ± 7.7	−25.7 ± 10.4
Multiple platform	+	−	49.0 ± 10.6	35.6 ± 9.6	−15.3 ± 6.5
Cuff platform	−	+	35.1 ± 7.0	35.9 ± 12.0	−15.0 ± 6.0
Home cage	−	−	34.9 ± 6.9	36.8 ± 9.7	+ 2.2 ± 2.8

tion. Following 72 h of treatment rats were placed in shuttleboxes for 15 min to assess the number of cage crossings and rearings. Before and after the experiment, animals were weighed. Results are given in table I.

An increase in the number of crossings and a decrease in the number of rearings were found as a result of PS deprivation. The multiple platform was mainly responsible for the increase in crossings, whereas the classical platform caused the decrease in rearings. Movement restriction resulted in a decrease in the number of crossings and in the number of rearings, for both of which effects the classical platform was responsible.

The findings suggest that PS deprivation leads to enhanced activity, an effect that is attenuated when movement restriction and/or the stress associated with it forms part of the deprivation treatment. In this respect, the multiple platform seems superior to the classical platform in depriving rats of PS. Now one has to explain why PS deprivation by means of the pendulum technique does not lead to enhanced activity. During the pendulum treatment, PS deprived rats are forced to move at regular times, whereas pendulum control rats can move freely. This means that the amount of activity is not controlled for. It might be possible that the increase of activity during treatment is responsible for a decrease in activity on subsequent testing, cancelling the increase of activity caused by PS deprivation. At the moment, however, no data are available to support this view.

In summary, two alternative PS deprivation techniques give rise to variable effects on general activity. The watertank technique and, even more pronounced, the modified watertank technique, leads to an increase in locomotor activity, whereas the pendulum technique does not produce a change in activity. It is felt that information regarding the variables responsible for these differential effects on behaviour is of importance for future research in this area.

References

1 Albert, I.; Cicala, G.A.; Siegel, J.: Psychophysiology *6:* 552–560 (1970).
2 Coenen, A.M.L.; Van Hulzen, Z.J.M.: in McConnell, Boer, Romijn, Van de Poll, Corner, Adaptive capabilities of the nervous system. Progress in brain research, vol. 53, pp. 325–330 (Elsevier, Amsterdam 1980).
3 Dement, W.: Science *131:* 1705–1707 (1960).
4 Fishbein, W.; Gutwein, B.M.: Behav. Biol. *19:* 425–464 (1977).
5 Hicks, R.A.; Moore, J.D.: Physiol. Behav. *22:* 689–692 (1979).
6 Jouvet, D.; Vimont, P.; Delorme, F.; Jouvet, M.: C.r. Séanc. Soc. Biol. *158:* 756–759 (1964).
7 Morden, B.; Mitchell, G.; Dement, W.: Brain Res. *5:* 339–349 (1967).
8 Ogilvie, R.D.; Broughton, R.J.: Psychophysiology *13:* 249–260 (1976).
9 Van Hulzen, Z.J.M.; Coenen, A.M.L.: Physiol. Behav. *25:* 807–811 (1980).
10 Vogel, G.W.: Archs gen. Psychiat. *32:* 749–761 (1975).

A.M.L. Coenen, MD, Department of Comparative and Physiological Psychology, University of Nijmegen, Montessorilaan 3, NL-6525 HR Nijmegen (The Netherlands)

Sleep Apneas in Alzheimer's Disease

S. Smirne, M. Franceschi, S.R. Bareggi[a], G. Comi, E. Mariani, M. Mastrangelo

Milan University, School of Medicine, Department of Neurology and Department of Pharmacology[a], Milan, Italy

It is well known that sometimes a sleep apnea syndrome occurs because of nervous system injuries of different type, such as those determined by infectious, vascular, neoplastic, traumatic and degenerative diseases. While studying different types of dementia, we observed in last years an unexpected high percentage of patients with Alzheimer's disease who presented sleep apneas.

Patients and Methods

Patients studied were 8 males and 15 females, aged 48–72 years (m ± SD = 60.9 ± 6.2), who presented dementia of Alzheimer's type since age 65 or earlier (Alzheimer's disease). The diagnosis was based on detailed anamnesis, neurological examination, neuropsychological and psychometric tests including aphasia, agnosia and apraxia tests, WAIS, WMS and a rating scale for dementia [2], routine EEG and either CT scan or PEG. The differential diagnosis of dementia of Alzheimer's type from multi-infarct dementia was made also on the basis of Hachinski's dementia score [6].

All the patients were studied by means of repeated all-night and daytime polysomnographies (EEG, EOG, chin and intercostal muscle EMG, thoracic and abdominal movements, nasal and buccal airflow) after at least 1 week withdrawal of psychoactive drugs.

In 13 patients homovanillic acid (HVA), 5-hydroxyindoleacetic acid (5-HIAA) and γ-aminobutyric acid (GABA) were determined in CSF obtained by lumbar puncture using procedures and methods described elsewhere [1].

Results and Conclusions

8 patients out of 23, 3 males and 5 females, presented a pathological number of sleep apneas with an 'apnea index' [5] ranging from 11 to 60 (m ± SE = 29.1 ± 7.2). Nevertheless, none of these apneic patients was obese or had anatomical abnormalities of the upper airways, and none showed evident excessive daytime somnolence (table I). In all the patients sleep ap-

Table I. Alzheimer's patients with sleep apneas

Patient	Sex	Age	Duration of disease years	IQ	Body weight Δ from mean, %	Upper airways	EDS	Apnea index
1	M	51	1	71	+ 3	normal	absent	60
2	F	72	6	57	− 26	normal	absent	58
3	F	53	2	55	− 14	normal	absent	39
4	F	67	3	58	+ 7	normal	absent	22
5	M	60	4	n.d.	− 8	normal	absent	16
6	F	62	2	62	+ 10	normal	absent	15
7	F	66	3	58	+ 5	normal	absent	12
8	M	69	3	58	− 12	normal	absent	11

EDS = Excessive daytime sleepiness; n.d. = not determinable.

neas were both obstructive and central, but a predominant type could be identified: obstructive apneas were predominant in 7 patients and central apneas only in 1. In all the patients apneas were repetitive either during non-REM and REM sleep, except in 1 patient who presented repetitive apneas only during non-REM sleep immediately successive to a REM period. No correlation was found between apnea index and duration or clinical severity of the disease.

Previously we observed in Alzheimer's patients that, as the dementing process proceeded, the sleep progressively lost some distinguishing characteristics, such as the phasic events and the slow waves of high amplitude of non-REM sleep as well as the desynchronized activity of REM sleep. Since generally in severely demented patients the waking EEG was slower than in the mildly demented ones, there was a final trend of wake and sleep EEG patterns to become similar [8]. In apneic Alzheimer's patients no correlation was found between apnea index and wake EEG slowing or disorganization of sleep EEG patterns. Moreover, no difference in clinical and neurophysiological characteristics was observed between apneic and non-apneic patients.

As to CSF biochemistry we previously reported an HVA concentration significantly lower in severely demented Alzheimer's patients than in mildly demented ones and in controls, and no significant change of 5-HIAA concentration determined by the disease, while GABA levels were homogeneously decreased in Alzheimer's patients independently from the severity of the disease, suggesting an age- rather than disease-related reduction [1].

In apneic Alzheimer's patients the CSF concentrations of HVA, 5-HIAA and GABA were not different from those of non-apneic patients.

From this study the following conclusions can be drawn: (1) A pathological number of sleep apneas occurs in nearly one third of Alzheimer's patients. (2) Sleep apneas are both obstructive and central, but predominantly obstructive apnea patients are prevalent. (3) Sleep apneas in Alzheimer's patients are not accompanied by evident excessive daytime somnolence. (4) Sleep apneas are unrelated to obesity or anatomical abnormalities of upper airways. (5) Apneic and non-apneic Alzheimer's patients show no difference in clinical, wake and sleep EEG and CSF biochemical patterns.

Neuropathological [7] and biochemical [4] findings suggesting the same impairment of cortex and brainstem neurons might explain the frequent observation of sleep apneas in Alzheimer's patients, but the recent report [3] of a rather high percentage of elderly subjects with sleep apneas and hypopneas has to be taken into consideration. Moreover, the possible role of sleep hypoventilation and ensuing hypoxia in the dementing process has to be further investigated.

References

1   Bareggi, S. R.; Franceschi, M.; Bonini, L.; Zecca, L.; Smirne, S.: (in press).
2   Blessed, G.; Slater, E.; Roth, M.: Br. J. Psychiat. *114:* 797–811 (1968).
3   Carskadon, M. A.; Brown, E. D.; Dement, W. C.: Sleep Res. (in press, 1980).
4   Davies, P.; Maloney, A. J. F.: Lancet *ii:* 1403 (1976).
5   Guilleminault, C.; Van den Hoed, J.; Mitler, M. M.: in Guilleminault, Dement, Sleep apnea syndromes, pp. 1–12 (Liss, New York 1978).
6   Hachinski, V. C.; Iliff, L. D.; Zilhka, E.; Du Boulay, G. H.; McAllister, V. L.; Marshall, J.; Ross Russel, R. W.; Symon, L.: Archs Neurol., Chicago *32:* 632–637 (1975).
7   Ishii, T.: Acta neuropath. *6:* 181–187 (1966).
8   Smirne, S.; Franceschi, M.; Bareggi, S. R.; Mariani, E.; Comola, M.: Sleep Res. (in press, 1980).

Dr. S. Smirne, Clinica Neurologica dell'Università, Ospedale S. Raffaele, I-20090 Milano-Segrate (Italy)

Data Compiled through the Study of Sleep Concerning the Prognosis of Petit Mal

J. Touchon, A. Besset, M. Billiard, P. Passouant

Service de Physiopathologie des Maladies Nerveuses, Centre Gui de Chauliac, Cliniques St. Eloi, Montpellier, France

It has been shown [1, 2] that a relationship exists between the fluctuations of alertness and petit mal electroclinical manifestations. It was therefore interesting to investigate the modifications of this relationship depending on the state of development of this disease.

Material and Methods
92 polygraphic recordings of nocturnal sleep were performed in 56 cases of petit mal epilepsy, 27 men and 29 women (mean age 14.3 ± 2, range 7–33). The average length of surveillance in the ward was 7 years ± 3, range 3–22. Three groups were defined with respect to course of disease: petit mal was well controlled in 31 subjects (55.4%), poorly controlled in 14 (25%) and variably controlled in 11 subjects (19.6%). Recordings for all subjects were performed after a cessation of treatment for at least 2 days prior to recording. Whenever possible relaxed wakefulness before sleep onset was recorded for 0.5 h and in all cases the recording was followed an hour later by spontaneous waking.

Results
The Existence of 3 c/s Paroxystic Spike and Wave Discharges in the Various States of Alertness (table I). The distribution of the cases is different in the 3 groups (table I). In the well-controlled group 3 c/s paroxystic spike and wave discharges exist, in the majority of cases, during stage I (93.54%), REM sleep (80.64%) and awakening (93.54%). On the other hand, in the poorly controlled group these discharges exist only for a maximum of 50% of cases in the same states of alertness. Results for the variably controlled group are intermediary between the two others.

Table I. Existence of 3 c/s paroxystic spike and wave discharges during the various states of alertness. Distribution of the cases according to the course of the disease

n	RW (%)	I (%)	II (%)	III–IV (%)	REM (%)	A (%)
WCG 31	18 (58.1)	29 (93.54)	12 (38.7)	0 (0)	25 (80.64)	29 (93.54)
VCG 11	6 (54.54)	7 (63.63)	4 (36.3)	0 (0%)	7 (63.63)	8 (72.72)
PCG 7	7 (50)	7 (50)	1 (7.14)	0 (0%)	6 (42.85)	3 (21.42)

RW = Relaxed wakefulness before sleep onset; I = stage I; II = stage II; III–IV = stages III and IV; REM = REM sleep; A = awakening; WCG = well-controlled group; VCG = variably controlled group; PCG = poorly controlled group.

Table II. Distribution of cases according to the existence of disorganized paroxystic spike and wave discharges (<3 c/s) and polymorphic epileptic patterns in stage I, REM and wakefulness

n	DPD (%)	PEP (%)
WCG 31	1 (3.2)	1 (3.2)
VCG 11	4 (36.3)	1 (9)
PCG 14	10 (71.4)	10 (71.4)

DPD = Disorganized paroxystic discharges <3 c/s; PEP = polymorphic epileptic patterns. Other abbreviations as in table I.

Table III. Average hourly frequency according to the various states of alertness

	RW	I	II	REM	An	Am	F
WCG	10.52 ± 2	12.09 ± 5	0.46 ± 0.2	7.41 ± 1.6	21.84 ± 7	26.52 ± 7	$p < 0.01$
VCG	14.58 ± 4	5.02 ± 1	1.59 ± 0.47	5.53 ± 1.7	3.71 ± 3	7.45 ± 2.3	$p < 0.005$
PCG	6.36 ± 3	5.38 ± 2	0.17 ± 0.08	5.33 ± 2	2.1 ± 0.8	2.59 ± 1	n.s.

An = awakening after sleep onset; Am = awakening in the morning. Other abbreviations as in table I.

Existence of Disorganised Paroxystic Spike and Wave Discharges (< 3 c/s) and Polymorphic Epileptic Patterns. The study concerns states of alertness for which these patterns are classically considered not to exist: states of wakefulness, stage I, REM sleep. As shown in table II these patterns are found to exist essentially in the poorly controlled group.

Hourly Frequency of 3 c/s Paroxystic Spike and Wave Discharges (duration > 3s) during the Various States of Alertness (table III). In the well-controlled group awakening stages are the highest point of the hourly frequency followed in decreasing manner by relaxed wakefulness before sleep onset – stage I, REM and stage II (table III). Such a graduation does not exist within the other groups. As indicated by a multivariate analysis, a relationship exists between awakening, REM sleep and the association relaxed awakening before sleep onset – stage I in the well-controlled group (F = 5.16 p<0.01). Such a correlation is not found in the other group.

Comment

Thus, 2 groups can be distinguished according to the ability of wakefulness and states bordering wakefulness (stage I and REM) to favor 3 c/s paroxystic spike and wave discharges. In the first case there exists a precise link between epileptic activity and states of alertness. The level of alertness gives to the epileptic pattern its form and its frequency. In the second case (poorly controlled group) this relationship does not exist or is not clearly apparent. One could suppose that in this case there exists a modification of the function, regulating alertness, and that this modification is responsible for the differences observed. Thus, the study of the relationship between states of alertness and paroxystic discharges in petit mal could be an interesting method in the determination of prognosis criteria for this kind of epilepsy.

References

1 Sato, S.; Dreifuss, F.E.; Penry, J.K.: Neurology, Minneap. *23:* 1335–1334 (1973).
2 Passouant, P.: in Koella, Levin, Sleep 1976. 3rd Eur. Congr. Sleep Res., Montpellier 1976, pp. 57–65 (Karger, Basel 1977).

J. Touchon, MD, Service de Physiopathologie des Maladies Nerveuses,
Centre Gui de Chauliac, Cliniques St. Eloi, F-34059 Montpellier (France)

G. Methodology of Sleep Studies. Automatic Analysis

Principles of the Automatic Pattern Recognition of Human Sleep Waveforms[1]

K. B. Campbell

Psychophysiology Section, MRC-APU, Cambridge, England, and
School of Psychology, University of Ottawa, Ottawa, Canada

The classification of more than a kilometer of a paper recording into the various stages of sleep is a time-consuming and often unreliable task. While human staging procedures do serve as fairly accurate summaries of 20 to 30-s epochs of EEG, they are accomplished at a tremendous loss of other potential data. Many other types of analysis that theoretically could be carried out are, in practice, ignored due to the immensity of the huge quantity of data. Over the past decade, various attempts at the computer analysis of sleep have been made. It is generally acknowledged that hybrid systems have been the most successful. Using a pre-processor developed in Canada [3] linked to a digital computer in Cambridge, such a system was constructed for stage classification purposes and to permit the detailed types of analyses that to date remain relatively unexploited by the human scorer.

Methods and Materials

The processing of the analogue signal is accomplished by alpha, delta and spindle hardware detectors operating on the EEG and a rapid eye movement (REM) detector on the EOG. The alpha and spindle detectors rely on phase-locked loop (PLL) circuitry [for details, see 1,

---

[1] Supported by research grant 183-77-1 ENV UK from the Environmental Research Programme of the Commission of European Communities, Brussels.

2]. Instead of using the traditional phase error signal of the PLL to determine the presence of alpha or spindle waveforms, the more reliable approach of a quadrature phase detector was substituted. Delta detection is based on zero-crossing and minimum amplitude (75 µV) techniques. The REM detector consists of a simple filter and a threshold discriminator. The output from each of the detectors is binary ('1' or '0') which is then fed into the digital interface of a minicomputer (Data General Eclipse S200) whose software was written to mimic the logic of standard sleep scoring procedures. For purposes other than staging, storage on the computer's 2.5 megabyte disc unit is required.

Results

To assure functioning of the spindle detector, a second automatic system relying on complex demodulation methods [5] was employed for comparison purposes. The two systems performed virtually identically being triggered by more than 80% of EEG waveforms judged as 'spindles' by human scorers [2].

With respect to the actual stage classification of sleep, three procedures were employed: (1) comparison with two human scorers; (2) comparison with a second automatic system, and (3) comparison of computer determined variables with established norms. The overall level of agreement between the hybrid system and human scoring, based on 8 nights of recordings from different subjects (aged 20 to 58), was 84% for all stages, ranging from a low of 61% for stage 1 to a high of 92% for stage 2. Evaluation using a second automatic system was made on 40 nights of recordings from an independent laboratory. Although the basis of operation of the *Kumar*-designed analyser [4] was quite different from the present one, inter-system agreement was approximately 0.80. Finally, the computer-derived values of a number of sleep parameters obtained on 60 sleep recordings from young subjects were remarkably similar to the norms established by *Williams et al.* [6].

The stage attributed to a particular epoch of sleep is recognized as a convenient but arbitrary means of managing a very large mass of data. To go beyond simple stage classification, software has been written to determine the evolution of the various waveforms (spindle, delta and REM) encountered in sleep recordings. The number of seconds of each is accumulated over a convenient interval – 15 min, typically – across the duration of the sleep period and the corresponding distributions are plotted. There is no concern whatsoever about arbitrary stages. Rather unexpected biorhythmic features have emerged as typified by figure 1 (plots of the usual sleep stages and the development of REM, spindle and delta activity during the night). While the 90-min cyclic oscillation of REM activity has often been reported,

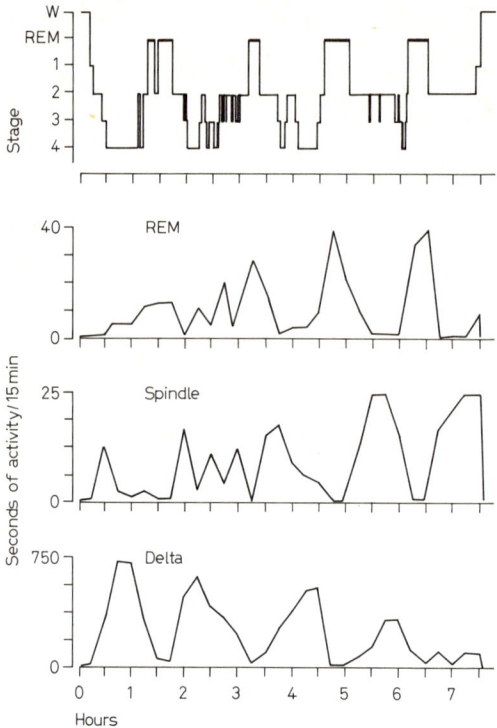

Fig. 1. Computer analysis of a 7.5-h sleep period of a typical subject. In the uppermost portion, traditional stage classification has been carried out. In the lower portions, the evolution of REM, spindle and delta activity. Note that the peaks of a particular waveform correspond to a predictable stage. However, the 90-min oscillation of delta and spindle waveforms is not apparent in the development of their respective stages (4 and 2).

both delta and spindle activity reveal similar biorhythmic features. Both REM and spindle transients markedly increase as the night progresses. However, one is out of phase with the other. Delta activity is characterized by a decayed oscillating waveform, being out of phase with REMs. In a separate study, it was found that the total amount of delta activity significantly decreases in older subjects. Interestingly, the relative manner in which delta evolves across the night was essentially identical for both younger and older age groups.

References

1 Broughton, R.; Healey, T.; Maru, J.; Green, D.; Pagurek, B.: Electroenceph. clin. Neurophysiol. *44:* 677–680 (1978).
2 Campbell, K.; Kumar, A.; Hofman, W.: Electroenceph. clin. Neurophysiol. *48:* 602–605 (1980).
3 Green, D.: A hybrid preprocessor for sleep staging using the EEG; thesis (M. Eng.), Ottawa.
4 Kumar, A.: Patt. Recog. *9:* 43–46 (1977).
5 Kumar, A.; Hofman, W.; Campbell, K.: Waking Sleeping *3:* 325–333 (1979).
6 Williams, R.L.; Karacan, I.; Hursch, C.: Electroencephalography (EEG) of human sleep (Wiley, New York 1974).

K.B. Campbell, PhD, School of Psychology, University of Ottawa, Ottawa K1N 6N5 (Canada)

Sleep 1980. 5th Eur. Congr. Sleep Res., Amsterdam 1980, pp. 452–455
(Karger, Basel 1981)

Estimation of the Circadian Sleep-Wake Cycle with a Doppler Radar[1]

G. Chouvet[a], P. Odet[b], J. P. Etienne[b], P. Chemarin[b], J. L. Valatx[a], J. F. Pujol[a]

[a] INSERM U. 171 et U. 52, Laboratoire de Médecine Expérimentale, Université Claude-Bernard, Lyon, and
[b] ICPI, Centre de Recherches Appliquées, Lyon, France

The initial purpose of this study was to develop an activity detector sensitive not only to gross locomotor activity (as the usual activity meters commercially available) but also sensitive to small movements of the animal, especially laboratory mouse, as observed during eating, drinking, grooming... in order to better approximate the sleep-wake cycle without actual polygraphic recordings.

A simple and cheap movement detector was developed around a microwave miniradar (DA 8525/11 – AEI Semiconductors) with an hyperfrequence (9.9 GHz) resonant cavity for low-power emission (10 mW) and a mixed diode to detect the Doppler effect. When there is a radial (relative to the emission vector) displacement of an obstacle echoing in the field of the oscillator, a signal whose Doppler frequency is proportional to the radial speed of displacement is collected on the reception diode and is amplified ($G = 2,500$). Its amplitude is proportional to the echoing surface, related to the reflection coefficient of the obstacle and inversely related to the square of the distance from the emitter. The voltage level at the radar output is in the microvolt range and explains the high gain of the band-pass amplifier, with cut-off frequencies of 0.6 and 60 Hz corresponding to radial speeds of 1–90 cm/sec with the emission frequency used. Activity was quantified in two ways: (1) Integration of the output voltage over 30-s periods by using the EMG channel of a sleep-classifier already described [2]. (2) As the output voltage and the Doppler frequency are not linearly related to movements, depending on many parameters of this one (speed, amplitude, echoing surface, reflection coefficient, etc.) a more natural way of activity measurements was developed around a microprocessor (INTEL 8748). With a threshold circuit, adjustable by the user, a microprogram cumu-

[1] This work was supported by INSERM (CRL No. 78.5.004.6) CNRS (L.A. 162) and DRET.

lates over a preset period (usually 60 s) the number of seconds containing at least one triggering. At the end of each period, the cumulative count is output on a serial peripheral device (RS232 C), thus giving an estimation of the 'active time' over this period. The 'active time' (number of seconds where at least one triggering has occurred) estimated over 1-min periods, was found more precisely related to the sleep-wakefulness data than simple integration or counts of all the triggerings (non-linearity with exact activity). Moreover, this method permits a multiplexing of several Doppler detectors on the same microprocessor.

An important point when using a microwave detector is the problem of received power at the level of the mouse and its biological effects. When using the radar without its cone of emission, in order to prevent an excessive concentration of microwaves over a narrow beam (but so decreasing the signal/noise ratio), we have verified that the emission level is included in a 60° cone, with half of the emitted power inside a half-angle of 30°. In these conditions, the power received by the animal is around 30 μW/cm^2 at 20 cm from the emitter and 4 μW/cm^2 at 40 cm. Even if this range is higher than the UK health safety level for humans (3 μW/cm^2), to our knowledge, all the data available on the biological effects of microwaves on rodents report that specific effects are observed only for a higher range, over 100 μW/cm^2 (see in *Adair and Adams* [1]).

With an emitter placed at 40 cm from the center of the cage, the noise level is approximatively of 40 mV peak to peak (after 2,500 amplification), the respiratory rhythm is easily detectable when the mouse is sleeping (200 mV p.p.) and levels over 1 V are always obtained for gross activity. Spectral density measurements were performed on the output by means of Fast Fourier Transform (FFT). During sleep, a low-energy level is concentrated in the 3- to 5-Hz range corresponding to mouse's breathing. For gross activity, the spectrum has an important peak around 1–3 Hz and decreases progressively over these frequencies. Except noise, no useful energy was usually observed over 5–10 Hz, contradicting the report of *Marsden and King* [3]. Some exceptions were noticed when the animal was scratching itself; in such cases, but for very brief episodes, output voltages around 15–20 Hz were observed. This fact favors the present method of counting after appropriate triggering (1 or more triggerings within 1 s) comparative to that of *Vanuytven et al.* [4] where each triggering is counted, thus resulting in very high counts in such cases.

The Doppler radar system is able to clearly differentiate the rest/activity cycles of two strains of mice (fig. 1), reflecting very accurately the sleep-wake circadian rhythms known on these strains. The data presented here clearly demonstrate the usefulness of a Doppler system for quantitative study of rest/activity circadian rhythms. It has the advantage of detecting all types of movements and not only gross locomotor activity, and may replace polygraphic recordings with efficacity when used in the sleep-wake cycle-related experiments.

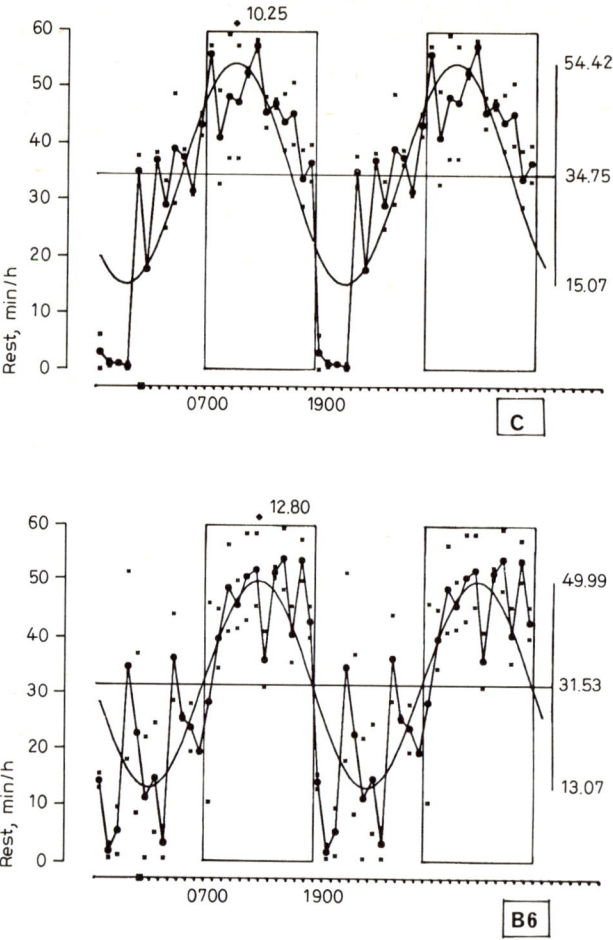

Fig. 1. Rest-activity circadian cycle of 2 mice of different strains (C57BL/6, B6, Balb/c, C). Abscissae: Legal time in hours (light, 0700–1900; dark, 1900–0700). Ordinates: The 'active time' was measured every 60-s epoch and punched out on a serial peripheral device for each mouse. The ordinates represent the mean hourly level of rest time (60 minus active time) computed on two successive 24-h periods (as indicated by crosses). The best-fitting sine rhythm around the daily mean level is plotted as a continuous line. The daily mean level, the amplitude, and acrophase of each of the rhythms are typical of the sleep-wake circadian rhythms known in these strains, especially acrophase near 10 h for Balb/c and 13 h for BL/6.

References

1 Adair, E.R.; Adams, B.W.: Science *207:* 1381–1383 (1980).
2 Chouvet, G.; Odet, P.; Valatx, J.L.; Pujol, J.F.: Waking Sleeping *4:* 9–31 (1980).
3 Marsden, C.A.; King, B.: Pharmacol. Biochem. Behav. *10:* 631–635 (1979).
4 Vanuytven, M.; Vermeire, J.; Niemegeers, J.: Psychopharmacology *64:* 333–336 (1979).

G. Chouvet, PhD, INSERM U. 171 et U. 52, Laboratoire de Médecine Expérimentale, Université Claude-Bernard, 8 Avenue Rockefeller, F-69005 Lyon (France)

Computer Analysis of the EEG in Infrahuman Species

A. F. Glatt

Biological Research Department, Pharma Division, Ciba-Geigy Ltd., Basel, Switzerland

Itil [1] and *Fink* [2] introduced computerized EEG analysis as a 'diagnostic' tool in human psychopharmacology. Researchers have made many attempts to use the EEG in animals as well to supplement the 'classical' screening methods. However, EEG (as any other) studies in animals are plagued with some very important drawbacks. Having no possibility to verbally 'command' a particular behavioral activity (or inactivity) we are at a loss when it comes to the (desired) comparison of particular drug effects in identical *states*, for instance, level of *general vigilance*. Still, such 'state-dependent' comparisons are of paramount importance since – as often has been experienced – the 'state' per se is liable to affect our indicators (namely the EEG pattern) to a far greater extent than our psycho- and neurotropic drugs do. Secondly, a number of drugs are liable either to eliminate completely the appearance of particular states, making state comparisons even more difficult.

It seems that there are two ways to get around, or at least reduce the impact of, these difficulties: (1) we can 'wait' and investigate the EEG pattern under placebo and drug conditions only when a particular state – namely REM sleep or 'approach' – occurs spontaneously, or (2) we can try to actively enforce the occurrence of a particular state by manipulations of the external conditions, by electrical stimulation of certain brain structures or by the administration of a second drug, well known and defined as being able to induce certain states.

Methods

Tif:RAI rats (average body weight of 200 g) were implanted under pentobarbital with epidural electrodes (4.0 mm lat. to midline and 5.5 mm behind the transversal suture). The electrodes were soldered to a 4-pole socket (Microtech Inc.) cemented to the skull. A cable leading to a slip-ring device (M. Gautschi, Gossau), was connected to the EEG machine (Grass 7B polygraph) and a Honeywell 101 taperecorder. Experimentation started 2 weeks after surgery. The EEG data (from tape) were digitized and analyzed by a PDP 11/34 computer using FFT or period-analysis techniques. All programs were developed by Dr. *H. Demiéville*, Mr. *B. Müller* and *R. Wenk* of Ciba-Geigy Laboratories, Basel and Summit (USA). EEG recordings were paralleled by visual inspection; typical behavioral patterns (namely grooming, exploration, resting, sleep) were noted and recorded on the EEG record for later ('time-locked') comparison with the (computerized) EEG.

Results

A three-dimensional display (over discrete time steps) of the power density spectra is not a convenient technique; it lacks clarity and does not allow 'first-gaze overview' for comparison with particular 'states'. We prefer – for continuous analysis and comparison with (continuously changing) 'states' – presentation of the 'mean power' along the time axis in a number of arbitrarily selected frequency bands; this display is akin to the one generated by the integration of the output of narrowly tuned analog filters. This presentation quite adequately reveals 'contingency' of EEG pattern and 'state'. The following additional analyses were performed: (1) During relatively long time periods (1 h or more) well adapted animals reveal highly constant amounts of waking (W), slow wave sleep (SWS) and paradoxical sleep (PS). Two 75-min periods of summated power spectra (selected from long-lasting recording periods before and after application of a placebo) are virtually identical (fig. 1a). (2) Pairs of shorter (summated) power spectra obtained during characteristic W, SWS, and PS periods (each pair separated by several hours) are again virtually identical (fig. 1b, d, f). (3) 'Arousal' was induced by the experimenter entering the animal room, once before, and – several hours later – once after placebo. Here too, the summated power spectra representing the 'stimulated state' are virtually identical (fig. 1c). Slight deviations may indicate 'biological' variations or deficiencies still inherent in our standardization procedures.

In addition we studied (again in a preliminary fashion) the effect of neuroleptics. Figure 1e shows the increase in power output at 8 and 16 Hz and the general enhancement in the beta frequencies. These effects on the EEG, obtained with haloperidol (0.1 mg/kg p.o.), were *not* paralleled by obvious changes in (spontaneous) behavior, except for some tendency towards ataxia. With higher doses the EEG effects became more pronounced and clear-cut behavioral changes became apparent. Additional experiments are in progress to investigate whether such EEG changes are specific for *all*

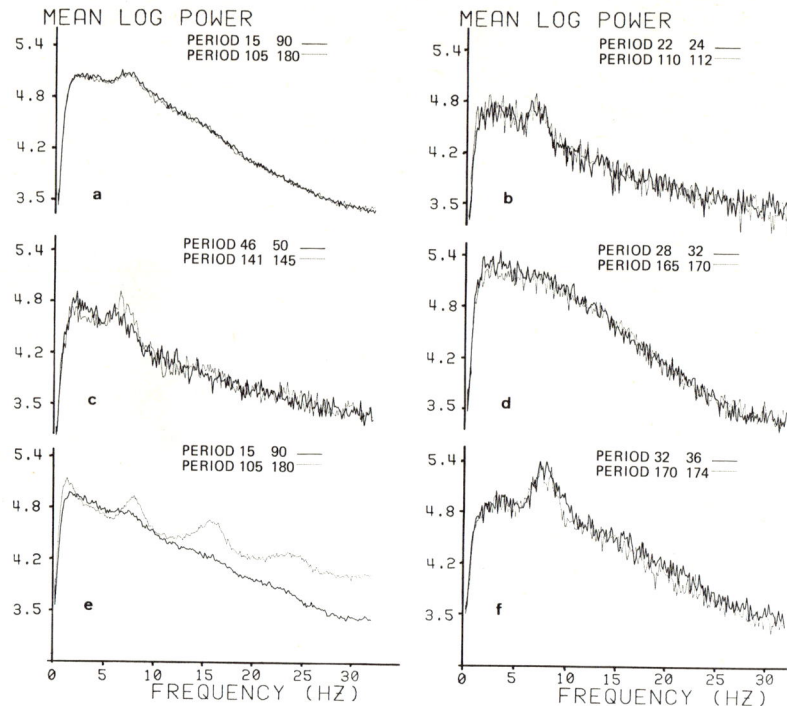

Fig. 1. Superposition of the mean log power spectrum of a control period (solid line) upon that of a subsequent placebo (a–d, f) or haloperidol (e) period (dotted line) in a rat.

neuroleptically active drugs and whether they are specific for rats only or are obtained in other species as well.

References

1 Itil, T.M.: in Itil, Psychotropic drugs and the human EEG. Mod. Probl. Pharmacopsychiat., vol. 8, pp. 44–75 (Karger, Basel 1974).
2 Fink, M.: in Rémond, EEG informatics. A didactic review of methods and applications of EEG data processing, pp. 301–318 (Elsevier, Amsterdam 1977).

Dr. A.F. Glatt, Biological Research Department, Pharma Division, Ciba-Geigy Ltd., CH-4000 Basel (Switzerland)

An Automatic Sleep Classifier for Cat[1]

G. Chouvet[a], C. Buda[b], M. Janin[b], P. Odet[c], J. F. Pujol[a]

INSERM[a] U. 171 and[b] U. 52, Laboratoire de Médecine Expérimentale, Université Claude-Bernard, Lyon, and
[c] ICPI, Centre d'Electronique Appliquée, Lyon, France

In recent years, there has been growing interest in automatic scoring of sleep records in laboratory animals: rat, dog and mouse. Although the majority of these systems are working on-line, sometimes multiplexed on several animals and are relatively inexpensive because of their hardware requirements, their adoption is only possible for a particular species with a given sleep stage classification. We report here the automatic scoring of vigilance states of the cat by using a powerful flexible system primarily built for laboratory rodents [1], combining an extreme hardware modularity and a modifiable truth table for sleep-stage decision.

Almost all the methods of automatic scoring are based on a simple hypothesis. Polygraphic signals carry some kind of information useful for visual sleep scoring and we suppose that a part of this information may be measured by a set of indices which summarize one epoch of the signals. This set of indices may be considered in a multidimensional space as the coordinates of a point featuring the polygraphic epoch. The basic assumption is that similar epochs, by visual polygraphic perception, will be close together in this space ('birds of a feather flock together'), in other words that points corresponding to the same state of vigilance will cluster together in the data space. With his own interpretation criteria, the user knows a priori the different clusters, and the problem is related to classification analysis. So, after the feature extraction which provides the coordinates of a polygraphic epoch, one must make a sleep-stage decision by discriminant procedures. However, even with very sophisticated discriminant functions or other decision rules, a poor feature extraction will always lead to a high error rate with the working classifier.

As simple comparators simulate in fact linear discriminant functions in the indices space, this approach is implementable on a relatively simple hardwired system primarily designed for rodents' sleep-stage scoring. In order to mimic visual scoring, and after selecting on the basis of computer simulations, implementable indices with a high discriminative power, four indices were selected for rodents: the F index (the output voltage of its associated integrator) is proportional to the ECoG energy in the theta band relative to the one in the delta band, the Z index provides a rough estimate of the ECoG mean frequency by ECoG zero-crossings counting, the D index is proportional to the dispersion or variability of ECoG amplitude and the M index is the usual integrated EMG. These four indices are measured by means of analog integrators, and their resetting is carried out by a central clock (30 s for the cat). A microprocessor-controlled system (8748 INTEL) performs at each clock pulse the A/D conversion of the integrator outputs, displays the numerical values on the front panel for adjustment purposes, and/or prints out these values on a serial peripheral device (RS 232C) for eventual numerical sleep monitoring. At the end of each integration period, these output voltages are compared to a preset reference voltage adjustable on the front panel by the user. These solid-state comparators divide the index space into 16 (2^4) subspaces. The logical outputs (4 bits) of the 4 comparators enter a hardwired boolean truth table, modifiable by microswitches, transforming the 4 bit-words into 3 bit-words, allowing 8 (2^3) different vigilance states. This final 3 bit-word is then passed to another microprocessor-controlled device in order to print out, after data compression, the sequence of vigilance states on a serial peripheral device for further off-line data analysis.

An important feature of this basic system is constituted by its flexibility and extreme hardware modularity (interchangeability of the integrators) which permit its use for cat sleep-stage scoring with only two leads with a very low error rate.

When using this standard system on the cat (fronto-occipital ECoG and neck EMG), wakefulness (W) and paradoxical sleep (PS) epochs are characterized by high Z and low D according to their polygraphic definition (low voltage fast activity), stage one of slow wave sleep (S1) by high Z and high D (low-voltage fast activity superimposed with spindles of slow waves) and stage two of slow wave sleep (S2) by low Z and high D (large continuous slow waves). The F index, even after trying several ratios of different frequency bands was found to be insufficient to discriminate with a low error rate W from PS. In the same way, the M index (integrated neck EMG) was not a good index for discriminating W from PS, especially because of periods of quiet wakefulness without muscle tone and/or episodic muscular twitches during some PS episodes. Moreover the eventual degradation of EMG electrodes especially on the cat, does not favor the use of EMG for automatic sleep scoring.

This problem may be overcome by using implanted bipolar electrodes in the lateral geniculate nucleus (LGN) on which high amplitude phasic potentials (PGO waves) appear selectively during (or just prior to, or during S2 stage) PS episodes, as already described in the literature. Replacing the M integrator by a D integrator in the hardwired system and using a LGN lead result in a very high D for PS (or S2 with PGO) and a low D for W, S1 and S2 without PGO spikes. So, the four classical sleep stages of the cat may be automatically scored when using three indices on two leads (ECoG and LGN) and an appropriate truth table (table I).

Table I.

Automatic	Visual[1]				Visual[2]		
	W	S1	S2	PS	Z (ECoG)	D (ECoG)	D (PGO)
W	96.4	0.9	0	0.3	1	0	0
S1	2.7	95.5	3.7	0	1	1	0
S2	0	2.8	96.3	0	0	1	0
PS	0.9	0.9	0	99.7	1	0	1

[1] Concordance matrix between visual and automatic scoring of 2,160 30-s epochs. The results are expressed in percentages, relatively to visual interpretation in each sleep stage.
[2] Simplified truth table used for cat. Two leads: ECoG and LGN (PGO spikes). The other logical combinations are very unusual and may be coded as 'unknown' state except for S2 with PGO (011).

The validity of this system has been tested extensively on 3 cats and a concordance matrix between visual and automatic scoring was computed on 3 × 12 h of recordings (table I). This matrix reflects especially the lability of the exact limits between W and S1 or S1 and S2. The PS epochs misclassified in W by the automatic classifier are due to the infrequent absence of PGO spikes during 30-s PS periods. However, these epochs, surrounded by PS epochs, may be easily coded as PS by an off-line data processing. The W or S1 epochs misclassified in PS are generally due to large artifacts on the LGN (and EMG) lead. As previously, they may be also rescored because of their context, during further off-line data processing. Nevertheless, the global concordance rate (around 95%) remains in the range of the routine interscorer agreement in visual interpretation.

Reference

1   Chouvet, G.; Odet, P.; Valatx, J.L.; Pujol, J.F.: Waking Sleeping *4:* 9–31 (1980).

G. Chouvet, PhD, INSERM U. 171 et U. 52, Laboratoire de Médecine Expérimentale, Université Claude-Bernard, 8 Avenue Rockefeller, F-69005 Lyon (France)

Author Index

Ackenheil, M. 9
Adrien, J. 290, 328
Åkerstedt, T. 16, 98, 190
Alihanka, J. 344

Ballenger, J.C. 23
Bareggi, S.R. 442
Bari, F. 272
Bastiaans, J. 86
Baylor, G.W. 354
Benedek, G. 272, 314
Benoit, O. 195, 206, 391, 431
Bergström, L. 414
Bernhard, J. 370
Bertrand, A. 389
Besset, A. 445
Billiard, M. 120, 124, 389, 400, 445
Blanchet, V. 220
Blois, R. 301
Boenke, R. 337
Bonnet, M. 357
Borbély, A.A. 40
Bouard, G. 391
Bour, H. 236
Bour, H.L. 321
Bowker, R.M. 304
Bruder, S. 198
Buda, C. 459
Bunney, W.E., Jr. 23

Burkhardt, R. 370
Burtschy, B. 51

Calasso, E. 361
Campbell, K.B. 448
Cavero, I. 297
Chemarin, P. 452
Chouvet, G. 51, 452, 459
Cipolli, C. 178, 361
Cluydts, R.J.G. 102
Coenen, A.M.L. 438
Comi, G. 442
Corner, M.A. 236, 318, 321
Cutler, N.R. 23

Dardenne, A. 133
Declerck, A.C. 324, 429
De Graaf, W. 380
De Koninck, J. 91
Delerm, B. 334
Depoortere, H. 283, 297
Diest, R. van 228

Eastman, C. 59
Ehrhart, J. 212
Elsenga, S. 2
Empson, J.A.C. 364, 367
Emptoz, H. 51
Endo, S. 311, 386
Eschenlauer, R. 212
Etienne, J.P. 452

Faeh, M. 370
Foret, J. 195
Foulkes, D. 174, 246
Franceschi, M. 442
Freixa i Baqué, E. 334

Gagneux, J.M. 220
Gaillard, J.-M. 292, 301
Garma, L. 391
George, B. 431
Gillberg, M. 16, 98
Glatt, A.F. 456
Gnirss, F. 307, 417, 421
Griefahn, B. 215
Groos, G.A. 42
Gros, E. 215

Haefely, W. 155
Haller, H. 337
Hasan, J. 344
Hayashi, Y. 386
Hearne, K.M.T. 364
Helojoki, M. 294
Heynick, F. 373
Hishikawa, Y. 128, 435
Hofman, W.F. 228
Hoofdakker, R.H. van den 2
Horáková, A. 394
Horne, J.A. 95, 348, 406, 408

Iijima, S. 128, 435

Author Index

Janin, M. 459
Jimerson, D.C. 23
Józsa, C. 314
Jurriëns, A.A. 217

Kafi, S. 292
Kahn, J.P. 328
Kaneda, H. 128
Kauth, H. 215
Kendel, K. 337
Kiesswetter, E. 198
Knauth, P. 198
Kobayashi, T. 311
Koella, W.P. 141, 158
Koukkou, M. 170
Kramer, M. 182, 357
Kumar, A. 211, 228

Landau-Ferey, J. 431
Lanfumey, L. 290
Lefèvre-Borg, F. 297
Lehmann, D. 170
Lienhard, J.P. 212
Liry, J.L. de 133
Lorenz, A. 9
Lustenhouwer, A.W. 429

Maccolini, S. 361
Malo, J.L. 397
Mariani, E. 442
Marias, J. 147
Mastrangelo, M. 442
Matsuo, R. 128
McQuarrie, E. 357
Michel, F.B. 389
Miettinen, M.V.J. 287
Milane, J. 389
Mirmiran, M. 236, 318, 351
Monday, J. 397
Monnier, M. 301
Monod, N. 403
Montplaisir, J. 133, 397
Moore, A.T. 340
Moser, U. 370
Muzet, A. 212

Nedopil, N. 426
Nevšímalová, S. 120, 394

Niemegeers, C.J.E. 279
Nogues, B. 403

Obál, F. 272, 314
Obál, F., Jr. 272, 314
Odet, P. 452, 459
Okuno, H. 311
Oswald, I. 142

Pani, R. 361
Pans-Raedts, M. 324
Parmeggiani, P.L. 331
Paroubková, D. 394
Partinen, M. 383
Passouant, P. 258, 389, 400, 445
Perez, E. 331
Perez, M.A. 424
Poelstra, P.A.M. 380
Porter, J.M. 406, 408
Post, R.M. 23
Pujol, J.F. 452, 459
Putkonen, P.T.S. 294, 383, 414

Romberg, H.P. 198
Romo, R. 424
Roth, B. 120, 394
Routhier, H.L. 51
Roy, J.C. 334
Rutenfranz, J. 198
Rüther, E. 9, 426

Ságová, V. 394
Saito, Y. 311
Salzarulo, P. 178, 361
Samson-Dollfus, D. 403
Seitz, R. 370
Shapiro, C.M. 340
Shimizu, T. 435
Sijben-Kiggen, R. 324
Smirne, S. 442
Soukos-Valavani, I. 377
Spiegel, R. 275
Sugita, Y. 128, 435

Schenker, J. 307, 421
Scherschlicht, R. 147

Schneeberger, J. 147
Schneider-Helmert, D. 85, 107, 307, 411, 417, 421
Schoenenberger, G.A. 417
Schulz, H. 72

Steinberg, R. 426
Steiner, M. 147
Strauch, I. 169, 370, 377

Tashiro, T. 128
Teshima, Y. 128, 435
Tilley, A.J. 364, 367
Tissot, R. 301
Torsvall, L. 190
Touchon, J. 400, 445
Tsuji, Y. 311
Tuozzi, G. 361

Uhde, T.W. 23

Valatx, J.L. 51, 452
Vallet, M. 220
Van Den Broeck, W.A.E. 279
Van den Dungen, H. 351
Van Hulzen, Z.J.M. 438
Van Luijtelaar, E.L.J.M. 438
Velasco, F. 424
Velasco, M. 424
Visser, P. 85, 102, 380

Wahlländer, B. 9
Walsh, J. 397
Watanabe, H. 386
Wauquier, A. 279, 324
Weber, U. 337
Wehr, T.A. 26, 64
Weitzman, E.D. 23
Wilkinson, R.T. 225
Williams, H.L. 231
Wirz-Justice, A. 26, 64

Yoshida, R. 386

Zamboni, G. 331
Zander, K.J. 9
Zulley, J. 202

3 1162 00234 3458

DATE DUE ✓			
DATE DE RETOUR			
~~JUL 1 9 1983~~			
~~APR 9 1989~~			
~~DEC 2 0 1992~~			

LOWE-MARTIN No. 1137